Advanced Dynamics of Rolling Elements

Pradeep K. Gupta

Advanced Dynamics of Rolling Elements

With 97 Figures

Springer-Verlag
Berlin Heidelberg New York Tokyo

Pradeep K. Gupta
PKG Incorporated
117 Southbury Road
Clifton Park, New York 12065
U.S.A.

Library of Congress Cataloging in Publication Data
Gupta, Pradeep K.
Advanced dynamics of rolling elements.
Bibliography: p.
Includes indexes.
1. Rolling contact. 2. Ball-bearings. 3. Roller bearings. I. Title.
TJ183.5.G87 1984 621.8′22 84-10567

Typeset by Asco Trade Typesetting Ltd., Hong Kong.
Printed and bound by R. R. Donnelley and Sons, Harrisonburg, Virginia.
Printed in the United States of America.

9 8 7 6 5 4 3 2 1

ISBN-13: 978-1-4612-9767-3 e-ISBN-13: 978-1-4612-5276-4
DOI: 10.1007/978-1-4612-5276-4

to Shelly

Preface

In any rotating machinery system, the bearing has traditionally been a critical member of the entire system, since it is the component that permits the relative motion between the stationary and moving parts. Depending on the application, a number of different bearing types have been used, such as oil-lubricated hydrodynamic bearings, gas bearings, magnetic suspensions, rolling element bearings, etc. Hydrodynamic bearings can provide any desired load support, but they are limited in stiffness and the associated power loss may be quite large. Gas bearings are used for high-precision applications where the supported loads are relatively light, bearing power losses are very low, and the rotating speeds generally high. For superprecision components where no frictional dissipation or bearing power loss can be tolerated, magnetic suspensions are employed; again, the load support requirements are very low. Rolling element bearings have been widely used for those applications that require greater bearing versatility, due to the requirements for high-load and high-stiffness characteristics, while allowing moderate power loss and permitting variable speeds. A study of the dynamic interaction of rolling elements is, therefore, the subject of this text.

Texts covering the analysis and design methodology of rolling elements are very limited. Notable works include *Analysis of Stresses and Deflections* (Jones, 1946, Vols. I and II), *Ball and Roller Bearings, Their Theory, Design and Application* (Eschmann, Hasbargen, and Weigand, 1958), *Ball and Roller Bearing Engineering* (Palmgren, 1959, 3rd ed.), *Advanced Bearing Technology* (Bisson and Anderson, 1965), and *Rolling Bearing Analysis* (Harris, 1966). Most of these texts were published at a time when the available computational means were very limited. Thus, all of this work was restricted to solving an equilibrium problem, which is a first approximation to the analysis of rolling bearings.

During the last decade, the operating speeds and temperatures that bearings must survive have increased greatly, due to the demand for higher efficiency. This situation has created considerable interest in the advancement of rolling bearing technology. With the advent of modern high-speed computers, the simple quasi-static equilibrium models of the 1960s have been replaced by advanced dynamic models capable of providing real-time performance simulations of rolling bearings. To a rolling bearing engineer, such an advancement offers design tools for improved designs against skidding, skewing, cage instabilities, lubricant traction behavior, and time-varying operating environments, as compared to the simple fatigue life and stiffness computations provided by the static models.

The principal motivation of the text is to document the vastly improved state of the art from an engineer's viewpoint. The text is a result of the research I have carried out under both government and industrial sponsorship over the past several years. The objectives are to

1. present a generalized formulation of a dynamic model to simulate rolling bearing performance under arbitrary operating conditions, and
2. provide the reader with a computer code to implement the analytical model in actual practice.

The dynamic model basically consists of the formulation of the coordinate frames and equations of motion of the bearing elements, and the geometric interactions which provide the foundation for modeling the applied force and moment vectors. These subjects are covered in Chapters 2 and 3, following the basic background presented in the first chapter. Specialized subjects particularly applicable to oil-lubricated rolling bearings are included in Chapters 4 and 5. Chapter 6 is devoted to numerical analysis, which logically precedes the presentation of the computer program framework and structure in Chapter 7. Typical results of the computer program and dynamic simulations of rolling bearings are presented in Chapter 8. In view of the very limited experimental data, a rigorous experimental validation of the analytical models has been difficult; however, I have attempted to present some comparisons between the analytical results and the recently available experimental data in Chapter 9. I expect that experimental validations and subsequent improvements of the analytical models will continue as more experimental data become available. Finally, in Chapter 10 I have briefly presented some general design guidelines from an engineer's viewpoint.

I have named the computer program developed in this text ADORE (*A*dvanced *D*ynamics *O*f *R*olling *E*lements). In view of the complexity and length of the code, I have made ADORE operational on some of the worldwide computer networks, such as the CYBERNET Services.* The intent is

*CYBERNET is a registered trademark of Control Data Corporation.

to provide the reader immediate access to the code. As may be expected, sophisticated computer codes are constantly updated as the technology behind the codes advances and the understanding of the fundamental model grows. Because ADORE is operational on a widely available computer network, I will be able to effectively maintain and support the code, and the user will have the benefit of the latest version of the program which may be substantially improved in comparison to the original program presented in this text. I greatly encourage interested readers to use ADORE on the network and, if possible, to supply me with comments and suggestions which will be crucial to its future improvement. In order to provide adequate instructions to the user while running ADORE on the network, I have included the source listing of the main program in one of the appendices. I hope this listing, which is strictly for reference purposes, will help the reader understand some of the finer nuances of ADORE and that, through example, it will inspire sophisticated readers to develop their own codes, which will eventually help advance the general technology behind ADORE.

The text is directed principally toward users and designers of advanced rolling bearing systems; in particular, engine manufacturers, manufacturers of inertial guidance systems, and the rolling bearing industry. It will also be very helpful to research staff both in government and in industrial laboratories.

It is my greatest pleasure to acknowledge Mechanical Technology Incorporated for the research opportunities and environment provided to me during my tenure there as a Senior Scientist (1971–82). In particular, I would like to mention the encouragement and support provided by Dr. Donald F. Wilcock, Dr. Jed A. Walowit, Mr. Oscar Pinkus, Mr. Wilbur Shapiro, and Dr. Jeffrey A. Asher. I am very grateful to Mrs. Terri Brandt and Mrs. Pat Marx for the outstanding typing of the manuscript. My thanks are also due to Ms. Donna Graham for her ingenious editing, and Miss Rose Ann Coons who drew all of the illustrations in the book. Finally, the encouragement I received from my wife Shelly and my daughters Neha and Priya has been crucial to the preparation of this text. The extensive time I spent on the computer, particularly during nights and holidays, has been a great sacrifice on their part and I am extremely thankful for their support.

Clifton Park, New York Pradeep K. Gupta

Contents

CHAPTER 1

Introduction

The development of a comprehensive dynamic model which provides a real-time simulation of the performance of a rolling bearing is substantiated as the primary objective of this text. The salient features of the quasi-static and dynamic models for the various types of rolling bearings are discussed. Also, the key interactions in a rolling bearing are identified and the nomenclature used throughout the text is presented.

The machine elements that support a rotating shaft (or the rotor) against a stationary housing (or the stator) of a rotating machinery system are commonly known as bearings, which, simply by the nature of their function, have proven to be the most critical elements governing overall system performance. Of the many bearing types, rolling bearings have been popular because of their low power consumption, high stiffness, and high load support capabilities. While the high-load capabilities have been effective for high-speed turbine applications, the high-stiffness and low-friction characteristics of rolling bearings have promoted their use in superprecision, high-speed spindles and guidance systems. In general, rolling bearing applications cover a wide range of operating loads and speeds.

Recognition of rolling contact fatigue as the classical failure mode in rolling bearings has fostered the continued development of high-endurance materials and has motivated vast improvements in manufacturing processes; however, bearing fatigue life remains a critical parameter. For prescribed materials and operating loads, the fatigue life of a rolling bearing rapidly decreases as the rotating speed increases, because the number of load cycles per unit time is proportionately increased. Having an even greater impact on the fatigue life of a rolling bearing is the contact load, which increases with the square of the speed because of centrifugal effects. Likewise, hoop stresses in the bearing elements increase with the square of the speed, which imposes an increased fracture toughness requirement for the materials used. The speed limitation in rolling bearings is commonly indicated by a design parameter called DN (the product of the shaft diameter in millimeters and the rotating speed in revolutions per minute) which is essentially an indicator of the order of the surface velocity in rolling bearings. The present design limit is about 2×10^6 DN, although some advanced designs up to 3×10^6 DN have been produced.

In addition to fatigue, the dynamics of the various elements in a rolling bearing become particularly important at high speed. Both operational and geometric parameters, including the manufacturing tolerances on the various dimensions, affect the dynamic stability of the bearing elements. Hence, the bearing may fail because of unstable motion of the bearing elements rather than because of fatigue, and fatigue failure may not be a consideration in the design process. In fact, the actual bearing life may be several orders of magnitude less than the expected fatigue life. For such cases, dynamically stable motion of each bearing element will often constitute acceptable bearing performance. Thus, a realistic analysis of the interaction between the various bearing elements and the associated dynamics of rolling bearings are the objectives of this text.

A generalized dynamic analysis of bearing performance basically consists of a formulation and integration of the differential equations of motion of each bearing element. In other words, a real-time simulation of bearing performance is obtained by solving a time transient or a propagation problem. The analytical foundations for such an advanced model and a computer program, ADORE (Advanced Dynamics of Rolling Elements), for obtaining the real-time performance simulation of rolling bearings are presented in subsequent chapters in this text. A brief review of the various elements in a rolling bearing, the comparison of the static and dynamic models used for rolling bearing analysis, and a discussion of the nomenclature used in this text are discussed in the following paragraphs.

1.1 Rolling Bearing Elements and Basic Interactions

A rolling bearing basically consists of three fundamental parts: a number of rolling elements, the outer and inner races, and the cage. All of these basic parts of a rolling bearing are referred to as bearing elements throughout this text. Depending on the type of bearing, the rolling elements may be balls, cylindrical rollers, spherical rollers, tapered rollers, or spherical tapered rollers. The geometry of the races and cage varies with the type of bearing.

Figure 1-1 presents a schematic drawing of the commonly used rolling bearing types, although in special applications the actual bearing may be either a modification or a combination of these basic configurations. The analysis and computer code to be presented are intended for analyzing all of the basic bearing types; however, since the computer code allows for only one row of rolling elements, the treatment of spherical roller bearings is somewhat limited.

The interaction between the bearing elements constitutes the basic formulation of the dynamic model and real-time performance simulation. The

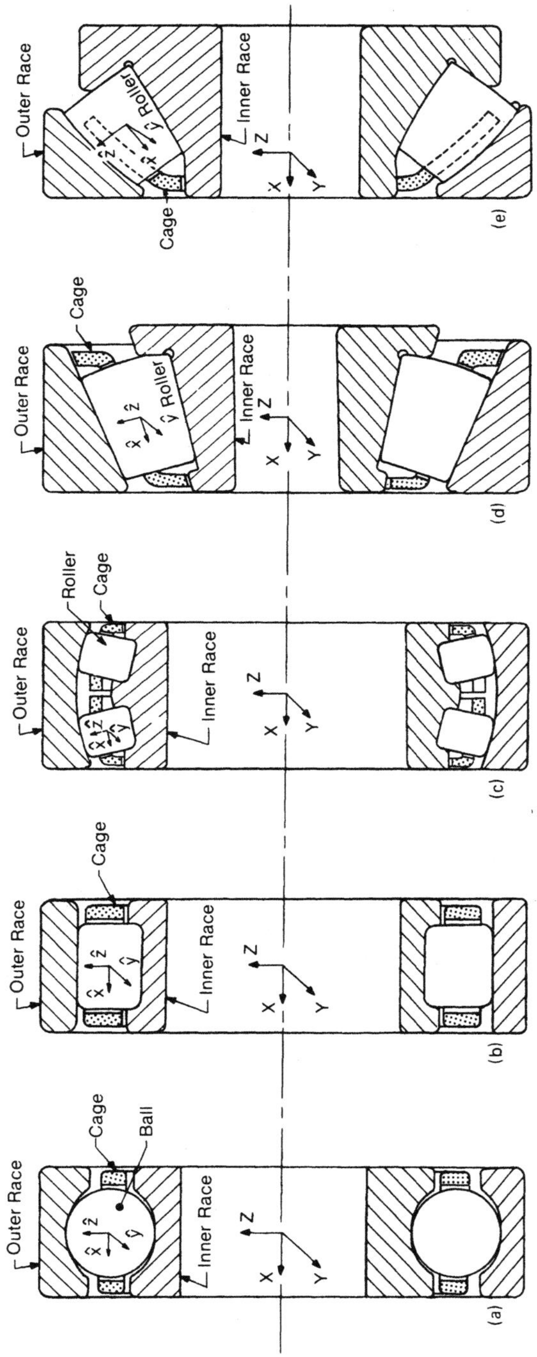

Figure 1-1 Schematic description of various types of rolling bearings. (a) Ball bearing; (b) cylindrical roller bearing; (c) spherical roller bearing; (d) tapered roller bearing; (e) spherical tapered roller bearing.

model is based on the following fundamental interactions:

- rolling element/race;
- rolling element/cage;
- cage/race;
- interaction between the rolling elements in the absence of cage.

Often, a number of secondary interactions must also be modeled for specific applications. Lubricant churning and drag effects, thermal and mechanical distortions due to bearing interaction with the external environment, and any geometric system constraints fall into this category.

1.2 Types of Analytical Models

There are two types of analytical formulations generally used for simulating rolling bearing performance. The static model is based on the static force and moment equilibrium equations, whereas the integration of the differential equations of motion constitutes the dynamic model. Since centrifugal forces and gyroscopic moments are very often included as applied forces in the equilibrium equations, the static models are generally referred to as quasi-static. Historically, both types of models have been used, depending on the performance parameters of interest.

1.2.1 Quasi-Static Model

For prescribed loads on the bearing, the force and moment equilibrium equations are written for each bearing element. These equations include the centrifugal forces and gyroscopic moments, together with the externally applied forces and moments. The resultant set of nonlinear algebraic equations is solved by a Newton–Raphson-type numerical method.

The rolling element angular velocities are computed from the prescribed angular velocities of the races, using certain kinematic hypotheses that constrain the orientation of the rolling element angular velocity vector. For a roller bearing, such a constraint is quite simple; it is assumed that pure rolling exists at both races and that the roller rotates about its axis of symmetry. In an angular contact ball bearing, however, the kinematics are somewhat more complicated due to the possible spin velocity (an angular velocity along an axis normal to the contact plane) of the balls. In this case, a "race control" hypothesis, which restricts the ball to spin either on the outer or the inner race, is commonly used. If no spin is allowed on the outer race, the hypothesis is called *outer race control*; if the ball is constrained from spinning on the inner race, the hypothesis is called *inner race control*. Whichever race gives the larger spin moment is the *controlling race*. In other words, the ball is allowed to spin on the race which gives the smaller spin moment for the known contact loads. At low speeds, lower spin torque on the outer race results in inner race control, whereas at high speeds, the greatly increased contact load at the outer race due to the centrifugal force on the ball results in outer race control.

Although such race control hypotheses provide the necessary condition for defining the ball angular velocity vector and solving the equilibrium equations, their practical validity is often questionable. For example, it has been shown experimentally that in well-lubricated ball bearings, the race control hypotheses are invalid and the ball spins on both races. An additional complication is that spin velocity is dependent on the lubricant behavior in the ball/race contact zone. The incorporation of realistic lubricant behavior in a quasi-static model is difficult due to the fact that most lubricants typically demonstrate a decrease in traction force with an increase in slip velocity, especially at high loads and high sliding speed. Such behavior results in a numerical convergence problem while solving the equilibrium equations. Hence, a realistic simulation of the lubricant behavior is one of the limitations of the quasi-static model.

When a cage is present in the bearing, frequent collisions of short duration between the rolling element and cage are imminent. In other words, the interaction is highly dynamic and the inertia of the cage and rolling element plays an important role. Also, since the orbital acceleration of the rolling element is dependent on the rolling-element-to-race traction force, lubrication effects are very important in modeling the rolling-element-to-cage contacts and the resulting motion of the cage. Compared to the bearing load supporting forces at the rolling-element-to-race contacts, these collision forces are quite small and may not greatly influence rolling element motion; however, since these dynamic forces are the only forces acting on the cage, they are significant in defining the cage motion. Thus, any treatment of cage behavior is outside the realm of the quasi-static models.

Similar to the cage effects, there are also dynamic effects associated with rolling elements that cannot be treated by equilibrium models. For example, roller skidding and skewing are highly dynamic phenomena, and the static equilibrium solutions may not provide any acceptable simulation of such behavior. Similarly, the influence of time-varying loads on the motion of both the rolling elements and the cage cannot be analyzed by the solution to the equilibrium problem alone.

In spite of the above limitations of the quasi-static models, they are still very useful in the design of rolling bearings since they provide a realistic load distribution on the rolling elements and well accepted predictions of fatigue life and bearing stiffness. Also, when doing a detailed dynamic simulation, the static equilibrium solutions very often provide favorable initial conditions for the integration of the differential equations of motion to determine the true steady-state behavior of the bearing.

The salient features of a quasi-static model may be briefly summarized as follows:

- The solution to a set of nonlinear algebraic equations of static equilibrium is obtained.

- Kinematic constraints are employed for computing angular velocities of rolling elements.
- Very effective for estimating the overall load distribution, bearing fatigue life, and stiffness.
- Since a single solution provides all of the performance parameters, the model requires only moderate computational effort.
- Convergence problems are experienced when attempting to simulate realistic lubricant traction behavior of a lubricant.
- None of the bearing elements accelerate and all velocities are assumed to be constant.
- No cage motion analysis may be considered.
- Treatment of rolling element instabilities such as roller skidding and skewing is very limited.
- No treatment for time-varying loads and race speeds.

1.2.2 Dynamic Model

In a dynamic model, the equilibrium equations of the quasi-static model are replaced by the differential equations of motion for each bearing element, and the model basically constitutes a propagation problem. The applied forces and moments are computed by the various interactions between the bearing elements, and the set of differential equations is integrated with the prescribed initial conditions. Due to the complexities in the computation of the applied forces, the integration is performed numerically. However, since all of the equaticns of motion are reducible to a set of first-order linear differential equations, conventional explicit or implicit algorithms are readily applicable. The integration is, therefore, quite straightforward, even when the number of equations is quite large. The initial conditions may be prescribed either from previously available solutions or from the quasi-static solution.

The dynamic model, in essence, provides a real-time simulation of bearing performance and it eliminates most of the problems associated with the equilibrium problem in the quasi-static model. For example, the angular velocities are now computed by integrating the corresponding accelerations, and no kinematic constraints of any nature are necessary. Also, unlike the quasi-static model, since no iterations are involved and every bearing element can accelerate arbitrarily, all lubrication effects can be treated in a very realistic fashion. Since all internal and external interactions may now vary arbitrarily with time, a realistic simulation of time-varying loads and speeds can be easily obtained. Similarly, the model allows for a complete treatment of all internal transient forces, such as those resulting from the rolling element/cage and race/cage interactions; therefore, all the instabilities associated with both the cage and the rolling elements can be simulated.

In addition to the simulation of all transient effects, the dynamic model provides an improved steady-state solution which may now be presented as a time-cycle solution rather than the one-point solution of the equilibrium problem. This provides a significant design strength to the model, where bearing performance must be free of all of the dynamic instabilities associated with the bearing elements. In summary, the general features of a dynamic model may be presented as follows:

- The model consists of the integration of the differential equations of motion of the bearing elements, thereby providing a real-time simulation of bearing performance.
- No kinematic constraints, such as race control, are now necessary.
- Any arbitrary lubricant behavior can be modeled since the dynamic model is free of the convergence problems associated with solving the equilibrium problem.
- The accelerations on the various elements are determined by the applied forces and moments, which are computed from the interactions between the bearing elements.
- All external interactions of the bearing may vary arbitrarily with time, and any variations in applied loads and speeds can be treated.
- A real-time simulation of all of the instabilities associated with either the rolling elements or the cage is obtained.
- Since the model constitutes a propagation (or a time transient) problem, it generally requires a substantially large computing effort in comparison to that required for the single solution of the equilibrium problem.

In spite of the increased computing effort, the dynamic model provides a significant reduction in the overall design costs. The real-time simulation capabilities of the dynamic model help replace some of the extensive experimental investigations. A parametric evaluation of the simulated bearing performance helps identify the range of critical design parameters over which experimental investigations are required for the design of advanced rolling bearing systems.

The design strengths of any analytical model have to be substantiated through experimental validation of the model's predictions, particularly when the model is highly complex. The comparison of the predictions of the computer code ADORE with the available experimental data, as presented later in the text, lends credence to its design strengths. It is expected that such validations of ADORE will continue as more experimental data becomes available, and that, ultimately, ADORE will be sufficiently strengthened so as to be widely accepted as a comprehensive analytical tool for the design and performance evaluation of rolling bearing systems.

1.3 Nomenclature

The analytical model presented in the text extensively uses a vector-matrix notation, and a large number of coordinate frames are employed. The general nomenclature is, therefore, divided into three parts: coordinate frames, vector transformations, and key symbols.

1.3.1 Coordinate Frames

Figure 1-1 shows the most fundamental coordinate frame (X, Y, Z) for the various bearing types. This coordinate frame, with its origin at the bearing center, is fixed in space, and it is referred to as the inertial coordinate frame. For the purpose of computing the quasi-static solution and the initial conditions for integrating the differential equations of motion, the applied loads on the bearing are prescribed in these coordinates. Thus the thrust load on a bearing is along the X axis, the radial load is along the Z axis, and the relative misalignment is indicated by a rotation about the Y axis.

The $(\hat{x}, \hat{y}, \hat{z})$ coordinate frame, shown in Figure 1-1, is fixed in the rolling element. This frame of reference is used for formulating the rotational motion of the rolling element. Similar to the $(\hat{x}, \hat{y}, \hat{z})$ system, body-fixed coordinate frames are also defined for the cage and the races. Further details of the various coordinate frames and the rules for vector transformation are discussed below.

1.3.2 Vector Transformations

The transformation of a vector from one coordinate frame to another is one of the most basic operations performed throughout this text. A vector is normally denoted by a boldface character and the transformation matrix from the coordinate frame u to v is written as $[T_{uv}]$. Additionally, all coordinate frames conform to the right-hand screw rule, and a superscript on a vector refers to the coordinate frame in which the components of the vector are represented. Thus, the transformation of a vector $\mathbf{r}$ from coordinate frame u to v is written as

$$\mathbf{r}^v = [T_{uv}]\mathbf{r}^u$$

Most of the transformation matrices used are orthogonal; hence, the transpose, denoted by $[T'_{uv}]$, is also the inverse. Symbolically,

$$[T_{vu}] = [T_{uv}^{-1}] = [T'_{uv}]$$

and

$$\mathbf{r}^u = [T'_{uv}]\mathbf{r}^v$$

In lieu of the generic symbols u and v used above, the following symbols denote some of the most commonly used coordinate frames:

a Azimuth frame with the x axis along the bearing axis, the z axis pointing radially outward, and the y axis determined by the right-hand screw rule.

b Coordinate frame fixed in the rolling element (ball or roller) along the principal axes of inertia, the x axis being the polar axis in the case of a roller. See frame $(\hat{x}, \hat{y}, \hat{z})$ in Figures 1-1 and 2-1.

c Coordinate frame fixed in the cage along its principal axes of inertia with the polar axis being along the x axis.

f Coordinate frame fixed in the race flange in the case of roller bearings, as defined in Figure 3-11.

i Inertial (space fixed) coordinate frame, as defined by the (X, Y, Z) system in Figures 1-1 and 2-1.

$\mathcal{c}$ A rolling element/race contact coordinate frame $(\bar{x}, \bar{y}, \bar{z})$ with $\bar{x}$ and $\bar{y}$ axes in the plane of contact ($\bar{x}$ along the major, and $\bar{y}$ along the minor axis of the contact ellipse in the case of elliptical point contact), and the $\bar{z}$ axis normal to the contact plane in accordance with the right-hand screw rule. See Figure 3-4.

p Pocket coordinate frame, fixed in the cage pocket with the x, y, z axes similar to those for the azimuth frame.

r Coordinate frame fixed in the race along its principal axis of inertia with the x axis oriented along the polar axis.

Normally, any two of the above characters will appear as subscripts to the matrix $[T]$ to represent the two coordinate frames between which the transformation applies. For example, $[T_{ib}]$ will denote a transformation from the inertial to the rolling element coordinate frame. For a vector, any one of the above characters may appear as a superscript and it defines the coordinate frame in which the vector is represented. Thus, the components of the vector $\mathbf{r}^f$ represent the x, y, and z components in the race flange coordinate frame.

1.3.3 List of Symbols

Although most of the symbols are defined in their respective place of use, the following are some of the most commonly used symbols in the text:

a	Semimajor axis of contact ellipse
α	Coefficient of linear thermal expansion
$\mathcal{A}$	Area
A, B, C, D	Parameters in traction slip equation
A_j	Polynomial coefficients for predictor
b	Semiminor axis of contact ellipse
B_j	Polynomial coefficients for predictor
$\mathcal{B}$	Coefficient matrix for explicit integration formula
C_D	Drag coefficient
d	Rolling element diameter
$\mathbf{e}$	Unit vector
E	Modulus of elasticity
$\mathcal{E}$	Elliptic integral

f	Race curvature factor defined as the ratio of the radius of curvature of race groove to nominal rolling element diameter
$\mathscr{f}$	Friction force at any point within the contact zone
$\hat{\mathscr{f}}$	Friction factor
$\mathscr{f}$	Integrated friction force at a contact
$\mathbf{F}$	Applied force vector
$g, \bar{g}$	Residuals in a nonlinear algebraic equation
G	Elasticity parameter
$\mathscr{G}, \mathscr{G}^*$	Coefficients of an explicit integration formula
$\mathbf{G}$	Applied moment vector
$\mathbf{i}, \mathbf{j}, \mathbf{k}$	Unit vectors along a coordinate frame
h	Lubricant film thickness
$\mathbf{h}$	Angular momentum vector
I_1, I_2, I_3	Principal moments of inertia
K	Thermal conductivity
ℓ	Contact length of a line contact
L	Roller length
$\mathscr{L}$	Length scale
m	Mass
M_c	Churning moment
p	Contact pressure at any point in the contact zone
p_H	Maximum Hertzian contact pressure
q	Contact load per unit length
Q	Contact load
$\mathscr{Q}$	Load scale
Q_m	Thermal parameter
$\mathbf{r}, \mathbf{R}$	Position vector
$\mathscr{r}$	Roller radius at any point along roller length
R	Equivalent radius for two interacting elements
$\bar{R}$	Radius of curvature of the deformed race and ball surfaces
$\mathscr{R}$	Radius of race groove center locus for ball bearing or radius of the raceway land (cylindrical surface) in the center for roller bearings
$\hat{R}$	Radius of cage at the guiding land
Re	Reynolds number
S	Cage half-width
$\mathscr{S}$	Race half-width for roller bearings
t	Time
T	Absolute temperature
$\mathscr{T}$	Time scale
Ta	Taylor's Number
$\mathbf{u}$	Surface velocity
U	Rolling velocity
$\bar{U}$	Speed parameter
$\mathbf{v}$	Mass center velocity
W	Dimensionless hydrodynamic load
$\bar{W}$	Load parameter

w	Thickness of cage (difference of outer and inner radius)
x, r, θ	Cylindrical coordinate frame
x, y, z	Coordinate frame
X, Y, Z	Coordinate frame
$\bar{x}, \bar{y}$	Dimensionless coordinates in the contact ellipse
$\mathbf{X}$	Generalized position vector
$\mathscr{X}$	Axial relative displacement of the races
$\mathbf{Y}$	Generalized derivative vector
$\mathscr{Y}$	Relative radial displacement of the races
α_1	Outer race contact angle
α_2	Inner race contact angle
α	Pressure–viscosity coefficient
β	Temperature–viscosity coefficient
δ	Contact deflection
Δ_{bp}	Cage pocket radial clearance
Δ_n	Time step size at the nth step
ε	Limit on the local truncation error per unit step size
η, ξ, λ	Transformation angles
$\bar{\xi}, \bar{\zeta}$	Dimensionless coordinates
κ	Traction coefficient
μ	Absolute viscosity
ν	Kinematic viscosity
ϕ	Angular position
ϕ_T	Thermal reduction factor
ϕ_S	Starvation factor
ρ	Density
$\bar{\rho}$	Curvature
σ	Poisson's ratio
τ	Shear stress
ψ	Azimuth angle
Θ	Relative race misalignment
Υ	Elasticity constant in point-contact load-deflection relation
$\boldsymbol{\omega}, \boldsymbol{\Omega}$	Angular velocity vectors

1.4 Summary

The objective of the text is stated as the development of a comprehensive dynamic model for the performance of a rolling bearing. The model will essentially consist of the formulation and integration of the differential equations of motion of each bearing element. Various types of rolling bearings are introduced and the primary interactions in a rolling bearing are identified. A brief comparison between the conventional quasi-static model and the fully dynamic model, developed in this text, is presented and the capabilities of these two general type of models is discussed. The final section in this chapter discusses the nomenclature, with particular emphasis on the various coordinate frames and vector transformations, used throughout the text.

CHAPTER 2

Equations of Motion and Coordinate Transformations

The coordinate frames, the relevant transformations, and the differential equations of motion for simulating the general motion of the bearing elements in a six-degree-of-freedom system are defined, and the various steps involved in the simulation process are schematically outlined.

The general steps in a real-time dynamic simulation of a system basically consist of defining the various coordinate frames, performing the relevant transformations, formulating the differential equations of motion, computing the applied forces and moments, and integrating the differential equations of motion with certain prescribed initial conditions. The objective of this chapter is to define the coordinate frames and transformations and formulate the equations of motion for each element in a rolling bearing.

2.1 Coordinate Frames and Transformations

The general motion of any bearing element can be completely described in a six-degree-of-freedom system. The three coordinates will locate the mass center of the element in an inertial coordinate frame, and the angular orientation will be completely defined by three angles. If a bearing element is now considered as a rigid body, then two sets of coordinate frames, as shown in Figure 2-1, must be defined:

1. An inertial coordinate frame (X, Y, Z), which is fixed in space.
2. A body-fixed coordinate frame $(\hat{x}, \hat{y}, \hat{z})$, which is fixed in the bearing element.

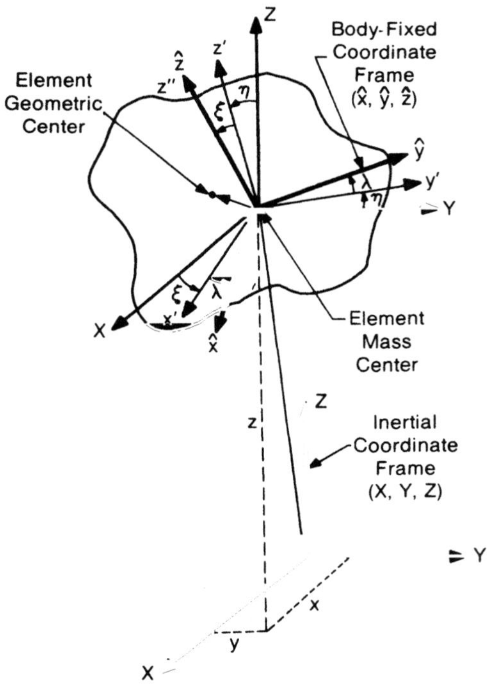

Figure 2-1 Generalized coordinate frames and degrees of freedom for any bearing element.

Generally, the body-fixed coordinate frame will be coincident with the principal axes of the element, and the inertial frame will be a convenient orthogonal set of coordinates with the origin at the bearing center. The geometry of the element can be described relative to the geometric center, which, in turn, can be located with respect to the mass center in the body-fixed coordinate frame.

The transformation from the inertial to the body-fixed frame can be accomplished by the three successive rotations shown in Figure 2-1:

$$\text{(i) } \eta\mathbf{I}; \qquad \text{(ii) } \xi\mathbf{j}' \qquad \text{(iii) } \lambda\hat{\mathbf{k}} \tag{2.1}$$

where $\mathbf{I}, \mathbf{J}, \mathbf{K}, \hat{\mathbf{i}}, \hat{\mathbf{j}}, \hat{\mathbf{k}}$, etc., are unit vectors along the corresponding X, Y, Z and $\hat{x}$, $\hat{y}$, $\hat{z}$ axes.

Using the superscript i for the inertial frame and b for the body-fixed frame, the vector transformation from the inertial to the body-fixed frame is written as

$$\mathbf{r}^b = [T_{ib}]\mathbf{r}^i \tag{2.2}$$

where $[T_{ib}]$ is the relevant transformation matrix. In terms of the three rotations in equation (2.1), the transformation matrix can be shown to be given by

$$[T_{ib}] = [T_{ib}(\eta, \xi, \lambda)] =$$

$$\begin{bmatrix} \cos\xi\cos\lambda & \begin{matrix}\cos\eta\sin\lambda \\ +\sin\eta\sin\xi\cos\lambda\end{matrix} & \begin{matrix}\sin\eta\sin\lambda \\ -\cos\eta\sin\xi\cos\lambda\end{matrix} \\ -\cos\xi\sin\lambda & \begin{matrix}\cos\eta\cos\lambda \\ -\sin\eta\sin\xi\sin\lambda\end{matrix} & \begin{matrix}\sin\eta\cos\lambda \\ +\cos\eta\sin\xi\sin\lambda\end{matrix} \\ \sin\xi & -\sin\eta\cos\xi & \cos\eta\cos\xi \end{bmatrix} \tag{2.3}$$

The angles (η, ξ, λ) are very similar to the classical Euler angles [88] and those used by Walters [94]. Also, the transformation matrix can readily be shown as orthogonal; thus, the inverse is the same as the transpose.

Similarly, it will be necessary to establish a relationship between the angular velocity vector $\boldsymbol{\omega}^b$ in the body-fixed frame and the time derivatives of the transformation angles (η, ξ, λ) because the angles will be used as fundamental variables later. Using simple geometry, an incremental rotation of the body $\delta\mathbf{n}$ can be written in terms of the three rotations described in (2.1) as

$$\delta\mathbf{n} = \Delta\eta\mathbf{I} + \Delta\xi\mathbf{j}' + \Delta\lambda\hat{\mathbf{k}} \tag{2.4}$$

where

$$\mathbf{j}' = \sin\lambda\,\hat{\mathbf{i}} + \cos\lambda\,\hat{\mathbf{j}} \tag{2.5}$$

and

$$\begin{aligned} I &= \cos\xi\,\mathbf{i}' + \sin\xi\,\hat{\mathbf{k}} \\ &= \cos\xi\,(\cos\lambda\,\hat{\mathbf{i}} - \sin\lambda\,\hat{\mathbf{j}}) + \sin\xi\,\hat{\mathbf{k}} \end{aligned} \tag{2.6}$$

The required transformation will be obtained by combining equations (2.4) through (2.6):

$$\begin{aligned} \delta\mathbf{n} = {} & (\Delta\eta\cos\xi\cos\lambda + \Delta\xi\sin\lambda)\hat{\mathbf{i}} + (-\Delta\eta\cos\xi\sin\lambda + \Delta\xi\cos\lambda)\hat{\mathbf{j}} \\ & + (\Delta\eta\sin\xi + \Delta\lambda)\hat{\mathbf{k}} \end{aligned} \tag{2.7}$$

or, in terms of time derivatives,

$$\boldsymbol{\omega}^b = [\mathscr{C}]\begin{Bmatrix}\dot{\eta} \\ \dot{\xi} \\ \dot{\lambda}\end{Bmatrix} \tag{2.8}$$

where the transformation matrix $[\mathscr{C}]$ is given by

$$[\mathscr{C}] = \begin{bmatrix} \cos\xi\cos\lambda & \sin\lambda & 0 \\ -\cos\xi\sin\lambda & \cos\lambda & 0 \\ \sin\xi & 0 & 1 \end{bmatrix} \tag{2.9}$$

The matrix $[\mathscr{C}]$ is not orthogonal; however, simple algebraic manipulation will show that the inverse is given by

$$[\mathscr{C}^{-1}] = \begin{bmatrix} \dfrac{\cos\lambda}{\cos\xi} & -\dfrac{\sin\lambda}{\cos\xi} & 0 \\ \sin\lambda & \cos\lambda & 0 \\ -\tan\xi\cos\lambda & \tan\xi\sin\lambda & 1 \end{bmatrix} \tag{2.10}$$

and the inverse of equation (2.8) is written as

$$\begin{Bmatrix} \dot{\eta} \\ \dot{\xi} \\ \dot{\lambda} \end{Bmatrix} = [\mathscr{C}^{-1}]\boldsymbol{\omega}^b \tag{2.11}$$

The acceleration components are obtained by straightforward differentiation of equation (2.11):

$$\begin{Bmatrix} \ddot{\eta} \\ \ddot{\xi} \\ \ddot{\lambda} \end{Bmatrix} = \frac{\partial[\mathscr{C}^{-1}]}{\partial t}\boldsymbol{\omega}^b + [\mathscr{C}^{-1}]\dot{\boldsymbol{\omega}}^b \tag{2.12}$$

where the derivative of $[\mathscr{C}^{-1}]$ can be obtained by differentiating equation (2.10) as

$$\frac{\partial[\mathscr{C}^{-1}]}{\partial t} = \begin{bmatrix} \begin{matrix} -\dot{\lambda}\sin\lambda\sec\xi \\ +\dot{\xi}\cos\lambda\sec\xi\tan\xi \end{matrix} & \begin{matrix} -\dot{\lambda}\cos\lambda\sec\xi \\ -\dot{\xi}\sin\lambda\sec\xi\tan\xi \end{matrix} & 0 \\ \dot{\lambda}\cos\lambda & -\dot{\lambda}\sin\lambda & 0 \\ \begin{matrix} \dot{\lambda}\sin\lambda\tan\xi \\ -\dot{\xi}\cos\lambda\sec^2\xi \end{matrix} & \begin{matrix} \dot{\lambda}\cos\lambda\tan\xi \\ +\dot{\xi}\sin\lambda\sec^2\xi \end{matrix} & 0 \end{bmatrix} \tag{2.13}$$

Since the angular velocities and acceleration may now be transformed into the time derivatives of the three angles (η, ξ, λ) defining the angular orientation of the body in space, all of the fundamental quantities associated with any element can be described by the following vector:

$$\mathbf{x} = \{x, r, \theta, \dot{x}, \dot{r}, \dot{\theta}, \eta, \xi, \lambda, \dot{\eta}, \dot{\xi}, \dot{\lambda}\}^T \tag{2.14}$$

where the components (x, r, θ), $(\dot{x}, \dot{r}, \dot{\theta})$, (η, ξ, λ), and $(\dot{\eta}, \dot{\xi}, \dot{\lambda})$ denote the mass center position, mass center velocity, angular position, and angular velocity, respectively. The vector $\mathbf{x}$ will hereafter be called the generalized solution vector.

Equation (2.14) can be further generalized to include all of the elements in the bearing. Thus, for n elements, the general solution vector is written as

$$\mathbf{X} = \begin{matrix} \{x_1, r_1, \theta_1, \dot{x}_1, \dot{r}_1, \dot{\theta}_1, \eta_1, \xi_1, \lambda_1, \dot{\eta}_1, \dot{\xi}_1, \dot{\lambda}_1, \\ \dots\dots\dots\dots\dots\dots\dots\dots\dots\dots, \\ x_n, r_n, \theta_n, \dot{x}_n, \dot{r}_n, \dot{\theta}_n, \eta_n, \xi_n, \lambda_n, \dot{\eta}_n, \dot{\xi}_n, \dot{\lambda}_n\}^T \end{matrix} \tag{2.15}$$

It will be shown later that the generalized equations of motion of any element can be expressed in terms of the above 12 fundamental quantities.

2.2 Equations of Motion

As described in the foregoing in terms of the coordinates and degrees of freedom, the general motion of any bearing element may be considered in two parts: the motion of the element's mass center, and the rotational motion of the element. The mass center motion will be described by Newton's law of motion, and the classical Euler equations of motion [88] will be the basis for the angular or rotational motion.

2.2.1 Mass Center Motion

The translatory motion of the mass center of any element in the bearing, described by Newton's law of motion, is

$$m\ddot{\mathbf{r}} = \mathbf{F} \tag{2.16}$$

where m is the element mass, $\ddot{\mathbf{r}}$ is the acceleration vector, and $\mathbf{F}$ is the applied force vector.
Equation (2.15) can be written in component form in either Cartesian or cylindrical coordinates as

$$\begin{aligned} m\ddot{x} &= F_x \\ m\ddot{y} &= F_y \\ m\ddot{z} &= F_z \end{aligned} \tag{2.17}$$

and

$$\begin{aligned} m\ddot{y} &= F_x \\ m\ddot{r} - mr\dot{\theta}^2 &= F_r \\ mr\ddot{\theta} + 2m\dot{r}\dot{\theta} &= F_\theta \end{aligned} \tag{2.18}$$

Normally, it is convenient to consider the mass center motion of a rolling element in a cylindrical frame, while a Cartesian system may be preferred for the races and cage of the bearing.

For the prescribed position and velocities, vector $\mathbf{x}$ of equation (2.14), the applied forces are computed from the various geometric interactions and the relevant lubrication considerations. Details of all such interactions will be dealt with in subsequent chapters where it will be shown that the geometry of the bearing elements and the applied operating conditions are significant inputs to the computation of the applied forces between the interacting elements.

2.2.2 Rotational Motion

If the bearing element is considered to rotate about a fixed point 0 (see Figure 2-2) with an angular velocity, $\boldsymbol{\omega}$, and if the triad, formed by the unit vectors $\mathbf{i}$, $\mathbf{j}$, $\mathbf{k}$ is assumed to rotate with an angular velocity $\boldsymbol{\Omega}$ then the

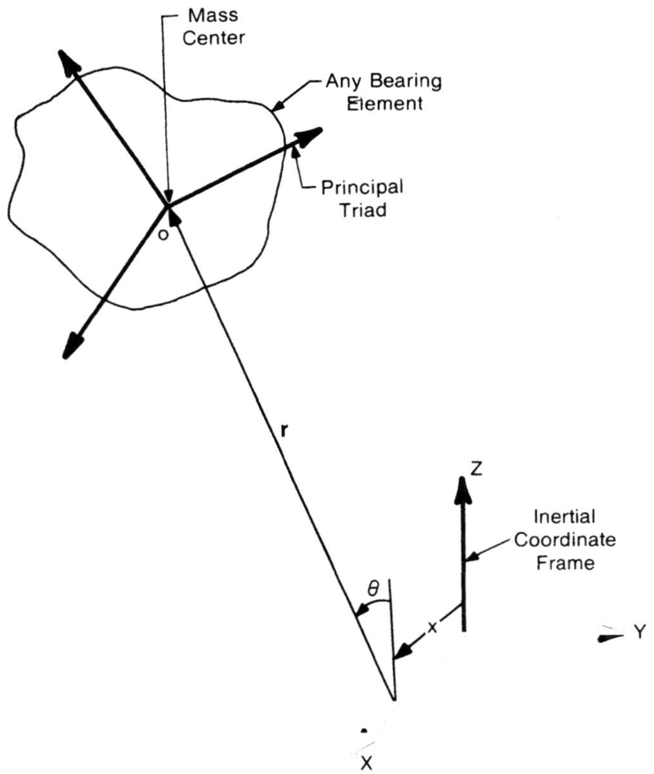

Figure 2-2 Principal triad and coordinates for rotational motion.

classical differential equation of motion [88] is written as

$$\dot{\mathbf{h}} = \frac{\delta h}{\delta t} + \boldsymbol{\Omega} \times \mathbf{h} = \mathbf{G} \tag{2.19}$$

where **h** is the angular momentum, **G** is the applied moment vector and

$$\dot{\mathbf{h}} = \frac{\delta \mathbf{h}}{\delta t} = I_1 \frac{d\omega_1}{dt}\mathbf{i} + I_2 \frac{d\omega_2}{dt}\mathbf{j} + I_3 \frac{d\omega_3}{dt}\mathbf{k} \tag{2.20}$$

where I_1, I_2, and I_3 are the principal moments of inertia of the bearing element.

Equations (2.17) and (2.18) can be combined to give the following component equations:

$$\begin{aligned} I_1\dot{\omega}_1 - I_2\omega_2\Omega_3 + I_3\omega_3\Omega_2 &= G_1 \\ I_2\dot{\omega}_2 - I_3\omega_3\Omega_1 + I_1\omega_1\Omega_3 &= G_2 \\ I_3\dot{\omega}_3 - I_1\omega_1\Omega_2 + I_2\omega_2\Omega_1 &= G_3 \end{aligned} \tag{2.21}$$

where the components G_1, G_2, G_3 are along the unit vectors **i**, **j**, **k**.

If the triad formed by the vectors **i, j, k** is assumed to be fixed in the body, the angular velocity vector $\mathbf{\Omega}$ equals $\boldsymbol{\omega}$, and equation (2.19) reduces to the classical Euler equations of motion:

$$\begin{aligned} I_1\dot{\omega}_1 - (I_2 - I_3)\omega_2\omega_3 &= G_1 \\ I_2\dot{\omega}_2 - (I_3 - I_1)\omega_3\omega_1 &= G_2 \\ I_3\dot{\omega}_3 - (I_1 - I_2)\omega_1\omega_2 &= G_3 \end{aligned} \tag{2.22}$$

It should be noted that the triad **i, j, k** is now fixed in the bearing element and it represents the principal triad.

Equation (2.22) applies to an arbitrary geometry of the bearing elements, allowing both balls and rollers to be treated with complete generality. In the case of perfectly spherical balls, however, an additional simplification can be made, because $I_1 = I_2 = I_3$ and any orthogonal coordinate system will represent the principal triad. Thus, if the ball has an orbital velocity $\dot{\theta}$, the rotational motion of the ball can be considered in a coordinate frame, which is neither fixed in space nor fixed in the ball, and it has an angular velocity $\mathbf{\Omega} = \{\dot{\theta}, 0, 0\}^T$. The x and z axes of such a coordinate frame will be parallel to the x and r axes, respectively, of the inertial frame, as shown in Figure 2-2. The y axis is then easily located by the right-hand screw rule. Using these simplifications, the equations for rotational motion of a spherical ball can be written as

$$\begin{aligned} I\dot{\omega}_1 &= G_1 \\ I\dot{\omega}_2 - I\omega_3\dot{\theta} &= G_2 \\ I\dot{\omega}_3 + I\omega_2\dot{\theta} &= G_3 \end{aligned} \tag{2.23}$$

where the applied moment vector **G** will now be in the coordinate frame defined above. For the prescribed positions and velocities of the interacting elements, the applied moment vector **G** is determined by the geometrical interaction and lubrication conditions to be discussed in subsequent chapters.

2.3 Moving Coordinate Frames

In most rolling bearing applications one of the races is fixed in space while the other race moves both in terms of rotation and translation. Thus the translatory motion can be considered in an inertial reference frame and the rotational motion for any bearing element can be treated in a coordinate frame fixed in the bearing element. In some advanced applications, however, the bearing as a whole moves in space and the coordinate frame, normally fixed in the stationary race, is no longer fixed in space. The translatory motion of all bearing elements now has to be considered in a moving reference frame.

Let the coordinate frame fixed in the bearing, normally regarded as the inertial frame, have an acceleration $\mathbf{a}_o$, angular velocity $\mathbf{\Omega}$, and an angular acceleration $\dot{\mathbf{\Omega}}$. Also, let the position of the mass center of a bearing element relative to the bearing coordinate frame be $\mathbf{r}$, the position of the origin of the bearing coordinate frame relative to a point fixed in space be $\mathbf{r}_o$, and the applied force on the bearing element be $\mathbf{F}$. The acceleration on the bearing element in the coordinate frame fixed in the bearing can now be written [88] as

$$m\mathbf{a} = \mathbf{F} - m\mathbf{a}_o - m\mathbf{a}_t - m\mathbf{a}_c \tag{2.24}$$

where m is the mass of the bearing element and the accelerations $\mathbf{a}_t$ and $\mathbf{a}_c$, generally called the acceleration of transport and the acceleration of Coriolis respectively, are given by

$$\mathbf{a}_t = \dot{\mathbf{\Omega}} \times \mathbf{R} + \mathbf{\Omega}(\mathbf{\Omega} \cdot \mathbf{R}) - \mathbf{R}\Omega^2 \tag{2.25}$$

and

$$\mathbf{a}_c = 2\mathbf{\Omega} \times \frac{\partial \mathbf{r}}{\partial t} \tag{2.26}$$

where

$$\mathbf{R} = \mathbf{r} + \mathbf{r}_o$$

Thus if the motion of the bearing coordinate is prescribed the accelerations of the various bearing elements can be determined in the coordinate frame fixed in the bearing.

2.4 General Motion Simulation

The formulation of the equations of motion, when combined with the analyses presented in subsequent chapters, becomes the basis for simulating the general motion of the bearing elements, and it provides the foundation for ADORE. The basic steps involved in obtaining the motion of the bearing elements are shown schematically in Figure 2-3. From the prescribed initial conditions, the general solution vector X is first set up for starting the integrating process. From the $(\dot{\eta}, \dot{\xi}, \dot{\lambda})$ components in the generalized solution vector at any instant of time, the angular velocity vector $\boldsymbol{\omega}$ of the various elements is computed using equation (2.8).

All position and velocity vectors, along with the prescribed constraints and geometry of the various elements, become inputs to the analyses for the computation of applied forces and moments. After transforming the applied force and moment vectors to the appropriate coordinate frames, equations (2.18) and (2.22) are used to obtain the mass center and angular acceleration vectors. It will be necessary to transform the angular accelerations into the second time derivatives of the angles (η, ξ, λ), using equation (2.12). The various accelerations, when assembled with the already known velocities, will result in the generalized derivative vector $\mathbf{Y}$, which may be written in a form similar to equation (2.15):

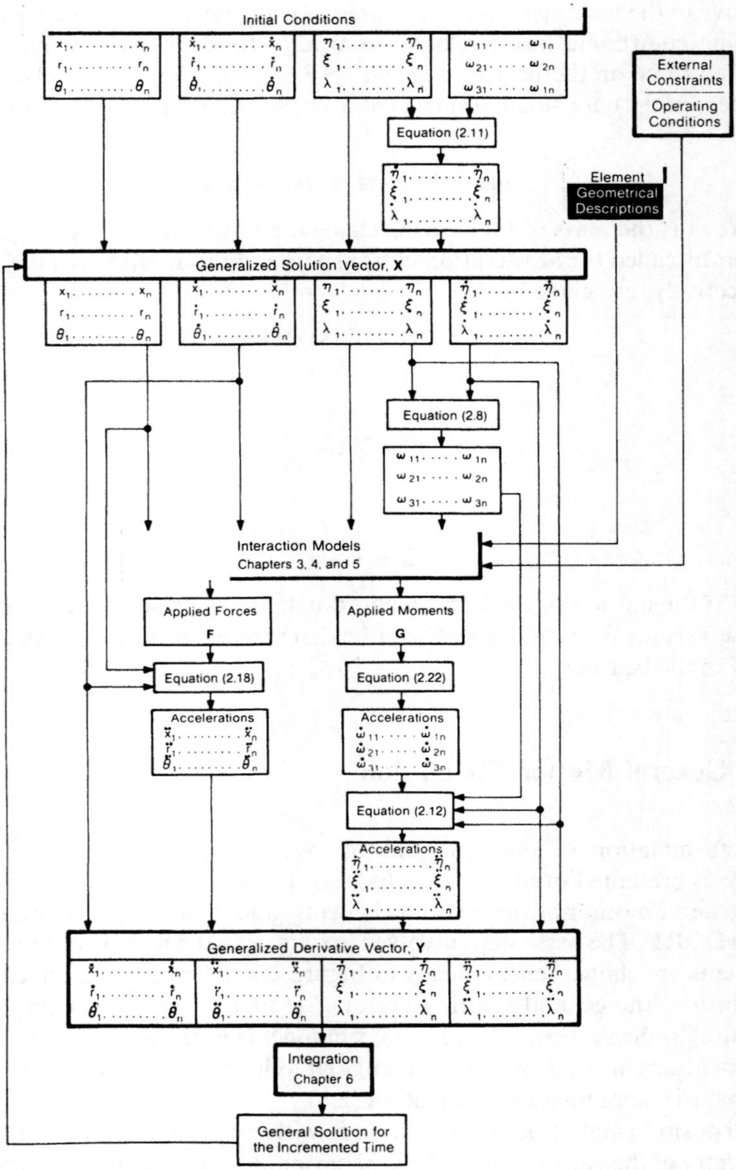

Figure 2-3 Schematic layout for the general motion simulation of bearing elements.

$$\mathbf{Y} = \begin{array}{l}\{\dot{x}_1, \dot{r}_1, \dot{\theta}_1, \ddot{x}_1, \ddot{r}_1, \ddot{\theta}_1, \dot{\eta}_1, \dot{\xi}_1, \dot{\lambda}_1, \ddot{\eta}_1, \ddot{\xi}_1, \ddot{\lambda}_1, \\ \dots\dots\dots\dots\dots\dots\dots\dots\dots\dots\dots\dots\dots\dots\dots\dots, \\ \dot{x}_n, \dot{r}_n, \dot{\theta}_n, \ddot{x}_n, \ddot{r}_n, \ddot{\theta}_n, \dot{\eta}_n, \dot{\xi}_n, \dot{\lambda}_n, \ddot{\eta}_n, \ddot{\xi}_n, \ddot{\lambda}_n\}^T\end{array} \tag{2.27}$$

By comparing the above equation with equation (2.15), it will be clear that

$$\mathbf{Y} = \dot{\mathbf{X}} \tag{2.28}$$

The generalized derivative vector can now be integrated, using one of the integration algorithms presented in Chapter 6, and solutions at the incremented time may be obtained. The entire process is repeated until the desired final time is reached.

2.5 Summary

The inertial and body-fixed coordinate frames are defined to formulate the general motion of the bearing elements in a six-degree-of-freedom system, three translational coordinates for the mass center and three rotations to define the angular orientation of the element with respect to its mass center. Based on the rules of transformation, an orthogonal transformation matrix is developed to establish the interrelationship between any two coordinate frames. The translational motion of the element's mass center is considered in an inertial coordinate frame, and the rotational motion of the element about its mass center is formulated in a body-fixed coordinate frame in order to preserve all six degrees of freedom in the general motion. Appropriate equations of motion are presented and the generalized solution and derivative vectors are formulated. Finally, an overview of the various steps involved in the simulation process are presented schematically.

CHAPTER 3

Geometric Interactions in Rolling Bearings

Models for simulating the interaction between the various elements in a rolling bearing are considered in terms of the contact geometry, the geometric interaction and the resulting normal and traction forces on the interacting elements. The generalized applied force and moment vectors as encountered in the differential equations of motion of the bearing elements are formulated.

The determination of the applied force and moment vectors is the most fundamental input to the generalized equations of motion formulated in the preceding chapter. These vectors, in turn, depend on the interaction between the various bearing elements and the interaction of the bearing with the environment. The development of a model for each such interaction is, therefore, the key to simulating the performance of the rolling bearing as a whole.

As discussed briefly in the first chapter, a rolling bearing consists of four basic elements: the rolling elements (balls, rollers, etc.), the outer race, the inner race, and the cage (see Figure 3-1). When the rolling elements are separated by a cage they cannot interact with each other. In the absence of a cage, however, interaction between the rolling elements occurs. The following interactions will, therefore, be fundamental to any rolling bearing configuration:

1. rolling element/race interactions;
2. rolling element/cage interactions;
3. cage/race interactions;
4. interactions between rolling elements;
5. external system interactions and constraints.

In generic terms an interaction model is basically a force–deformation relation, as shown schematically in Figure 3-2. The inputs to the model are the prescribed position and velocity vectors, and the outputs are the force and moment vectors acting on the interacting elements. Consider two inter-

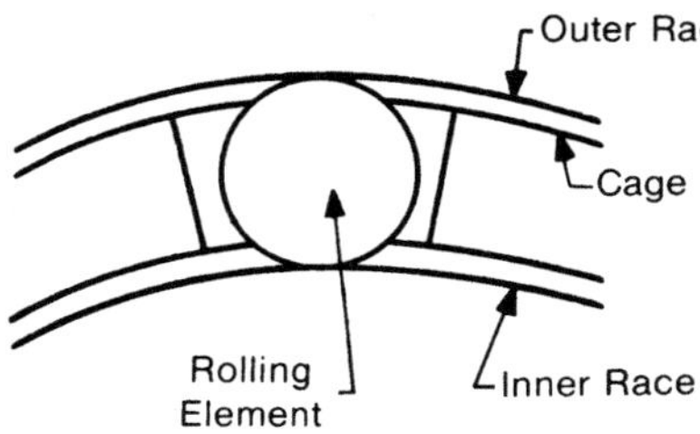

Figure 3-1 Elements of a rolling bearing.

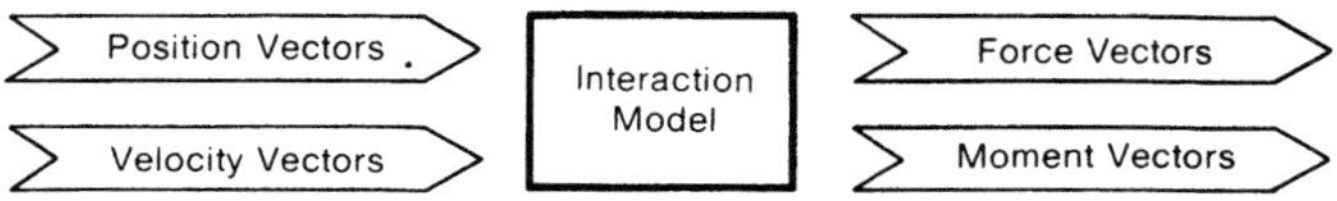

Figure 3-2 The input/output of an interaction model.

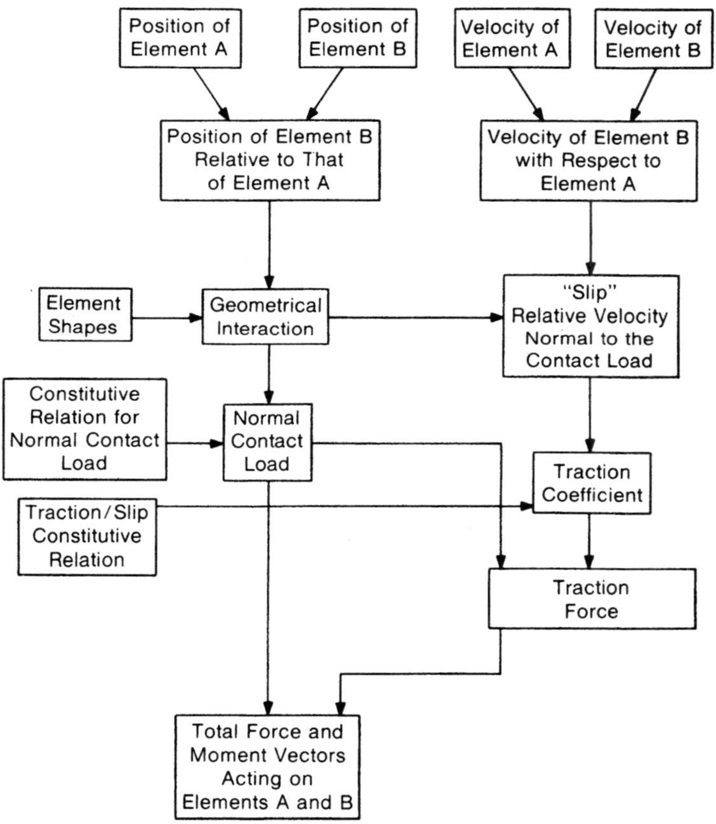

Figure 3-3 A generic description of the interaction model for two interacting bearing elements.

acting elements, A and B, with prescribed positions and velocities. As shown schematically in Figure 3-3, the position vectors help to locate one of the elements relative to the other. This relative position, with the knowledge of the geometry of the elements, gives the geometric interaction between the elements. A constitutive equation is then used to compute the normal contact load. Similarly, the absolute velocities of the two elements are subtracted to obtain the relative velocity. For the already determined geometric interaction and contact geometry, the relative velocity can be transformed into "slip," which is defined as the relative sliding velocity normal to the contact load. A traction–slip constitutive equation or model is now used to determine the traction coefficient, which is similar to a friction coefficient in that it is the ratio of the traction (or friction) force to the normal contact load. The traction coefficient, when multiplied by the normal load, gives the traction force. Both the normal and traction forces can now be added vectorially to determine the net force vector acting between the interacting elements. The moment vector acting on the element is determined by the cross product of the vector that locates the point of interaction with respect to the element's mass center and the net force vector.

3.1 Rolling Element/Race Interactions

Since the geometry of the bearing plays an important role in formulating the interaction between various elements, the considerations in ball and roller bearings will be discussed separately.

3.1.1 Ball/Race Interactions

The generic approach outlined above is applied to the ball/race interaction in an angular-contact bearing to determine the geometry of contact, the normal contact load, and the tractive forces and moments.

Geometric Considerations. The geometry of ball/race interaction is shown schematically in Figure 3-4. For simplicity, the diagram is drawn in the x–z plane, and the y axis is normal to the plane of the diagram. The position vector $\mathbf{r}_r$ locates the mass center of the race R_M relative to the inertial frame (X, Y, Z). Vector $\mathbf{r}_{rg}$ locates the race geometric center R_G. Similarly, the mass center of the ball B_M is located by $\mathbf{r}_b$ relative to the inertial frame, and $\mathbf{r}_{bg}$ locates the ball geometric center relative to the mass center.* The vectors $\mathbf{r}_{rg}$ and $\mathbf{r}_{bg}$ will generally be prescribed in the race-fixed frame (x, y, z) and

*For most practical problems the mass and geometric centers may be coincident, which simplifies the geometry to some extent. The purpose of defining the geometric center differently from the mass center is to preserve the generality and capability of simulating advanced configurations.

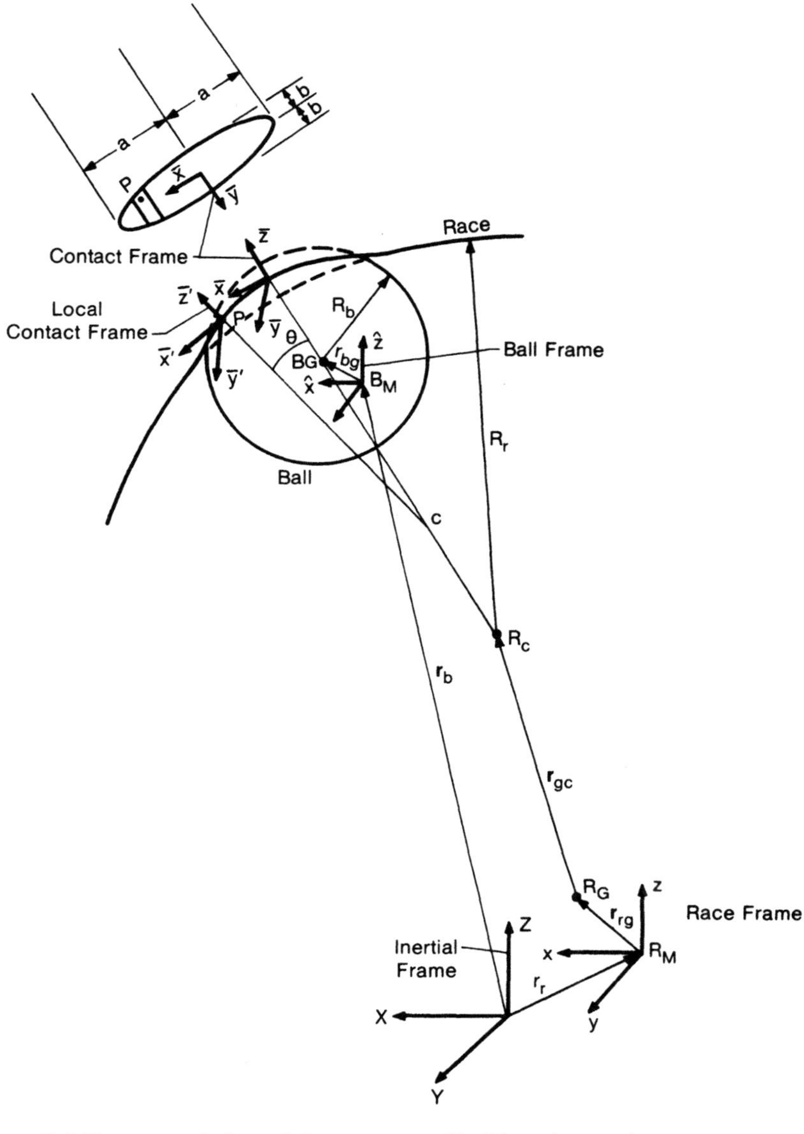

Figure 3-4 Exaggerated view of the geometry of ball/race interaction.

the ball-fixed frame $(\hat{x}, \hat{y}, \hat{z})$, respectively. Using the notations discussed in the preceding chapter let the relevant transformations be denoted as

$$\text{Transformation from inertial to race frame:} \quad [T_{ir}(\eta_r, \xi_r, \lambda_r)]$$

$$\text{Transformation from inertial to ball frame:} \quad [T_{ib}(\eta_b, \xi_b, \lambda_b)]$$

An azimuth frame (x_a, y_a, z_a) is defined such that z_a is parallel to the radial component of $\mathbf{r}_b$, x_a is parallel to the inertial X axis, and y_a is determined by the right-hand screw rule. Since the diagram in Figure 3-4 is drawn in the x–z plane, this coordinate frame is not shown in the figure. However, the transformation, which will depend only on the ball azimuth angle ψ, may be defined as

Transformation from inertial to ball azimuth frame: $[T_{ia}(\psi, 0, 0)]$

Now the interaction between the ball and the race can be determined by locating the geometric center of the ball B_G relative to the race curvature center R_c. In terms of the various position vectors shown in Figure 3-4 this relative position vector $\mathbf{r}_{br}$ will be given by the equation

$$\mathbf{r}_{br} = \mathbf{r}_b + [T'_{ib}]\mathbf{r}_{bg} - (\mathbf{r}_r + [T'_{ir}](\mathbf{r}_{rg} + \mathbf{r}_{gc})) \tag{3.1}$$

Everything on the right-hand side of the above equation is known except for the vector $\mathbf{r}_{gc}$, for which it will be necessary to determine a race azimuth angle ϕ, measured in the race-fixed coordinate frame, such that B_G, R_c, and R_G lie in one plane. The vector $\mathbf{r}_{brg}$, locating B_G relative to R_G, is written as

$$\mathbf{r}_{brg} = \mathbf{r}_b + [T'_{ib}]\mathbf{r}_{bg} - \{\mathbf{r}_r + [T'_{ir}]\mathbf{r}_{rg}\} \tag{3.2}$$

When $\mathbf{r}_{brg}$ is in the inertial frame, an angle ϕ' may be defined such that it denotes the angle between the inertial Z axis and the plane containing vector $\mathbf{r}_{brg}$, and a line parallel to the inertial X passing through R_G. Symbolically,

$$\phi' = \arctan\left(\frac{-r^i_{brg_2}}{r^i_{brg_3}}\right) \tag{3.3a}$$

If the race is assumed to be symmetric about the X axis, then the race-fixed coordinate frame may not be necessary and $\phi = \phi'$. However, if the generalities mentioned above are preserved, then ϕ is determined in the race-fixed frame such that $\mathbf{r}_{gc}$ and $\mathbf{r}_{brg}$ lie in the same plane. Now, if $\mathscr{R}$ is the radius of the circle representing the locus of the race groove centers, then

$$\mathbf{r}^r_{gc} = \begin{Bmatrix} 0 \\ -\mathscr{R}\sin\phi \\ \mathscr{R}\cos\phi \end{Bmatrix}$$

In order that $\mathbf{r}_{gc}$ and $\mathbf{r}_{brg}$ are coplanar, the y component of the vector

$$[T(\phi', 0, 0)][T'_{ir}]\begin{Bmatrix} 0 \\ \mathscr{R}\sin\phi \\ \mathscr{R}\cos\phi \end{Bmatrix}$$

must vanish. Therefore,

$$\phi = \arctan\left(\frac{T_{23}}{T_{22}}\right) \tag{3.3b}$$

where T_{22} and T_{23} are the relevant components of the matrix obtained by the product $[T(\phi', 0, 0)][T'_{ir}]$.

The computation of the ball azimuth ψ is quite straightforward once the orbital position θ, locating the ball mass center, is known from prescribed $\mathbf{r}_b \equiv (x, r, \theta)$ in the cylindrical coordinates.

$$\psi = \phi - \arctan\left(\frac{-r_2}{r_3}\right) \tag{3.4}$$

where r_2 and r_3 are the y and z components of the vector $[T'_{ib}]\mathbf{r}^b_{bg}$.

Once ϕ is known, $\mathbf{r}_{br}$ may be determined in the inertial frame using equation (3.1); and when ψ is known, it may be referenced in the ball azimuth frame as

$$\mathbf{r}^a_{br} = [T(\psi, 0, 0)]\mathbf{r}^i_{br} \tag{3.5}$$

Clearly, if there is any interaction between the ball and race, the contact load will be acting along $\mathbf{r}_{br}$ and, therefore, it will be convenient to define a unit vector in the ball azimuth frame as

$$\mathbf{e}^a_{br} = \frac{\mathbf{r}^a_{br}}{|\mathbf{r}_{br}|}$$

The contact angles at the ball/race contact, as shown in Figure 3-5, are defined in terms of the components of $\mathbf{e}^a_{br}$.

$$\alpha_1 = \arctan\left(\frac{e^a_{br_1}}{e^a_{br_3}}\right)$$

and

$$\alpha_2 = \arctan\left(\frac{-e^a_{br_2}}{\sqrt{(e^a_{br_1})^2 + (e^a_{br_3})^2}}\right) \tag{3.7}$$

The transformation from the azimuth frame (x_a, y_a, z_a) to the contact frame $(\bar{x}, \bar{y}, \bar{z})$ is obtained in terms of the following two sequential rotations:

(i) $\alpha_1 \mathbf{J}$, (ii) $\alpha_2 \mathbf{i}$

where $\mathbf{J}$ and $\mathbf{i}$ are unit vectors along the y_a and $\bar{x}$ axes, respectively.
After carrying out some straightforward algebra, the relevant transformation matrix can be shown to be

$$[T_{ac}] = \begin{bmatrix} \cos\alpha_1 & 0 & -\sin\alpha_1 \\ \sin\alpha_1 \sin\alpha_2 & \cos\alpha_2 & \cos\alpha_1 \sin\alpha_2 \\ \sin\alpha_1 \cos\alpha_2 & -\sin\alpha_2 & \cos\alpha_1 \cos\alpha_2 \end{bmatrix} \tag{3.8}$$

Normal Contact Load. Once the ball geometric center is located relative to the race curvature center by equation (3.5), the geometric interaction, or contact deflection, δ, may be written as

$$\delta = |\mathbf{r}_{br}| - (f - 0.5)d \tag{3.9}$$

where d is the ball diameter and f is the race curvature factor, which is defined as the ratio of the radius of curvature of race groove to the ball diameter.

A negative δ will indicate no contact between the ball and race. When a

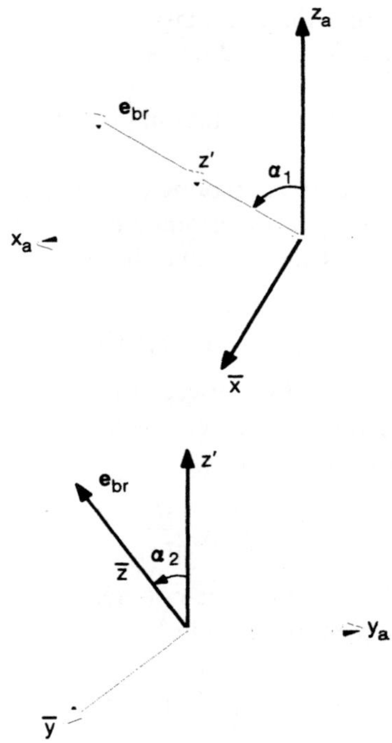

Figure 3-5 Definition of ball/race contact angles.

positive interaction exists, the contact stress, load, and contact geometry can be computed from the classical Hertzian point-contact solutions summarized in Appendix I.

Tractive Forces. The computation of tractive forces in the ball/race contact zone is generally quite complicated because of the large variations in local slip, and hence in the traction coefficient, from point to point in the contact ellipse. Since for most bearings the contact ellipse is quite narrow ($a \gg b$ in Figure 3-4), the variation in slip along the $\bar{y}$ axis may be neglected. Thus, the contact ellipse is divided in several elementary strips, as shown in Figure 3-4, and the slip velocity over the elementary strip is determined by that at the point lying on the major axis of the ellipse. An attractive feature of such a simplification is that the integration required for the computation of the total tractive forces and moments is necessary only along the $\bar{x}$ axis. It is generally agreed that for oil-lubricated elastohydrodynamic conditions (to be discussed in Chapter 4) such an assumption is quite realistic, especially in view of the assumptions in the traction/slip models. However, for a dry contact or for solid-lubricated conditions the point-to-point variation in the traction force must be considered and a generalized two-dimensional integration must be performed. In either case, for the computation of local slip, it will be necessary to determine the position vector locating any point P (see

Figure 3-4) within the contact ellipse relative to the ball and race mass centers. If the point P is denoted by normalized coordinates ($x = \bar{x}/a$, $y = \bar{y}/b$), then it is located with respect to the ball geometric center by the position vector

$$\mathbf{r}_p^c = \left\{ \begin{matrix} xa \\ yb \\ \sqrt{\bar{R}^2 - (xa)^2} - \sqrt{\bar{R}^2 - a^2} + \sqrt{\frac{1}{4}d^2 - a^2} \end{matrix} \right\} \tag{3.10}$$

where $\bar{R}$ is the radius of curvature (the center of curvature being point C in Figure 3-4) of the deformed pressure surface, defined by

$$\bar{R} = \frac{2fd}{2f + 1} \tag{3.11}$$

f is the race curvature factor, and d is the ball diameter. The curvature of the deformed surface in the y direction has been assumed to be negligible. This will be the case when $a \gg b$.

Let the race have a translational velocity $\mathbf{v}_r$ and an angular velocity $\boldsymbol{\omega}_r$. Likewise, let the ball have a translational velocity $\mathbf{v}_b$ and an angular velocity $\boldsymbol{\omega}_b$. Generally, the angular velocities are specified in the body-fixed frames while the translational velocities are known in the inertial frame. The ball velocity can be conveniently defined in the cylindrical coordinate frame in terms of components $\dot{x}$, $\dot{r}$ and $\dot{\theta}$. Then, the local slip velocity $\mathbf{u}_s$ of the race relative to the ball at point P is given in the contact frame in terms of the race and ball velocities, $\mathbf{u}_r$ and $\mathbf{u}_b$, respectively, as

$$\mathbf{u}_r^c = [T_{ac}][T_{ia}]\left[\mathbf{v}_r^i + \left([T'_{ir}]\boldsymbol{\omega}_r^r - \left\{ \begin{matrix} \dot{\theta} \\ 0 \\ 0 \end{matrix} \right\}\right) \times \left(\{\mathbf{R}_p^i + [T'_{ir}]\mathbf{r}_{rg}^r\}\right)\right] \tag{3.12}$$

$$\mathbf{u}_b^c = [T_{ac}]\left[\left\{ \begin{matrix} \dot{x} \\ 0 \\ \dot{r} \end{matrix} \right\} + [T_{ia}]([T'_{ib}]\boldsymbol{\omega}_b^b \times \{[T'_{ia}][T'_{ac}]\mathbf{r}_p^c + [T'_{ib}]\mathbf{r}_{bg}^b\})\right] \tag{3.13}$$

$$\mathbf{u}_s^c = \mathbf{u}_r^c - \mathbf{u}_b^c \tag{3.14}$$

where the vector $\mathbf{R}_p$ is computed (see Figure 3-4) as

$$\mathbf{R}_p = \mathbf{r}_p + \mathbf{r}_{bg} + \mathbf{r}_b - \mathbf{r}_r \tag{3.15}$$

When the contact half-width a spreads over a considerable angle across the ball surface (see Figure 3-4) transformation of the slip velocity in the local contact frame is necessary in order to determine the slip tangential to the contact surface. This transformation is defined in terms of the angle θ (see Figure 3-4) as

$$\mathbf{u}_s^l = [T_{cl}(0, \theta, 0)]\mathbf{u}_s^c \tag{3.16}$$

The components u_{s_1} and u_{s_2} will now determine the local slip normal to and along the rolling direction, respectively. The corresponding traction coefficients, defined as the ratio of shear to normal surface stress, can be determined from the prescribed traction/slip constitutive relation or by a "traction model." The details of such models are discussed in Chapter 4; for the present let the traction coefficients be denoted as

$$\begin{aligned} \kappa_1 &= \kappa_1(u_{s_1}, u_{s_2}) \\ \kappa_2 &= \kappa_2(u_{s_1}, u_{s_2}) \end{aligned} \tag{3.17}$$

Since the normal contact pressure is defined in the contact frame it will be convenient to define a traction coefficient vector $\boldsymbol{\kappa}$ corresponding to equation (3.17). Such a vector may then be transformed in the contact frame by the equation

$$\boldsymbol{\kappa}^c = [T'(0, \theta, 0)] \begin{Bmatrix} \kappa_1 \\ \kappa_2 \\ 0 \end{Bmatrix} \tag{3.18}$$

where θ is angle between the nominal and local contact frames as shown in Figure 3-4.

Net Forces and Moments. The total forces and moments on the interacting ball and race may be determined by first defining the incremental forces and moments at any point P in the contact zone (see Figure 3-4) and then integrating over the contact ellipse. Since the normal contact pressure is known and the traction coefficient given by equation (3.18), the incremental normal and tractive forces at the contact point P may be written as

$$d\mathbf{q}^c = -p_h ab\sqrt{1 - \bar{x}^2 - \bar{y}^2} \begin{Bmatrix} 0 \\ 0 \\ 1 \end{Bmatrix} d\bar{y}\, d\bar{x}$$

$$d\boldsymbol{f}^c = p_h ab\sqrt{1 - \bar{x}^2 - \bar{y}^2}\, \boldsymbol{\kappa}^c\, d\bar{y}\, d\bar{x}$$

or

$$d\mathbf{F}^c = d\mathbf{q}^c + d\boldsymbol{f}^c = p_h ab\sqrt{1 - \bar{x}^2 - \bar{y}^2} \left(\boldsymbol{\kappa}^c - \begin{Bmatrix} 0 \\ 0 \\ 1 \end{Bmatrix} \right) d\bar{y}\, d\bar{x} \tag{3.19}$$

where p_h is the Hertz contact pressure determined from the Hertz point-contact elasticity theory outlined in Appendix I.

Using equations (3.10) and (3.19) and knowing the relative positions of the ball and the race, the moment acting about the mass center of the ball and the race due to the incremental force $d\mathbf{F}$ can be written in terms of the following vector equations:

$$\begin{aligned} d\mathbf{G}_b^c &= (\mathbf{r}_p^c + \mathbf{r}_{bg}^c) \times d\mathbf{F}^c \\ d\mathbf{G}_r^c &= -(\mathbf{R}_p^c + \mathbf{r}_{rg}^c) \times d\mathbf{F}^c \end{aligned} \tag{3.20}$$

where $\mathbf{r}_p$ and $\mathbf{R}_p$ are given by equations (3.10) and (3.15) respectively; $\mathbf{r}_{bg}$ and $\mathbf{r}_{rg}$ are the position vectors locating the ball and race geometric centers relative to their respective mass centers; and the incremental vectors $d\mathbf{G}_b$ and $d\mathbf{G}_r$ denote the moments acting on the ball and race respectively. Since the moments in the body-fixed frame and the forces in the inertial frame will normally be required, a corresponding transformation of equation (3.19) and (3.20) will be necessary.

Once the incremental forces and moments are determined by equations (3.19) and (3.20), an integration over the contact ellipse will determine the total forces and moments acting on the ball and race. Depending on the degree of approximation acceptable, two cases may be considered.

Case 1: No Variation in Traction Along the $\bar{y}$ Axis. In this case, the traction coefficient is assumed to be constant along the $\bar{y}$ axis, and it is therefore only determined for $\bar{y} = 0$ in equation (3.10). With this simplification, the incremental force in equation (3.19) can be written as

$$\begin{aligned} \iint d\mathbf{F}^c &= \int_{-1}^{1} p_h ab \left(\boldsymbol{\kappa}^c - \begin{Bmatrix} 0 \\ 0 \\ 1 \end{Bmatrix} \right) 2 \int_0^{(1-\bar{x}^2)1/2} \sqrt{1 - \bar{x}^2 - \bar{y}^2}\, d\bar{y}\, d\bar{x} \\ &= \frac{\pi}{2} p_h ab \int_{-1}^{1} (1 - \bar{x}^2) \left(\boldsymbol{\kappa}^c - \begin{Bmatrix} 0 \\ 0 \\ 1 \end{Bmatrix} \right) d\bar{x} \end{aligned} \tag{3.21}$$

Equation (3.21) can be easily integrated numerically using the conventional Gaussian Quadrature formulae. Treatment of discontinuities in components of $\boldsymbol{\kappa}$ will require some care. When due to curvature of the ball/race interaction surface there are one or two points of pure rolling, the slip component changes sign; Figure 3-6 shows a typical variation. Because of this change in sign there is a corresponding discontinuity in $\boldsymbol{\kappa}$ and a subdivision of the domain $-1 \leq \bar{x} \leq 1$ into the two or three regions separated by the discontinuities becomes convenient. Numerical integration is then performed separately in each region and the results are summed. The location of the points of discontinuity are found by numerically solving the nonlinear algebraic equation

$$u_{s_2}^c = 0 \tag{3.22}$$

where $u_{s_2}^c$ is defined by equation (3.14).

Case 2: Two-Dimensional Variation in Traction. In this case, equations (3.19) and (3.20) are integrate numerically in two dimensions. In terms of the scaled variable (x, y), the elliptical contact zone is transformed into a unit circle, as shown in Figure 3-7. Furthermore, if the polar coordinates (r, θ) are used, the incremental force $d\mathscr{f}$ from equation (3.19) may be written as

$$d\mathscr{f}^c = \sqrt{1 - r^2}\mathscr{f}\ (r\cos\theta, r\sin\theta)\, ds\, dr \tag{3.23}$$

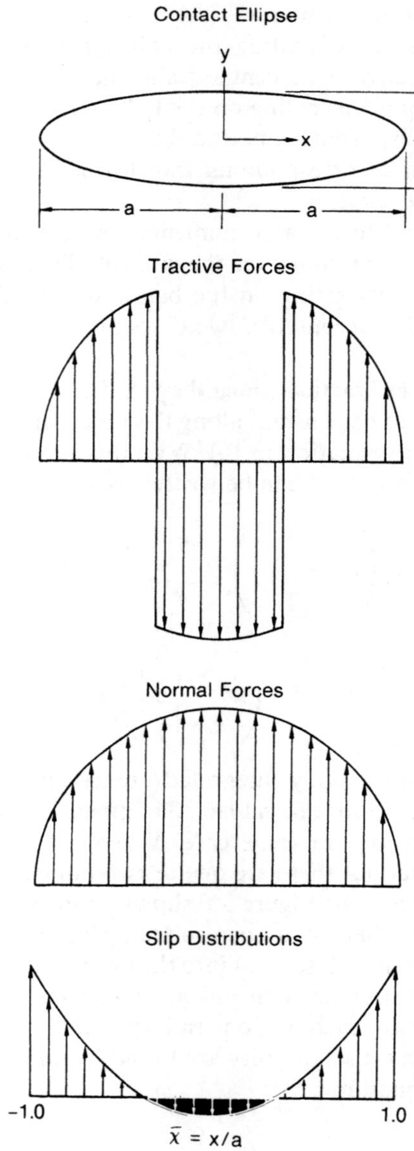

Figure 3-6 Typical variation of ball/race slip velocity in the rolling direction.

where $ds = rd\theta$ and

$$f^c = p_h ab \left[\boldsymbol{\kappa} - \begin{Bmatrix} 0 \\ 0 \\ 1 \end{Bmatrix} \right] .$$

The term $\sqrt{1 - r^2}$, resulting from the Hertzian elasticity solution, cannot be

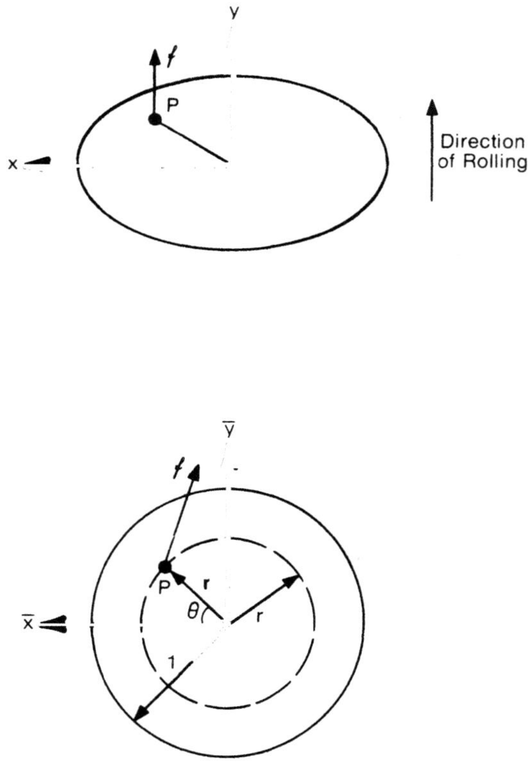

Figure 3-7 Elliptical contact zone transformed into a unit circle.

approximated by a polynomial, especially near the boundary, as $r \to 1$. Therefore, the conventional Gaussian Quadrature formulae [1] for a two-dimensional integration cannot be used economically. However, a modification in terms of the Chebyshev polynomial is possible.

Consider the double integral of equation (3.23) over the unit contact circle:

$$\iint d\mathscr{f}^c = \mathscr{f}^c = \int_0^1 \sqrt{1-r^2} \int_0^{2\pi r} \mathscr{f}(r\cos\theta, r\sin\theta)\, ds\, dr \tag{3.24}$$

The inner integral in (3.24) can be replaced by a summation and $\mathscr{f}^c$ may be written as

$$\mathscr{f}^c = \int \sqrt{1-r^2}\, \mathscr{f}_r\, dr \tag{3.25}$$

where

$$\mathscr{f}_r = \frac{\pi r}{m} \sum_{i=1}^{2m} \mathscr{f}\left(r\cos\frac{\pi i}{m}, r\sin\frac{\pi i}{m}\right) \tag{3.26}$$

and m is the number of equally spaced points on the circle of radius r and f is a functional notation of the resulting function of r.

Since f can be approximated by a polynomial, equation (3.25) represents a formula similar to the classical Chebyshev integration formula except that the limits of integration are 0 to 1 as compared to the conventional limits of -1 to 1.

By examining the numerical accuracy resulting from varying orders of polynomial representation in equation (3.25) it is found that a fifth-order approximation is adequate for the present problem. Also, m is arbitrarily selected as 72. This will give a point on the circle at every 5° and, hence, the variation of slip along the circle can be treated with adequate accuracy. Thus, the final integration algorithm, written in terms of Chebyshev weights and the roots of the polynomial, is

$$f^c = \sum_{j=1}^{n} w_j \left[\frac{\pi r_j}{m} \sum_{i=1}^{2m} f\left(r_j \cos\frac{\pi i}{m}, r \sin\frac{\pi i}{m} \right) \right] \tag{3.27}$$

where m and n are selected to be 72 and 5, respectively, and the roots r_j and weights w_j are computed to be as follows:

j	r_j	w_j
	0.0445594672	0.112314749
2	0.218694082	0.220987601
3	0.474310986	0.239278553
4	0.735889118	0.161581619
5	0.929067427	0.0512356414

Note that f in equation (3.27) is the local traction vector, which is related to the local slip vector via the traction slip relation.

3.1.2 Roller/Race Interactions

Similar to the ball/race contact the interaction between a roller and race can be simulated in terms of the geometry of contact and the resulting force and moment vectors as defined by appropriate constitutive relations.

Geometrical Considerations. Figure 3-8 schematically describes an exaggerated view of the geometric interaction between the roller and a race. For simplicity, the diagram is drawn in the x–z plane and the y axis is normal to the plane of the diagram. Also, both the race and the roller are assumed to be conical in order to include tapered roller bearings. For cylindrical roller bearings the cone angle may be set to zero. The position vector $\mathbf{r}_r$ locates the mass center of the race R_M relative to the inertial frame (X, Y, Z). Vector $\mathbf{r}_{rg}$ locates the race geometric center R_G relative to the mass center, and vector $\mathbf{r}_{gc}$ locates the center of race land R_l relative to the race geometric center R_G. Similarly, the mass center of the roller B_M is located by $\mathbf{r}_b$ relative to the inertial frame and $\mathbf{r}_{bg}$ locates the roller geometric center relative to the mass

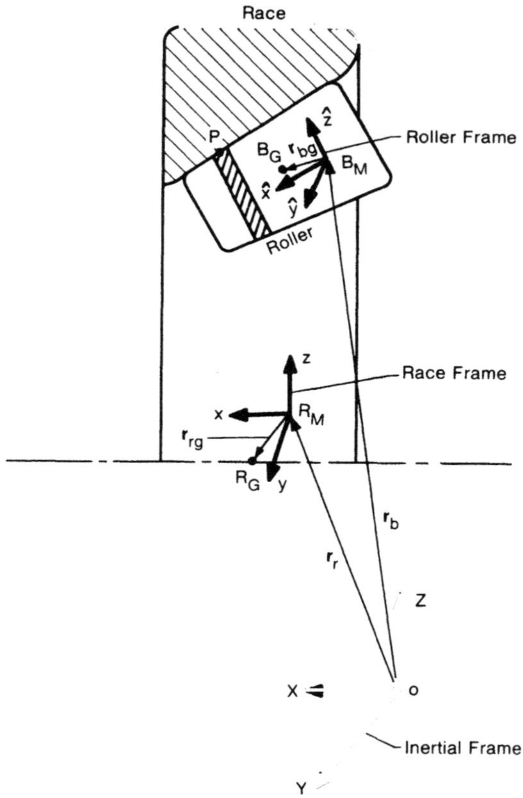

Figure 3-8 Exaggerated view of roller/race interaction.

center. The vectors $\mathbf{r}_{rg}$ and $\mathbf{r}_{bg}$, respectively, will generally be prescribed in the race-fixed frame (x, y, z) and the roller-fixed frame $(\hat{x}, \hat{y}, \hat{z})$. Also, the relevant transformations will be

$$\text{Transformation from inertial to race frame:} \quad [T_{ir}(\eta_r, \xi_r, \lambda_r)]$$

$$\text{Transformation from inertial to roller frame:} \quad [T_{ib}(\eta_b, \xi_b, \lambda_b)]$$

Since the radius of the roller can vary arbitrary along the roller axis, the roller/race interaction is somewhat more complicated than the ball/race contact. The general procedure will be to divide the roller into several elementary disks, as shown in Figure 3-8, and compute the interaction of the race with each disk independently. These local interactions can then be integrated to obtain the total roller/race interaction.

The interaction at point P (see Figure 3-4) on the roller surface can be determined by first locating the roller geometric center B_G relative to the race geometric center R_G and then locating the center of the elementary disk, containing the point P, with respect to the race geometric center. The vector

$\mathbf{r}_{br}$, locating the roller center with respect to the race center, can be simply written in terms of the various position vectors defined above as

$$\mathbf{r}_{br}^{r} = [T_{ir}]\{\mathbf{r}_{b}^{i} + [T_{ib}']\mathbf{r}_{bg}^{b} - \mathbf{r}_{r}^{i}\} - \mathbf{r}_{rg}^{r} \tag{3.28}$$

Now let the axial position of the elementary disk relative to the roller center be x in the roller frame. The center of the elementary disk can then be located relative to the race center by the vector $\mathbf{r}_{bl}$ as

$$\mathbf{r}_{bl}^{r} = \mathbf{r}_{br}^{r} + [T_{br}]\begin{Bmatrix} x \\ 0 \\ 0 \end{Bmatrix} \tag{3.29}$$

An azimuth angle ψ, locating the disk center relative to the race center, can now be defined as

$$\tan\psi = -\frac{r_{bl_2}^{r}}{r_{bl_3}^{r}} = -\frac{r_{br_2}^{r} + T_{br_{21}}x}{r_{br_3}^{r} + T_{br_{31}}x} \tag{3.30}$$

and the transformation from the race to azimuth frame may be written as

$$[T_{ra}] = [T_{ra}(\psi, 0, 0)]$$

The position of the point P relative to the race center can now be written as

$$\mathbf{r}_{b\ell p}^{a} = [T_{ra}]\left\{\mathbf{r}_{br}^{r} + [T_{br}]\begin{Bmatrix} x \\ -\imath\sin\phi \\ \imath\cos\phi \end{Bmatrix}\right\} \tag{3.31}$$

where the radius of the roller at the axial position x is $\imath$ and ϕ is the angle about the roller axis, locating the angular position of point P on the roller surface. In general $\imath$ can be a function of both x and ϕ.

The necessary condition for determining ϕ is written as

$$r_{b\ell p_2}^{a} = r_{br_2}^{a} + T_{21}x - T_{22}\imath\sin\phi + T_{23}\imath\cos\phi = 0 \tag{3.32}$$

where

$$[T] = [T_{ra}][T_{br}]$$

If $\imath$ is a function of ϕ, then the above equation must be solved numerically. In the case of axially symmetric roller ($\imath$ independent of ϕ), however, it can be easily shown that

$$\phi = \tan^{-1}\left(\frac{T_{23}}{T_{22}}\right) + \sin^{-1}\left(\frac{r_{br_2}^{a} + T_{21}x}{\imath\sqrt{T_{22}^2 + T_{23}^2}}\right) \tag{3.33}$$

Once the angle ϕ is known, the geometric interaction δ for the elementary disk is written as

$$\delta = (r_{b\ell p_3}^{a} - \mathscr{R})\cos\tilde{\alpha} \tag{3.34}$$

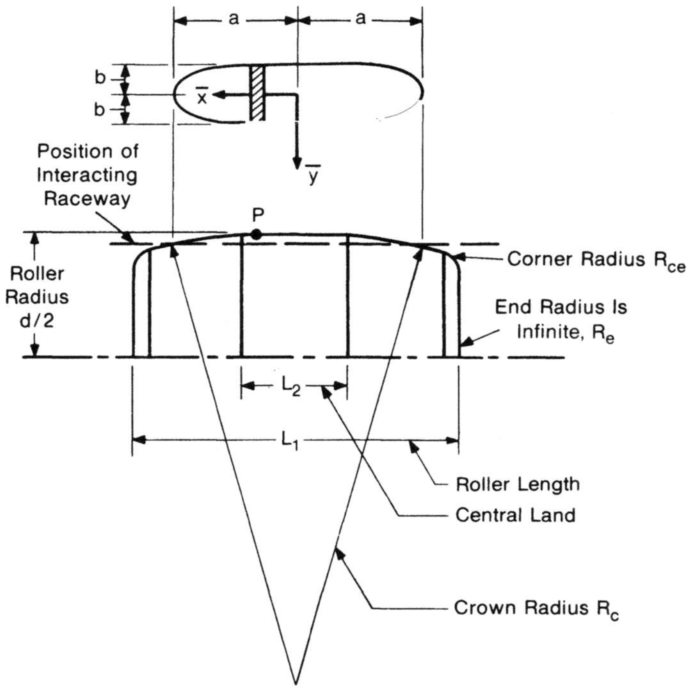

Figure 3-9 Roller/race contact geometry.

where $\tilde{\alpha}$ is the contact angle or the semicone angle of race and $\mathscr{R}$ is the radius of the race surface at the point P. In general the radius $\mathscr{R}$ may also vary arbitrarily and it can be written as

$$\mathscr{R} = \mathscr{R}(r^a_{b\ell p_1}, \psi) \tag{3.35}$$

A negative value of δ will indicate no contact, while a positive value will signify a contact. The unit vector along the geometric interaction, or the contact load, is given by

$$\mathbf{e}^r = [T_{ra}][T_{ac'}]\begin{Bmatrix} 0 \\ 0 \\ 1 \end{Bmatrix} \tag{3.36}$$

where the transformation from the azimuth to contact frame is simply written as

$$[T_{ac}] = [T_{ac}(0, \tilde{\alpha}, 0)]$$

For cylindrical roller bearings, the typical geometry of a roller is shown in Figure 3-9. The roller has a central land where the radius is constant and a crown radius which essentially reduces the roller radius external to the central flat zone; this reduction is often referred to as crown drop. Clearly the effective radius at any point P on the roller surface can be easily defined. Similarly, more complicated geometries such as out-of-round rollers, tapered rollers, etc., can be treated in a straightforward manner once the radius of the roller surface is defined.

In order to determine the total roller/race interaction it will be necessary to integrate the local interaction over the contact length. Therefore, the first step will be to determine the length of the contact zone. For an arbitrary roller and race profile this is simply done by finding the points of zero geometric interaction, which will essentially be given by the roots of the equation

$$(r^{a}_{b\ell p_3} - \mathscr{R}) = 0 \tag{3.37}$$

When the contact extends to the end of the roller, the extremity of the contact will be defined by the roller length and a solution to the above equation will not be necessary. In general, depending on the roller and race geometry and any misalignment in the bearing, equation (3.37) may have one or two roots within the roller length. Of course, if the contact is perfectly aligned and the roller land extends to its length, the length of the contact zone will be equal to the roller length.

Once the length of the contact zone is determined it will be a simple matter to divide the contact zone into a definite number of elementary disks and determine the roller/race interaction. For the purpose of integrating the local interactions a trapezoidal rule is found to be adequate, although other quadrature schemes may also be used.

It must be remembered that all of the above analysis applies to a line contact type roller/race interaction. For a spherical roller, where the entire surface of the roller is crowned and the race surface is also curved, the analysis will be almost identical to that present for a ball bearing.

Normal Contact Load. The load deflection relation for a line contact is not uniquely defined, as in the case of point contact. The deflection can be computed only in terms of strains corresponding to finite lengths of the two interacting bodies. One such relation [81] is of the form

$$\delta = \frac{2Q}{\pi\ell}\left[\frac{1-\sigma_1^2}{E_1}\left(\ln\frac{2d_1}{b} - \frac{\sigma_1}{2(1-\sigma_1)}\right) + \frac{1-\sigma_2^2}{E_2}\left(\ln\frac{2d_2}{b} - \frac{\sigma_2}{2(1-\sigma_2)}\right)\right] \tag{3.38}$$

where the contact half-width

$$b = \left(\frac{4Q}{\pi\ell\Sigma\bar{\rho}}\left|\frac{1-\sigma_1^2}{E_1} + \frac{1-\sigma_2^2}{E_2}\right|^{1/2}\right) \tag{3.39}$$

where σ_1, E_1, σ_2, E_2 are Poisson's ratio and the elastic modulus for the two interacting bodies; $\Sigma\bar{\rho}$ is the sum of the principal curvatures; Q is the normal contact load; and ℓ is the contact length. The determination of lengths d_1

and d_2 is rather arbitrary and, hence, application of (3.38) becomes a little difficult.

In view of this difficulty, Lundberg [71] has determined an empirical equation based on actual load deflection measurements in line contact. This relation is

$$\delta = \frac{2Q}{\pi \ell}\left[\frac{1-\sigma_1^2}{E_1} + \frac{1-\sigma_2^2}{E_2}\right]\left[\ln\frac{l}{b} + 1.1932\right] \tag{3.40}$$

where the contact half-width b is given by equation (3.39).

For a given Q, the computation of δ is straightforward from equation (3.40); however, when δ is prescribed, some type of iterative procedure will be necessary in order to compute Q.

Palmgren [76] gives a relation which is free of the above difficulty:

$$\delta = 0.39\left[\frac{4(1-v_1^2)}{E_1} + \frac{4(1-v_2^2)}{E_2}\right]^{0.90}\frac{Q^{0.90}}{\ell^{0.80}} \tag{3.41}$$

In terms of results, the above two relations are fairly close. Hence, if equation (3.40) is to be used, then the initial guess in the iterative scheme can be derived from equation (3.41) and, under such conditions, the iteration will converge rather rapidly.

Tractive Forces and Moments. For the purposes of computing the traction, it will be necessary to define the position vector locating the center of the elementary disk in the contact zone relative to the roller geometric center and the local slip velocities. Since the geometric interaction δ is always very small compared to both the roller and race radii, it is reasonable to assume that roller and race deform equally. The required position vector may then be written as

$$\mathbf{r}_p^b = \begin{Bmatrix} x \\ -(\imath - \delta/2)\sin\phi \\ (\imath - \delta/2)\cos\phi \end{Bmatrix} \tag{3.42}$$

In order to compute the local slip velocities, let the race have a translation velocity $\mathbf{v}_r$ and an angular velocity $\boldsymbol{\omega}_r$. Likewise, let the roller have a translational velocity $\mathbf{v}_b$ and an angular velocity $\boldsymbol{\omega}_b$. Generally, the angular velocities are specified in the body-fixed frames while the translational velocities are measured in the inertial frame. The roller velocity can be conveniently measured in the cylindrical coordinate frame in terms of components $\dot{x}$, $\dot{r}$, and $\dot{\theta}$. Then the local slip velocity $\mathbf{u}_s$ of the race relative to the roller at point P (see Figure 3-9) is given in the contact frame in terms of the race and roller velocities, $\mathbf{u}_r$ and $\mathbf{u}_b$ respectively, as

$$\mathbf{u}_r^c = [T_{ac}][T_{ia}]\left[\mathbf{v}_r^i + \left([T_{ir}']\boldsymbol{\omega}_r^r - \begin{Bmatrix} \dot{\theta} \\ 0 \\ 0 \end{Bmatrix}\right) \times (\mathbf{R}_p^i + [T_{ir}']\mathbf{r}_{rg}^r)\right] \tag{3.43}$$

$$\mathbf{u}_b^c = [T_{ac}]\left\{\left\{\begin{matrix}\dot{x}\\0\\\dot{r}\end{matrix}\right\} + [T_{ia}]\{[T'_{ib}]\boldsymbol{\omega}_b^b \times ([T'_{ia}][T'_{ac}]\mathbf{r}_p^c + [T'_{ib}]\mathbf{r}_{bg}^b)\}\right\} \tag{3.44}$$

$$\mathbf{u}_s^c = \mathbf{u}_r^c - \mathbf{u}_b^c \tag{3.45}$$

where the vector $\mathbf{R}_p$, which locates the center of the elementary disk with respect to the race geometric center, is computed as

$$\mathbf{R}_p = \mathbf{r}_p + \mathbf{r}_{bg} + \mathbf{r}_b - \mathbf{r}_r \tag{3.46}$$

The slip velocities normal to and along the rolling direction are given by the components $u_{s_1}^c$ and $u_{s_2}^c$, respectively.

Since in a roller bearing the slip will be primarily along the rolling direction (i.e., along the y axis) the tractive force will act along the y axis. The relevant traction coefficient κ, which will be determined by the appropriate traction model (to be described in Chapter 4), for the present may be denoted as

$$\kappa_2 = \kappa_2(u_2) \tag{3.47}$$

where the subscript 2 is used to denote the y direction. The tractive force, F_t, is now defined as

$$F_t = \kappa Q \tag{3.48}$$

where the normal load Q on the elementary disk is already known.

Both the normal and tractive forces acting on the roller can be combined in a general force vector, written conveniently in the contact frame as

$$\mathbf{F}^c = \left\{\begin{matrix}0\\\kappa Q\\-Q\end{matrix}\right\}$$

The force on the race will be $-\mathbf{F}$, while the moments on the roller and race are defined as

$$\mathbf{G}_b = (\mathbf{r}_p + \mathbf{r}_{bg}) \times \mathbf{F}$$

and

$$\mathbf{G}_r = -(\mathbf{R}_p + \mathbf{r}_{rg}) \times \mathbf{F}$$

All vectors must be transformed to the appropriate coordinate frames. As required by the equations of motion, the forces are transformed to the inertial frame, while the moments are transformed to a body-fixed frame.

It should be remembered that all of the above forces and moments are on a given elementary disk in the contact zone, and an integration over the contact zone will give the total force and moment vectors. As discussed earlier a simple trapezoidal rule is found to be adequate for computing the total roller/race interaction.

3.1.3 Roller/Race-Flange Interactions

In a roller bearing subjected to slight misalignment, the loads between the roller and raceways may be somewhat asymmetric and, hence, the resulting tractions develop a moment about the transverse axis of the roller. Such a moment will lead to angular rotation of the roller about its transverse axis. This highly undesirable type of motion is generally called skewing. When the roller skews, the corners on the roller could interact with the cage and the flanges on the race. Generally, the axial clearance between the roller end and cage is quite large compared to that between the roller end and race flange. Hence, the cage interaction can be neglected, but it is essential that the roller end and race-flange interaction be modeled in a realistic fashion in order to simulate roller skewing and the resulting bearing behavior. This is especially true for a tapered roller bearing where there is a continued interaction between the roller end and the race flange. Unfortunately, local geometry and kinematics greatly complicate the problem and development of a realistic model becomes extremely difficult. An acceptable model should not only simulate the problem correctly, but it should also be simple enough to be rapidly executable on a digital computer, since computer time is a key factor in sophisticated programs, such as the one under consideration.

A first-order interaction model for the roller-end and race-flange contact is derived on the basis of the following two assumptions:

1. If the roller end is not contacting the race flange, there is no contact force. In other words, any hydrodynamic action is neglected.
2. When the race flange is contacting at the roller corner, which will have a finite radius of curvature, the contact load is approximated by an effective Hertzian-type contact model and the tractive force is determined by a prescribed traction–slip relation.

It is, therefore, clear that the objective is to treat metal-to-metal contact only. Basically, there will be two parts to such a treatment; e.g., the contact geometry and the load deflection relationship. Both of these parts will be discussed in this section.

Geometrical Considerations. The roller geometric center was located with respect to the race geometric center in the preceding section. For the roller-corner and flange interaction the center of curvature of the roller corner is located with respect to a point on the race flange. Thus, an additional transformation from the race-azimuth to the flange frame is introduced:

Transformation from race-azimuth to race-flange frame: $[T_{rf}(0, \tilde{\theta}, 0)]$

where the angle $\tilde{\theta}$ will depend on the particular flange under consideration, as shown, in an exaggerated fashion, in Figure 3-10.

Note that the race azimuth angle will now be defined by the relevant center of curvature of the roller corner and not by the center of the roller. Consider

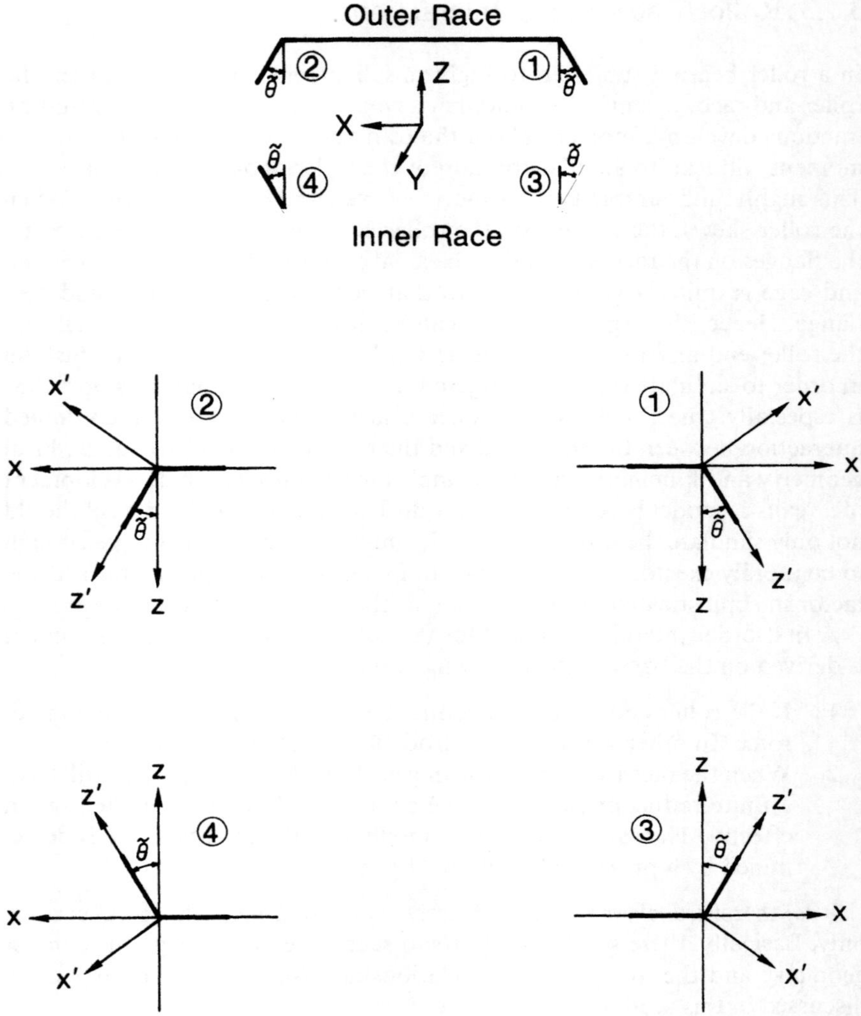

Figure 3-10 Sign convention for race flange angles. (X_a, Y_a, Z_a) is the face aximuth frame and (X', Y', Z') is the flange frame. The Y axis is not shown since the figure is drawn in the X–Z plane.

a point O on the locus of the center of curvature of the roller corner, as shown in Figure 3-11. Since the locus will be a circle, the position vector locating O with respect to the roller geometric center in the roller frame is given as

$$\mathbf{r}_e^b = \begin{Bmatrix} \pm \ell \\ -R\sin\phi \\ R\cos\phi \end{Bmatrix} \tag{3.49}$$

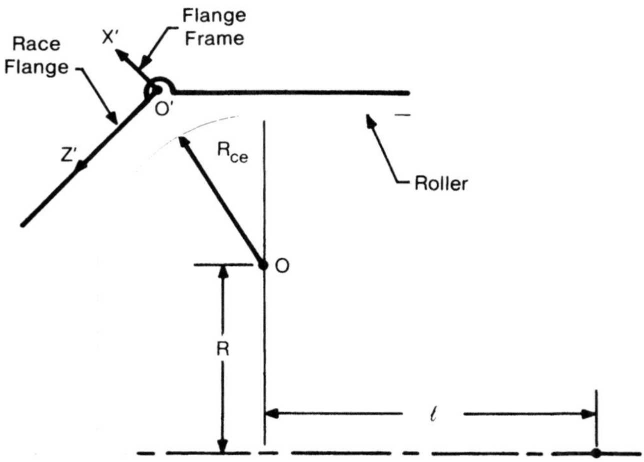

Figure 3-11 Exaggerated view of roller corner and race flange interaction.

where ϕ is the angular coordinate about the X axis in the roller fixed frame, and R is the radius of the circle which is the locus of the center of curvature of the roller corner as shown in Figure 3-11.

It will be necessary to write $\mathbf{r}_e$ in the race frame, and ultimately in the flange frame, and, hence, we recall the transformation from the roller to race frame from the preceding section:

$$[T_{br}] = [T_{ir}][T'_{ib}] \tag{3.50}$$

Also, a vector $\mathbf{r}_{brg}$ locating the roller geometric center with respect to the race geometric center can be written in terms of the various vectors introduced in Figure 3-8:

$$\mathbf{r}^r_{brg} = [T_{ir}]\{\mathbf{r}^i_b + [T_{ib}]\mathbf{r}^b_{bg}\} - \{[T_{ir}]\mathbf{r}^i_r + \mathbf{r}^r_{rg}\} \tag{3.51}$$

Now for any selected value of ϕ in (3.49), the race-azimuth angle ψ and the transformation from the race-to-race azimuth frame $[T_{ra}(\psi, 0, 0)]$, is defined such that the vector

$$\mathbf{r}_{er} = \mathbf{r}_e + \mathbf{r}_{brg}$$

when written in the race-azimuth frame has only x and z components. Hence,

$$\mathbf{r}^r_{er} = \mathbf{r}^r_{brg} + [T_{br}]\mathbf{r}^b_e \tag{3.52}$$

and

$$\mathbf{r}^a_{er} = \begin{Bmatrix} r^a_{er_1} \\ 0 \\ r^a_{er_3} \end{Bmatrix}$$

where

$$r^a_{er_1} = r^r_{er_1}$$

and

$$r^a_{er_3} = \sqrt{(r^r_{er_2})^2 + (r^r_{er_3})^2}$$

The vector $\mathbf{r}_{er}$ locates the roller corner curvature center with respect to the race geometric center. As shown in Figure 3-11, let O' be a reference point on the flange. The roller curvature center can now be referenced relative to O' as

$$\mathbf{r}^a_{ef} = \mathbf{r}^a_{er} - \begin{Bmatrix} \mathscr{S} \\ O \\ \mathscr{R} \end{Bmatrix} \tag{3.53}$$

where $\mathscr{S}$ is the race half-width and $\mathscr{R}$ is the radius of the race land.

When the above is written in the flange frame,

$$\mathbf{r}^f_{ef} = [T_{rf}]\mathbf{r}^a_{ef} \tag{3.54}$$

The point of maximum interaction will be determined by maximizing the x component of $\mathbf{r}^f_{ef}$, by varying ϕ in equation (3.49). Hence,

$$\frac{\partial}{\partial \phi}(r^f_{ef_1}) = 0$$

After carrying out some routine algebra, it may be shown that the above equation may be written as

$$T_{rf_{11}}[-T_{br_{12}}\mathscr{R}\cos\phi - T_{br_{13}}\mathscr{R}\sin\phi]$$
$$+ \frac{T_{rf_{13}}}{r^a_{er_3}}\left\{\sum_{i=2}^{3} (r^r_{b_i} \pm T_{br_{i1}}\ell - T_{br_{i2}}\mathscr{R}\sin\phi + T_{br_{i3}}\mathscr{R}\cos\phi)\right. \tag{3.55}$$
$$\left.\times (-T_{br_{i2}}\mathscr{R}\cos\phi - T_{br_{i3}}\mathscr{R}\sin\phi)\right\} = 0$$

The above is solved for ϕ by standard bisection. Note that there will be two possible roots, one denoting the maximum and the other defining a minimum. The interest here is, of course, in the maximum.

For a positive geometric interaction, the elastic deflection δ will be defined as

$$\delta = (r^f_{ef_1})_{\max} - R_{ce} \tag{3.56}$$

If δ is negative, then there will be no interaction between the roller corner and the race flange.

Normal Contact Load. If contact exists between the roller corner and race flange, then the contact load can be determined by using the classical point-contact Hertzian solution (see Appendix I). The deflection δ is computed as described above; the material properties are prescribed; and the only additional parameters needed are the effective radii of curvature of the interacting bodies, which may be defined as follows:

$$\text{Race flange} \begin{cases} R_1 = \infty \\ R_2 = \mathscr{R}, \end{cases}$$

$$\text{Roller corner} \begin{cases} R_1 = R_{ce} \\ R_2 = (R_{ce} + R)/\cos\theta_s \end{cases}$$

where θ_s is the skew angle. This angle is included in the transformation T_{ib} used above and in the preceding section.

Tractive Forces and Moments. The computation of tractive force will be very similar to the procedure described for the roller/race interaction presented in the preceding section. The contact point is located relative to the roller and race mass centers and the slip velocities are computed. So, if $\mathbf{r}_p$ defines the contact point relative to the roller geometric center, and $\mathbf{R}_p$ defines it relative to the race geometric center, it will be straightforward to show that

$$\mathbf{r}_p^f = [T_{rf}][T_{ra}][T_{br}]\mathbf{r}_e^b + \begin{Bmatrix} 0 \\ 0 \\ R_{ce} \end{Bmatrix} \tag{3.57}$$

and

$$\mathbf{R}_p^f = \mathbf{r}_p^f + [T_{rf}][T_{ra}]\mathbf{r}_{brg} \tag{3.58}$$

The local velocities at the contact point are then

$$\mathbf{u}_b = \mathbf{v}_b + \boldsymbol{\omega}_b \times (\mathbf{r}_p + \mathbf{r}_{bg})$$

$$\mathbf{u}_r = \mathbf{v}_r + \boldsymbol{\omega}_r \times (\mathbf{R}_p + \mathbf{r}_{rg})$$

the effective slip,

$$u_s = (u_{r_2} - u_{b_2}) \tag{3.59}$$

and the traction coefficient,

$$\kappa = \kappa(u_s)\frac{u_s}{|u_s|} \tag{3.60}$$

The total force acting on the roller is now deined in the flange frame as

$$\mathbf{F}_b^f = \begin{Bmatrix} 0 \\ \kappa Q \\ -Q \end{Bmatrix} \tag{3.61}$$

where Q is the normal contact load, as discussed above. The load of the race will be

$$\mathbf{F}_r = -\mathbf{F}_b \tag{3.62}$$

and the moments of the roller and raceway will be

$$\mathbf{G}_b = (\mathbf{r}_p + \mathbf{r}_{bg}) \times \mathbf{F}_b \tag{3.63}$$

and

$$\mathbf{G}_r = (\mathbf{R}_p + \mathbf{r}_{rg}) \times \mathbf{F}_r \tag{3.64}$$

It must be remembered that all vectors must be transformed to the appropriate frames. Generally, the forces will be transformed to the inertial frame and the moments to the body-fixed frame of reference.

3.2 Rolling Element/Cage Interactions

Similar to the rolling element/race interactions, the interactions between rolling element and cage can also be considered in terms of the contact geometry and the constitutive load deflection relations. In general, the rolling element/cage contacts consist of (i) frequent collisions lasting for very short time intervals, and (ii) hydrodynamic forces resulting from a lubricant film at the interface. There may also be a region where the interaction may be elastohydrodynamic in nature, but the time for this interaction will be extremely short due to the relatively high approach velocity leading to an ultimate collision. Primarily due to such dynamic complexities, a few assumptions in modeling the rolling element/cage contact are necessary.

1. The cage is assumed to be a rigid rotor and any gross bending for flexural deformation is ignored.
2. When the rolling element is not in contact with the cage, the interaction is assumed to be purely hydrodynamic and no elastic deformations resulting from the hydrodynamic forces are considered.
3. If the geometric interaction implies a contact, the local deformations are considered to be elastic.

With the above assumptions, the rolling element/cage interaction can be considered in three basic parts: (i) the determination of geometric interaction or minimum clearance between interacting elements, (ii) postulation of hydrodynamic models for prescribed clearances, and (iii) postulation of dry contact models in the case of contact. The geometric interaction or the minimum film is determined by the prescribed position vectors and geometry of the elements. Once a minimum clearance, the geometry of the elements, and the velocity vectors are known, the hydrodynamic interaction is determined by solving the Reynolds equation in the contact zone. If there is contact, either a Hertzian point-contact or a line-contact type of elastic model can be used to compute the normal forces; and with a prescribed traction model and velocity vectors, the tractive forces can be readily determined. The net forces and moments can then be determined by proper transformations and vector products as for the rolling element/race interaction. However, as will be seen below, an average interaction will be determined for each contact and the corresponding hydrodynamic or dry contact force will be determined. Thus, variation within the contact zone, as computed for the rolling element/race interaction, will not be considered. This is reasonable in view of the fundamental assumptions outlined above.

3.2.1 Geometric and Kinematic Considerations

For high-speed angular-contact ball bearings and cylindrical roller bearings, the cage is usually guided on one of the races. Since in some low-speed applications, the cage may ride on the rolling elements themselves, four distinct interactions (e.g., ball/cage with race guidance, ball/cage with ball guidance, roller/cage with race guidance, and roller/cage with roller guidance) must be considered.

Ball/Cage Interactions with Race Guidance. Race guided cages are most widely used in high-speed engine bearings. In the following analysis, the cage pockets are assumed to be cylindrical and the balls spherical. For a specified geometry, the relative position of the ball within the pocket is computed in terms of the position vectors locating the ball and cage in the inertial frame.

As shown in Figure 3-12, let $\mathbf{r}_b^i$, $\mathbf{r}_c^i$, and $\mathbf{r}_p^c$, respectively, be the position vectors locating the geometric center of the ball in the inertial frame, the geometric center of the cage in the inertial frame, and the center of the pocket in the cage frame. The transformations from the inertial to cage frame and the cage frame to pocket frame are defined by the matrices $[T_{ic}]$ and $[T_{cp}]$. The vector $\mathbf{r}_{bp}^p$ locating the center of ball relative to the center of pocket, in the pocket frame is written as

$$\mathbf{r}_{bp}^p = [T_{cp}]\{[T_{ic}]\mathbf{r}_c^i + \mathbf{r}_p^c - [T_{ic}]r_b^i\} \tag{3.65}$$

The minimum clearance or the maximum interference δ_{bp} between the ball and pocket wall will then be

$$\delta_{bp} = \Delta_{bp} - \sqrt{(r_{bp_1}^p)^2 + (r_{bp_2}^p)^2} \tag{3.66}$$

where Δ_{bp} is the radial-ball-to-pocket clearance. Also, the unit vector normal to the plane of contact, $\mathbf{e}_{bp}^p$, will have only x and y components in the pocket frame:

$$\mathbf{e}_{bp}^p = \frac{1}{\sqrt{(r_{bp_1}^p)^2 + (r_{bp_2}^p)^2}}\begin{Bmatrix} r_{bp_1}^p \\ r_{bp_2}^p \\ 0 \end{Bmatrix} \tag{3.67}$$

For the computation of hydrodynamic forces, the relevant angular velocity will be the ball angular velocity $\boldsymbol{\omega}_b^p$ transformed in the pocket frame. However, in the case of metallic contact, the tractive force will be determined by the local slip vector, which is to be computed from the specified velocities of the ball and cage. In the event that the mass center is different from the geometric center, a body-fixed coordinate with the origin at the mass center

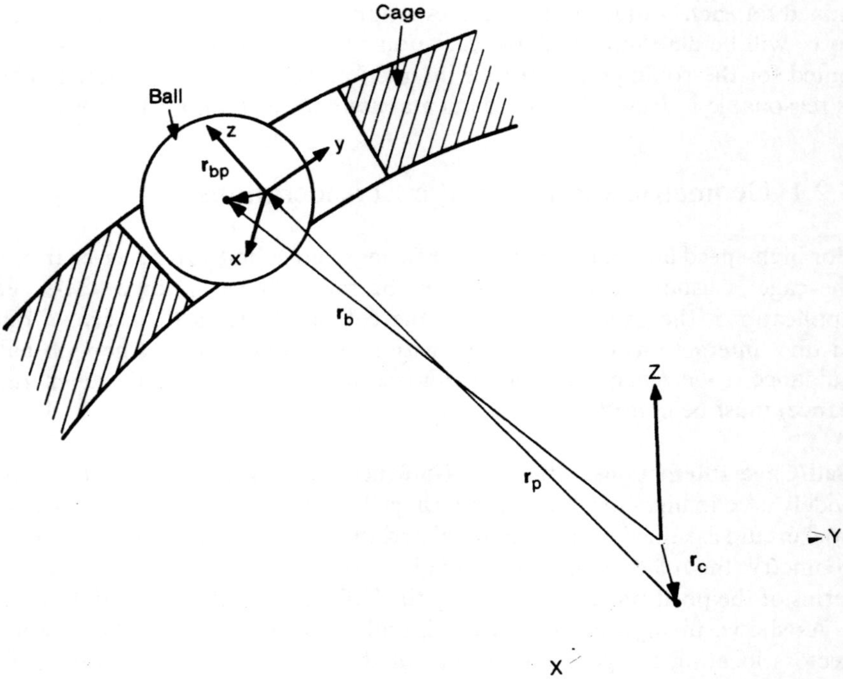

Figure 3-12 Schematic description of the various position vectors establishing the position of a rolling element in the cage pocket.

is generally considered, and the velocities are described in terms of the velocity of the mass center and the rotation of the body about the mass center. It will, therefore, be convenient to define $\mathbf{r}^c_{cg}$ and $\mathbf{r}^b_{bg}$ as the body-fixed vectors locating the geometric center relative to the mass center for the cage and ball respectively. Now, if the mass centers of the cage and ball are respectively located by $\mathbf{r}_{cm}$ and $\mathbf{r}_{bm}$, then

$$\mathbf{r}^i_c = \mathbf{r}^i_{cm} + [T'_{ic}]\mathbf{r}^c_{cg}$$

and

$$\mathbf{r}^i_b = \mathbf{r}^i_{bm} + [T'_{ib}]\mathbf{r}^b_{bg} \tag{3.68}$$

where $[T'_{ib}]$ is defined as the transformation matrix between the inertial and ball frames.

When the cage and ball velocities, $\dot{\mathbf{r}}_{cm}$ and $\dot{\mathbf{r}}_{bm}$, and the corresponding angular velocities, $\boldsymbol{\omega}_c$ and $\boldsymbol{\omega}_b$, are prescribed, the velocity of the ball relative to the cage and the point of contact may be written, once the vectors locating the contact point relative to the cage mass center $\mathbf{r}_{cc}$ and relative to the ball mass center $\mathbf{r}_{bc}$ are determined as

$$\mathbf{r}^c_{cc} = [T'_{cp}](\tfrac{1}{2}d\mathbf{e}^p_{bp} + \mathbf{r}^p_{bp}) + \mathbf{r}^c_p + \mathbf{r}^c_{cg} \tag{3.69}$$

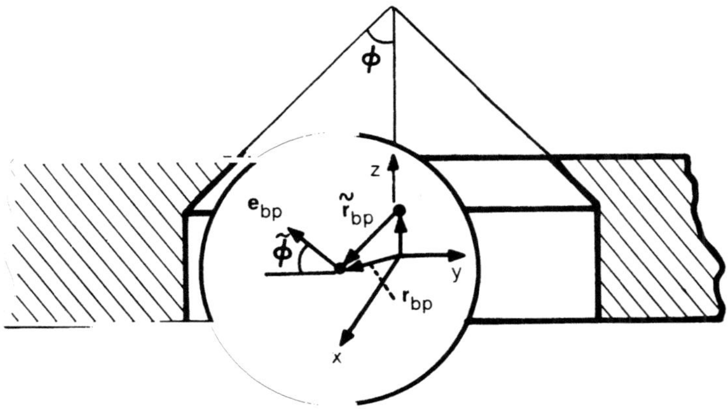

Figure 3-13 Exaggerated pocket geometry for a ball riding cage.

and

$$\mathbf{r}_{bc}^{b} = [T_{ib}][T_{ic}][T_{cp}'](\tfrac{1}{2}d\mathbf{e}_{bp}^{p}) + \mathbf{r}_{bg}^{b} \tag{3.70}$$

where d is the ball diameter and $[T_{ib}]$ is the transformation from the inertial to ball frame.

The velocity of the ball relative to the cage, $\mathbf{u}_{bc}$, is now straightforward:

$$\begin{aligned}\mathbf{u}_{bc}^{p} = {} & [T_{cp}][T_{ic}](\dot{\mathbf{r}}_{bm}^{i} - \dot{\mathbf{r}}_{cm}^{i}) - [T_{cp}](\boldsymbol{\omega}_{c}^{c} \times \mathbf{r}_{cc}^{c}) \\ & + [T_{cp}][T_{ic}][T_{ib}'](\boldsymbol{\omega}_{b}^{b} \times \mathbf{r}_{bc}^{b})\end{aligned} \tag{3.71}$$

In the event of Coulomb-type friction, the direction of friction force will be determined by the direction of relative velocity in the plane of contact, which may be readily determined by equations (3.71) and (3.67).

Ball/Cage Interactions with Ball Guidance. When the cage rides on the ball, the pocket geometry could be spherical or cylindrical with conical ends. In the case of a spherical pocket, $\mathbf{r}_{bp}$ will be given by equation (3.65) and the relevant interaction is determined by subtracting the argument of this vector from the given pocket clearance. For cylindrical pockets with conical ends, the ball geometric center is located with respect to the center of the base of the cone which provides the guiding surface. Since the geometry of the pocket is completely known, the vector $\tilde{\mathbf{r}}_{bp}$ can be readily determined, as shown in Figure 3-13.

In order to determine the magnitude of interaction with the conical surface, it will be convenient to transform $\tilde{\mathbf{r}}_{bp}^{p}$ into cylindrical coordinates (r, θ, z) such that

$$\begin{aligned} r &= \sqrt{(\tilde{\mathbf{r}}_{bp_1}^{p})^2 + (\tilde{\mathbf{r}}_{bp_2}^{p})^2} \\ \theta &= \arctan\left(\frac{\tilde{\mathbf{r}}_{bp_2}^{p}}{\tilde{\mathbf{r}}_{bp_1}^{p}}\right) \\ z &= \tilde{\mathbf{r}}_{bp_3}^{p} \end{aligned} \tag{3.72}$$

A unit vector, normal to the conical surface, from the center of the ball is given by

$$\mathbf{e}_{bp}^{p} = \begin{Bmatrix} \cos\tilde{\phi}\cos(\theta + \theta') \\ \cos\tilde{\phi}\sin(\theta + \theta') \\ \sin\phi \end{Bmatrix} \tag{3.73}$$

where $\tilde{\phi}$ is the cone angle and the angle θ' is measured from the axis along θ. Thus $\theta' = 0$ will result in a unit vector in the direction of maximum interaction.

Now the interaction δ_{bp} is given by

$$\delta_{bp}(\theta') = (\tfrac{1}{2}d + \Delta_{bp})\cos\phi - \tilde{\mathbf{r}}_{bp}^{p} \cdot \mathbf{e}_{bp}^{p} - \tfrac{1}{2}d \tag{3.74}$$

It should be noted that since $\mathbf{e}_{bp}^{p}$ is a function of θ', the interaction will depend on θ', and it will be necessary to determine the range of θ' for which δ_{bp} is negative. This range will also determine the length of the contact zone.

Also, by straightforward geometry, the relative position of the ball center and the length of the conical surface will determine if the vector $\mathbf{e}_{bp}$ falls on the given surface. The interaction analysis will be performed only when the vector $\mathbf{e}_{bp}$ falls on the given cage surface. Once the point of contact has been established the relative slip velocity will be defined by an equation similar to (3.71).

Unlike the exaggerated view shown in Figure 3-13, the opening at the conical end of the pocket will be smaller than the ball diameter. Hence, for a certain value of $\tilde{\mathbf{r}}_{bp}^{p}$ the ball may contact the cage surface all around the cone, i.e., the range of θ' may be 2π. In geometric terms this will be a configuration for maximum interaction at the guiding surface.

Roller/Cage Interactions with Race Guidance. For the roller/cage interaction, aside from determining the vector $\mathbf{r}_{bp}$ (using Equation 3.65) which locates the center of the roller with respect to the pocket center, it will also be necessary to determine orientation of the roller relative to the pocket. For this purpose, a roller-fixed coordinate frame, along its principal axis, will be defined and, similar to the case of a ball bearing, the transformation from the roller to pocket frame will be denoted by $[T_{bp}]$.

The geometry of the pockets in this case will be assumed to be rectangular, as shown in Figure 3-14. The point of maximum interaction will be determined by examining the vector that locates the edge of the roller relative to the pocket wall. In the pocket frame, the length of the roller, ℓ, and an angle, ϕ, will determine the position vector $\mathbf{r}_{e}^{p}$ of any point on the edge with respect to the pocket center. Symbolically,

$$\mathbf{r}_{e}^{p} = \mathbf{r}_{bp}^{p} + [T_{bp}] \begin{Bmatrix} \ell \\ \tfrac{1}{2}d\cos\phi \\ \tfrac{1}{2}d\cos\phi \end{Bmatrix} \tag{3.75}$$

Two types of interaction may be possible, namely, interaction with the

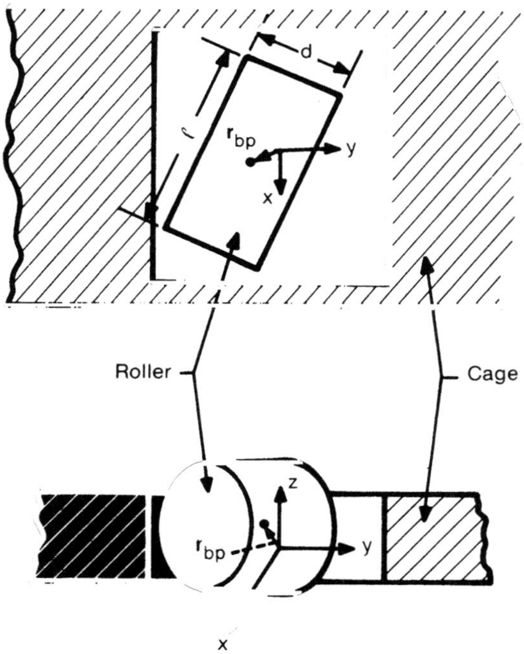

Figure 3-14 Roller and cage/pocket interaction.

cylindrical surface or with the end of the roller. For the point of maximum interaction along the cylindrical surface, r_{e_2} should be a maximum, or

$$\frac{\partial r_{e_2}^p}{\partial \phi} = 0$$

which gives

$$\phi = \arctan\left(\frac{T_{bp_{23}}}{T_{bp_{22}}}\right) \tag{3.76}$$

Similarly, for the end of the roller r_{e_1} must be a maximum, or

$$\frac{\partial r_{e_1}^p}{\partial \phi} = 0$$

which gives

$$\phi = \arctan\left(\frac{T_{bp_{1,3}}}{bp_{1,2}}\right) \tag{3.77}$$

In either case, once ϕ is determined, the magnitude of interaction δ_{bp} is written as the difference of the relevant half-width of the pocket and the corresponding component of $\mathbf{r}_e^p$. The unit vector normal to the plane of contact will clearly be normal to the relevant pocket wall, i.e., the unit vector will either be along the y axis or the x axis. Having determined the contact point, the relative sliding speed is computed by an equation similar to (3.71). It should be noted that both ends of the rollers must be analyzed separately.

Roller/Cage Interactions with Roller Guidance. When the cage rides on the rollers, the pocket is assumed to be cylindrical so that the roller may not be pushed out of the pocket ends. The assumption of such a geometry greatly simplifies the situation, since all geometric and kinematic consideration for this case will be identical to those presented for the cage/race interaction, the details of which will be described in section 3.3. Thus, once the vector $\mathbf{r}_{bp}^{p}$, locating the center of the roller with respect to the pocket center, is computed in accordance with equation (3.65), the remainder of the analysis will be identical to the one described in section 3.3.

3.2.2 Hydrodynamic Models

From most of the available literature, it can be established that contact forces resulting from ball/cage, roller/cage, or race/cage contacts are small in comparison to the load suporting forces at the ball/race or roller/race contacts. Furthermore, the hydrodynamic forces generated between the rolling element and cage interfaces are small compared to the forces resulting from metallic contact. In ultrahigh-speed applications, however, the magnitude of the hydrodynamic forces may be significant.

Concerning the hydrodynamics at the ball/cage interface, the two ball angular velocity components $\omega_{\hat{x}}$ and $\omega_{\hat{z}}$ will be significant, as shown in Figure 3-15. It is clear that the Reynolds equation will have to be solved in two dimensions in order to allow for both velocities. Such a solution will indeed be somewhat tedious. With the objective of an approximate treatment, however, the Reynolds equation may be solved along the two coordinates independently and the two solutions could be superimposed. Thus, for the component $\omega_{\hat{x}}$, the problem is similar to that of a "long bearing," where

$$\frac{\partial p}{\partial y} \ll \frac{\partial p}{\partial x} \tag{3.78}$$

and the Reynolds equation is written as

$$\frac{\partial}{\partial x}\left(h^3 \frac{\partial p}{\partial x}\right) = 6U\mu \frac{\partial h}{\partial x} \tag{3.79}$$

where p is the local pressure; h is the film thickness; $U = (\omega_{\hat{x}} d/2)$ is the relative sliding velocity; and μ is the viscosity of the lubricant.

In accordance with the coordinate system shown in Figure 3-15, let the exit be denoted by $x = x_0$, where the pressure gradient will be zero, and let the corresponding film thickness be h_o. Also, let the pressure be zero at the

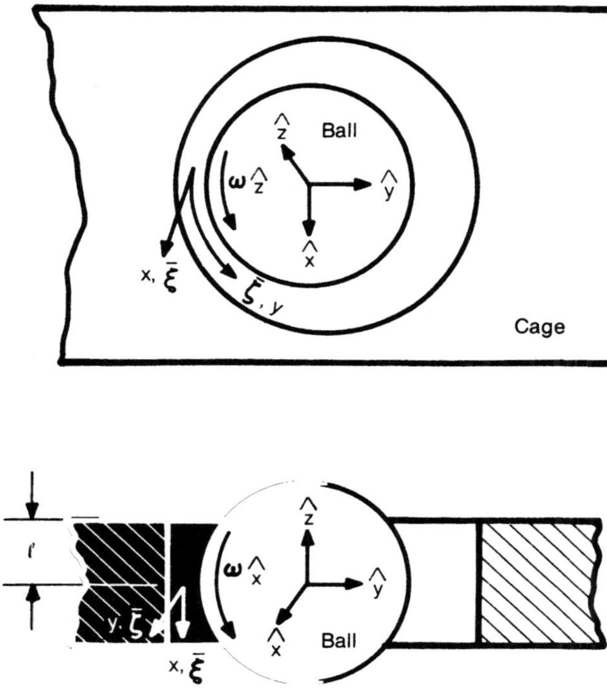

Figure 3-15 Schematic of ball/cage hydrodynamic interaction.

entrance edge of the cage located at $x = \ell$. Then, the integrated form of equation (3.79) may be written as

$$p(x) = 6\mu U \int_{\ell}^{x} \frac{h(x') - h_o}{h(x')^3} dx'$$

If the equivalent radii $[R = R_1 R_2/(R_2 - R_1)]$ along the x and y directions are R_x and R_y respectively, the film thickness at $x = 0$ is h_m, and the film thickness profile is approximated to be parabolic, then in terms of the following dimensionless coordinates

$$\bar{\xi} = \frac{x}{\ell}; \qquad \bar{\zeta} = \frac{y}{\ell}\sqrt{\frac{R_x}{R_y}}; \qquad P_1 = P\frac{\ell^3}{24\mu U R_x^2}; \qquad \mathscr{L} = \frac{\ell}{\sqrt{2R_x h_m}} \tag{3.80}$$

The integrated form of the Reynolds equation may be written as

$$P_1(\xi) = \int_{-1}^{\bar{\xi}} \frac{\bar{\xi}'^2 - \bar{\xi}_0^2}{(1/\mathscr{L}^2 + \bar{\zeta}^2 + \bar{\xi}'^2)^3} d\bar{\xi}' \tag{3.81}$$

The exit coordinate $\bar{\xi}_0$ is determined by the boundary condition $P_1(\bar{\xi}_0) = 0$. Thus, for any given value of $\bar{\zeta}$, equation (3.81) may be solved for $\bar{\xi}_0$ by setting $P_1(\bar{\xi}_0) = 0$. The resultant algebraic equation will be nonlinear, but the solution is straightforward by the conventional bisection method.

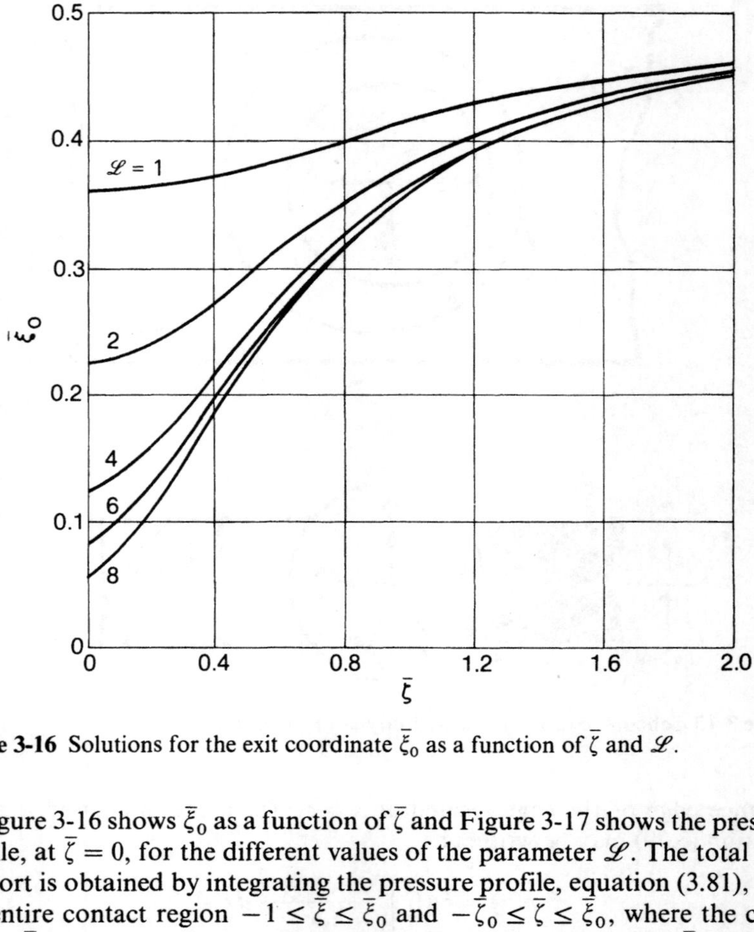

Figure 3-16 Solutions for the exit coordinate $\bar{\xi}_0$ as a function of $\bar{\zeta}$ and $\mathscr{L}$.

Figure 3-16 shows $\bar{\xi}_0$ as a function of $\bar{\zeta}$ and Figure 3-17 shows the pressure profile, at $\bar{\zeta} = 0$, for the different values of the parameter $\mathscr{L}$. The total load support is obtained by integrating the pressure profile, equation (3.81), over the entire contact region $-1 \le \bar{\xi} \le \bar{\xi}_0$ and $-\bar{\zeta}_0 \le \bar{\zeta} \le \bar{\xi}_0$, where the coordinate $\bar{\zeta}_0$ indicates the point of maximum film. Integration over $\bar{\xi}$ will give

$$\int_{-1}^{\bar{\xi}_0} P_1(\bar{\xi}) d\bar{\xi} = \int_{-1}^{\bar{\xi}_0} \int_{-1}^{\bar{\xi}} \frac{\bar{\xi}'^2 - \bar{\xi}_0^2}{(1/\mathscr{L}^2 + \bar{\zeta}^2 + \bar{\xi}'^2)^3} d\bar{\xi}' \, d\bar{\xi}$$

By switching the limits of integration and carrying out some routine algebra, it can be shown that

$$\int_{-1}^{\bar{\xi}_0} P_1(\bar{\xi}) d\bar{\xi} = \left[-\bar{\xi}_0^3 \left\{ \frac{\bar{\xi}'}{2a} \left(\frac{1}{2R^2} + \frac{3}{4aR} \right) + \frac{3}{8a^2\sqrt{a}} \tan^{-1} \frac{\bar{\xi}'}{\sqrt{a}} \right\} - \frac{\bar{\xi}_0^2}{4R^2} \right.$$
$$\left. + \bar{\xi}_0 \left\{ -\frac{\bar{\xi}'}{4R^2} + \frac{\bar{\xi}'}{8aR} + \frac{1}{8a\sqrt{a}} \tan^{-1} \frac{\bar{\xi}'}{\sqrt{a}} \right\} + \left(\bar{\xi}'^2 + \frac{a}{2} \right) \frac{1}{2R^2} \right]_{-1}^{\bar{\xi}_0} \tag{3.82}$$

where $R = 1/\mathscr{L}^2 + \bar{\zeta}^2 + \bar{\xi}'^2$ and $a = 1/\mathscr{L}^2 + \bar{\zeta}^2$.

As seen earlier in Figure 3-16, $\bar{\xi}_0$ is a function of $\bar{\zeta}$. Hence, the above equation will have to be integrated numerically over $\bar{\zeta}$ to compute the total load support. Also, from the given geometry and lubrication condition the

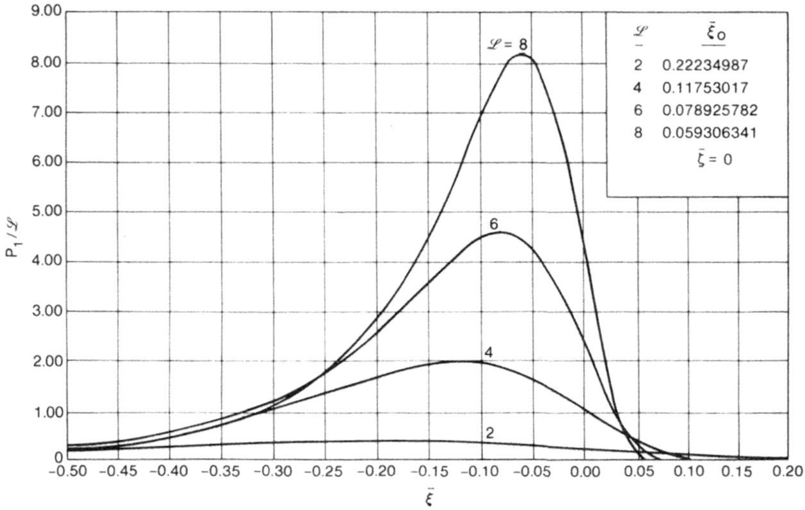

Figure 3-17 Typical pressure profiles for the "long bearing" case.

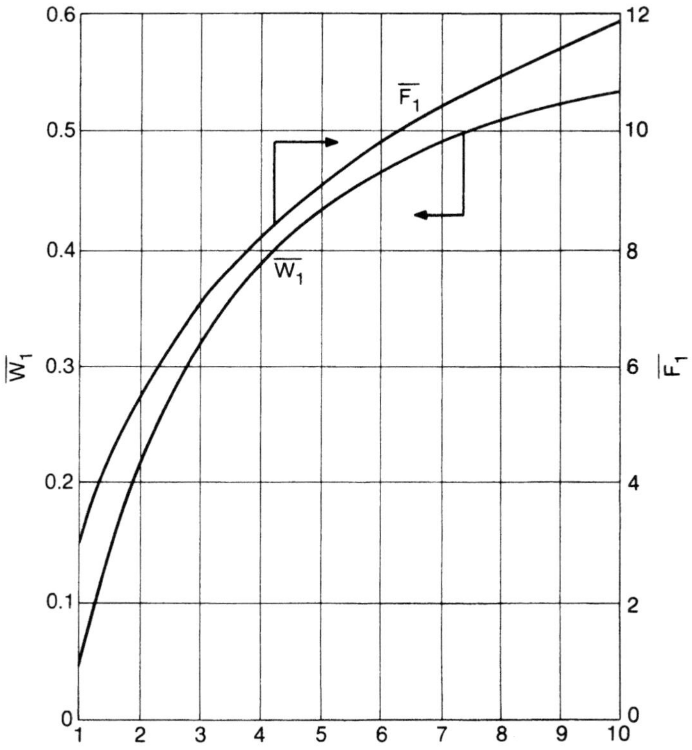

Figure 3-18 Normal ($\overline{W}_1$) and friction ($\overline{F}_1$) loads in the case of a "long bearing."

maximum film coordinate $\bar{\zeta}_0$ will be determined. In terms of the dimensional load W_1, the dimensionless load $\overline{W}_1$ is defined as

$$\overline{W}_1 = \frac{W_1}{24\mu U R_x}\sqrt{\frac{2h_m}{R_y}} \tag{3.83}$$

For a typical configuration, with $\bar{\zeta}_0 = 3$, the dimensionless load $\overline{W}_1$ is plotted as a function of the parameter $\mathscr{L}$ in Figure 3-18.

The net traction is obtained by approximating the shear stress τ_{xy} as

$$\tau_{xy} = \mu\frac{U}{h} \tag{3.84}$$

by integrating over the contact region, and by assuming that the lubricant does not contribute to any traction in the separation zone. Some straightforward algebra will show that the dimensionless tractive force $\bar{F}_1$ is given by

$$\bar{F}_1 = \frac{F_1}{2\mu U\sqrt{R_x R_y}} = \int_{-\bar{\zeta}_0}^{\bar{\zeta}_0} \frac{1}{\sqrt{1/\mathscr{L}^2 + \bar{\zeta}^2}}\left[\tan^{-1}\frac{\bar{\xi}_0}{\sqrt{1/\mathscr{L}^2 + \bar{\zeta}^2}} + \tan^{-1}\frac{1}{\sqrt{1/\mathscr{L}^2 + \bar{\zeta}^2}}\right]d\bar{\zeta}, \tag{3.85}$$

where F_1 is the dimensional tractive force. Once again, since $\bar{\xi}_0$ is a function of $\bar{\zeta}$, the above equation is integrated numerically. The results for $\bar{\zeta}_0 = 3$ are shown in Figure 3-18.

In the case of rotation ($\omega_{\hat{z}}$) the problem is somewhat similar to that of a "short bearing" where the Reynolds equation is written as

$$\frac{\partial}{\partial y}\left(h^3\frac{\partial p}{\partial y}\right) = 6V\mu\frac{\partial h}{\partial x} \tag{3.86}$$

where $V = \frac{1}{2}d\omega_{\hat{z}}$ is the relative sliding velocity along the y direction. The solution in this case will be somewhat simpler. It can be shown that in terms of the dimensionless variables, defined earlier by equations (3.80), the

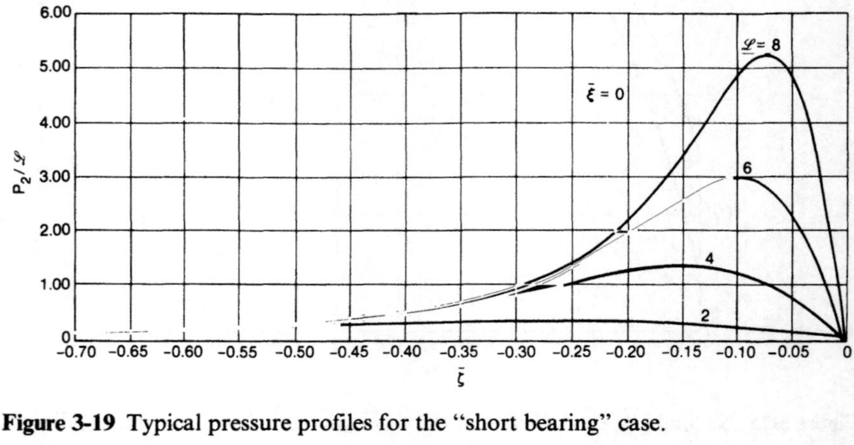

Figure 3-19 Typical pressure profiles for the "short bearing" case.

pressure profile may be written as

$$P_2(\bar{\zeta}) = \frac{\bar{\zeta}}{4}\left[\frac{1}{(1/\mathscr{L}^2 + \bar{\zeta}^2 + 1)^2} - \frac{1}{(1/\mathscr{L}^2 + \bar{\zeta}^2 + \bar{\xi}^2)^2}\right] \tag{3.87}$$

where

$$P_2 = \frac{pR_y\ell^3}{48\mu VR_x^3}\sqrt{\frac{R_x}{R_y}}$$

It may be clearly seen that since $\bar{\xi} \leq 1$, the pressure will be negative for $\bar{\zeta} > 0$. Thus the effective load support zone is terminated at $\bar{\zeta} = 0$. Typical pressure profiles for $\xi = 0$, are shown in Figure 3-19.

Similar to equation (3.82), integration of pressure and shear stress can be shown as

$$\overline{W}_2 = \frac{W_2\sqrt{2R_xh_m}}{48\mu VR_x^2}$$

$$= \frac{1}{2}\int_{-\bar{\zeta}_0}^{0} \bar{\zeta}\left\{\frac{1}{(1/\mathscr{L}^2 + \zeta^2 + 1)^2} - \frac{1}{2(1/\mathscr{L}^2 + \bar{\zeta}^2)^{3/2}}\tan^{-1}\frac{1}{\sqrt{1/\mathscr{L}^2 + \bar{\zeta}^2}} - \frac{1}{2(1/\mathscr{L}^2 + \zeta^2)(1/\mathscr{L}^2 + \zeta^2 + 1)}\right\} d\zeta \tag{3.88}$$

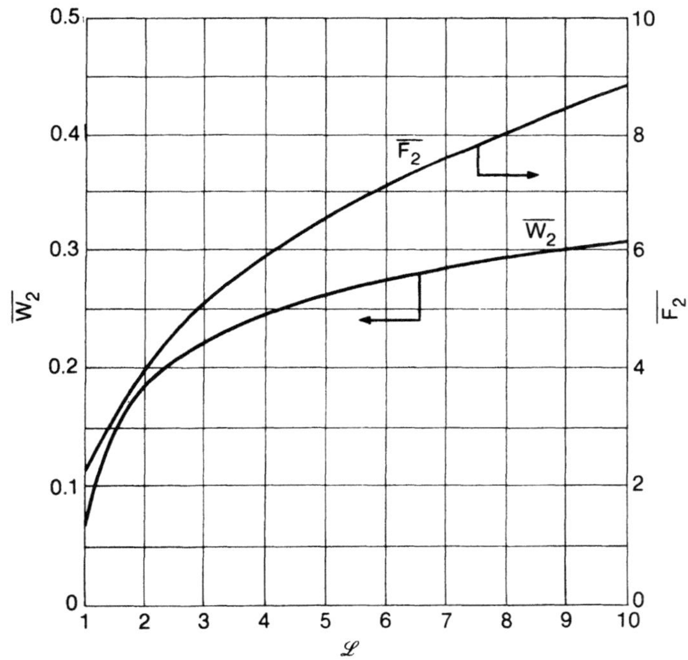

Figure 3-20 Normal ($\overline{W}_2$) and friction ($\overline{F}_2$) loads for the "short bearing" case.

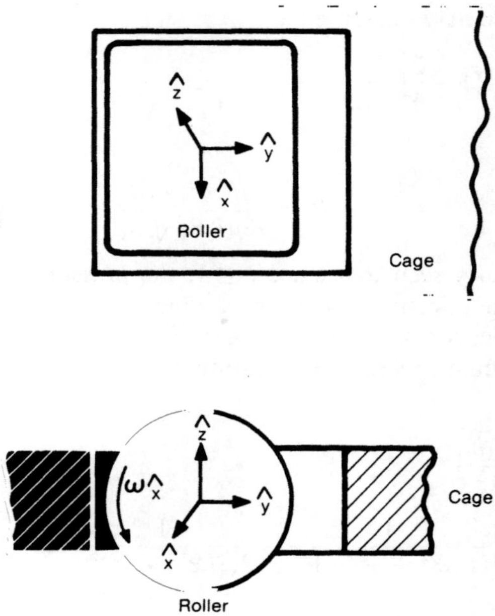

Figure 3-21 Hydrodynamic interaction between the cage and a roller.

and

$$\bar{F}_2 = \frac{F_2}{2\mu V\sqrt{R_x R_y}} = 2\int_{-\bar{\zeta}_0}^{0} \frac{1}{\sqrt{1/\mathscr{L}^2 + \zeta^2}} \tan^{-1} \frac{1}{\sqrt{1/\mathscr{L}^2 + \bar{\zeta}^2}} d\zeta \quad (3.89)$$

The results of the above expression, with $\bar{\zeta}_0 = 3$, are shown in Figure 3-20 as a function of the geometrical parameter $\mathscr{L}$.

Ball/cage hydrodynamic interaction for a ball riding cage can also be approximated by the above treatment, at least for the pockets with conical ends, where significant hydrodynamic forces will arise only in the cylindrical region of the pocket. When the pockets are spherical, the problem is somewhat more complicated. However, for the purpose of order-of-magnitude estimation of any hydrodynamic forces, the above treatment may not be unreasonable.

Roller/cage hydrodynamic interaction is considered only for the cylindrical surface and any forces at the roller ends are neglected. Thus the "long bearing" assumption, discussed above, will be valid and equations (3.82) through (3.85) will be relevant. As shown in Figure 3-21, there will be only one component of the roller angular velocity which will be responsible for any hydrodynamic effect. Also, the film thickness will vary only in the $\bar{\xi}$ direction. Thus, the pressure profile can be determined only at $\bar{\zeta} = 0$ and integration of that pressure along $\bar{\xi}$ will determine the load per unit length of the roller. Likewise, consideration of tractions will also be similar.

For the case of the roller riding cage with cylindrical pockets, the hydrodynamic analysis is more straightforward than the above approach. The interaction can be readily simulated by a conventional partial arc journal bearing, for which the solutions are very well-known [79].

3.2.3 Dry Contact Models

A metallic contact between the rolling element and the cage will generally result in large forces and severe interactions. Depending on the magnitude of the forces, the overall geometry of the cage may also change; and in such an event, the analysis of the interactions will indeed be very tedious. In the present investigation, it will be assumed that the metallic contact causes only local elastic deformations and that the cage is otherwise rigid; in other words, no gross bending or flexural deformation will be allowed. Thus, once the relative position or magnitude of interaction δ is known from the geometric considerations, the contact force is estimated by conventional Hertzian-type contact analysis. Clearly, this will be quite straightforward in the case of a point contact (see Appendix I) which will be relevant to the ball/cage interaction. In the case of a roller/cage contact, however, a load–deflection relation must be postulated, since the classical elasticity theory does not provide such a relation. Such relations will be similar to those described by equations (3.38) through (3.41) for the roller/race interaction.

3.3 Race/Cage Interactions

The assumptions outlined in the beginning of the preceding section also apply here, since the race/cage interaction is basically similar to that discussed for the rolling element/cage interaction.

3.3.1 Geometric and Kinematic Considerations

As shown in Figure 3-22, the geometrical considerations involve determination of vector $\mathbf{r}_{cp}$, locating the geometric center of the cage relative to the center of the race guiding surface, and the computation of the magnitude of the interaction δ_{cp}. Since the position vector that locates the cage center $\mathbf{r}_c$ is known at any instant of time, the desired vector $\mathbf{r}_{cp}$ is written as

$$\mathbf{r}_{cp}^r = [T_{ir}]\{\mathbf{r}_c^i - \mathbf{r}_r^i\} \tag{3.90}$$

where $[T_{ir}]$ is the transformation matrix from the inertial to the race coordinate frame.

In order to compute the magnitude of the interaction, it will be necessary to consider a point on the edge of the cage with respect to the cage center. A general description of such a point is given in terms of a vector $\mathbf{r}_p^c$, whereby

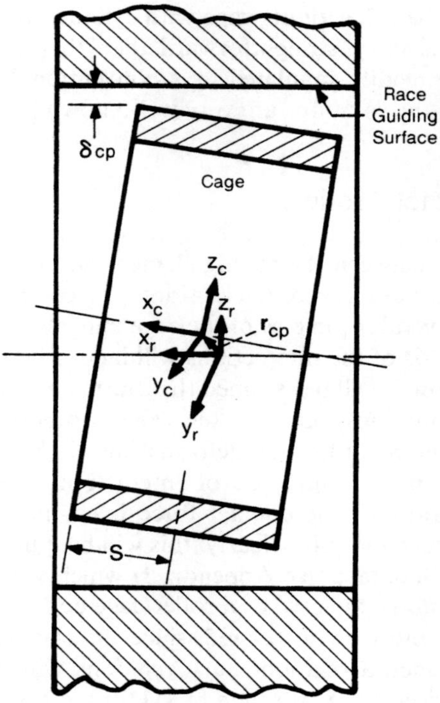

Figure 3-22 Exaggerated view of the race/cage interaction.

$$\mathbf{r}_p^c = \begin{Bmatrix} S \\ \hat{R}\cos\phi \\ \hat{R}\sin\phi \end{Bmatrix} \tag{3.91}$$

where S is the cage half-width, $\hat{R}$ is the radius of the cage guiding surface, and ϕ denotes the angle by which the point is located circumferentially.

If the transformation from the cage to race frame is denoted as $[T_{cr}]$, then the vector locating the point on the cage relative to race center is

$$\mathbf{r}^r = \mathbf{r}_{cp}^r + [T_{cr}]\mathbf{r}_p^c \tag{3.92}$$

For the point of minimum clearance, the projection of $\mathbf{r}^r$ in the y–z plane must be maximum for outer race guidance and minimum for inner race guidance. In either case,

$$\frac{\partial}{\partial\phi}[(r_2^r)^2 + (r_3^r)^2] = 0 \tag{3.93}$$

A combination of equations (3.90) through (3.93) will give an algebraic equation in which ϕ will have to be solved by standard bisection. Once $\mathbf{r}^r$ is known, the required interaction δ_{cp} can be determined with the prescribed cage/race clearance, Δ_{cr}. The unit vector normal to the plane of contact will be along the projection of $\mathbf{r}^r$ in the y–z plane. The determination of relative velocities of sliding at the point of contact will be identical to the approach discussed earlier [equations (3.68)–(3.71)] for ball/cage contacts.

3.3.2 Hydrodynamic Models

The hydrodynamics of the race/cage contact is very simply simulated by the well-known "short journal bearing" solution [79]. Since the geometry is indeed close to or may be identical to that of a conventional journal bearing, no modification to the available solution will be needed. Misaligned contacts are also treated within these existing solutions.

3.3.3 Dry Contact Models

The metallic interaction will be that of a line contact configuration and an appropriate load deflection relation, as considered in equations (3.38) through (3.41), will be required to compute the normal load. Equation (3.91) may be used for such a relationship because of the similitude to the roller cage interaction. Misaligned contacts are treated by methods similar to the ones used in analyzing misaligned roller bearing contacts [47]. Once the normal load is known, the friction force is estimated by a specified coefficient of friction, or a traction–slip relationship.

3.4 Interactions Between Rolling Elements

In a cageless bearing, commonly known as a full complement bearing, the interaction between adjacent rolling elements must be considered. This interaction can sometimes be fairly severe, especially when the operating speeds are high and the bearing elements experience considerable skid. Due to the counterrotation of the elements, the sliding speed between two interacting rolling elements will be very high and, hence, the frictional interaction can be quite severe. Likewise, the relative approach velocities will define the severity of collision and the resulting normal contact forces.

Similar to the geometric interactions presented in the preceding sections, the contact between two rolling elements can also be modeled in terms of the geometric interaction and the resulting normal and tractive forces. Since the constitutive relations will be identical to those already discussed for point and line contacts, it will only be necessary here to present the geometric interaction and the relative sliding speeds at the points of contact in ball and roller bearings.

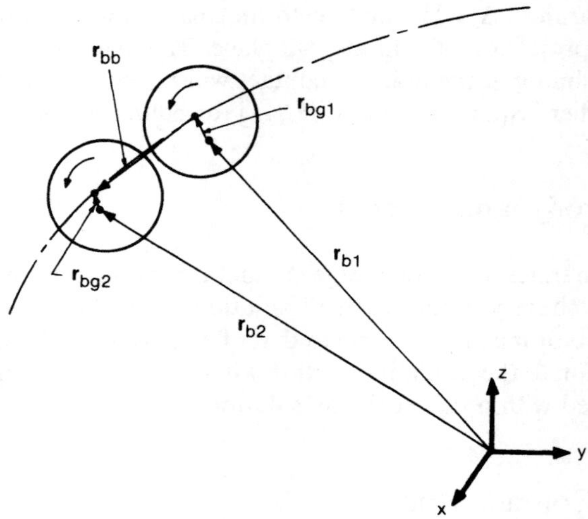

Figure 3-23 Interaction between two balls.

3.4.1 Ball Bearings

Figure 3-23 shows the various position vectors locating the two adjacent balls in an inertial reference frame. The relative position vector $\mathbf{r}_{bb}$ can be expressed as

$$\mathbf{r}_{bb} = \mathbf{r}_{b_2} + [T'_{ib_2}]\mathbf{r}_{bg_2} - \mathbf{r}_{b_1} - [T'_{ib_1}]\mathbf{r}_{bg_1} \tag{3.94}$$

If d_1 and d_2 are the diameters of ball 1 and ball 2, respectively, then the maximum geometric interaction is defined as

$$\delta = |\mathbf{r}_{bb}| - \tfrac{1}{2}(d_1 + d_2) \tag{3.95}$$

A positive δ will indicate no contact, while a negative value will represent a point contact with a maximum geometric interaction of $|\delta|$. The classical Hertzian theory (see Appendix I) can be used to determine the resulting normal contact force.

For a prescribed traction model, it will be necessary to determine the relative tangential slip, which is derived from the velocities of the interacting balls and the position vector locating the point of interaction. Also, the moments acting on the interacting balls can be determined by a vector product to the contact force and the position vector locating the contact point relative to the mass center. Now, from equation (3.94) a unit vector along the normal contact force is

$$\mathbf{e} = \mathbf{r}_{bb}/|\mathbf{r}_{bb}| \tag{3.96}$$

Thus, if Q is the normal contact load, as determined by the Hertzian theory for the geometric interaction of $|\delta|$, then the normal contact forces on balls 1 and 2 are, respectively, written as

$$\mathbf{F}_{1n} = -Q\mathbf{e}$$

and (3.97)

$$\mathbf{F}_{2n} = Q\mathbf{e}$$

The position vectors locating the point of contact relative to the mass center can also be written in terms of the unit vector $\mathbf{e}$, whereby

$$\mathbf{r}_{p1} = \tfrac{1}{2}d_1\mathbf{e}$$

and (3.98)

$$\mathbf{r}_{p2} = -\tfrac{1}{2}d_2\mathbf{e}$$

where $\mathbf{r}_{p1}$ and $\mathbf{r}_{p2}$ are the two vectors in the inertial frame, assuming that the geometric interaction $|\delta| \ll \tfrac{1}{2}(d_1 + d_2)$.

If $\mathbf{U}_1$, $\mathbf{U}_2$ and $\boldsymbol{\omega}_1$, $\boldsymbol{\omega}_2$ are, respectively, the mass center and the angular velocities of the two balls, then the relative velocity of ball 2 relative to ball 1 may be written as

$$\mathbf{U} = \mathbf{U}_2 - \mathbf{U}_1 + (\boldsymbol{\omega}_2 \times \mathbf{r}_{p2} - \boldsymbol{\omega}_1 \times \mathbf{r}_{p1}) \tag{3.99}$$

In general, the relative mass center velocity $(\mathbf{U}_2 - \mathbf{U}_1)$ may have a component tangential to the contact plane. However, this component will normally be very small compared to the surface velocities resulting from the ball angular velocities. Therefore, the relative slip vector $\mathbf{U}_s$ can be written as

$$\mathbf{U}_s = \boldsymbol{\omega}_2 \times \mathbf{r}_{p2} - \boldsymbol{\omega}_1 \times \mathbf{r}_{p1} \tag{3.100}$$

A traction coefficient κ corresponding to the absolute slip velocity $|\mathbf{U}_s|$ can be determined from the prescribed traction model, and thus the traction forces acting on balls 1 and 2 are respectively written as

$$\mathbf{F}_{1t} = \kappa Q\mathbf{e}_s$$

and (3.101)

$$\mathbf{F}_{2t} = -\kappa Q\mathbf{e}_s$$

where $\mathbf{e}_s$ is a unit vector $\mathbf{U}_s/|\mathbf{U}_s|$. Equations (3.97) and (3.101) may be added to determine the total contact force vector. A vector product of the position vectors, defined in equation (3.98), and these force vectors will give the required moment vectors.

3.4.2 Roller Bearings

The geometry of interaction between two rollers is much more complicated than that between the balls, because other than the variation of radius along the roller axis, the axes of two interacting rollers may be neither parallel nor coplanar. However, the angles between the two axes will generally be very small, and the geometry of interaction will be very similar to that between a roller and the inner race. Hence, if the radius of the inner race is made identical to that of the roller, then the analysis presented in Section 3.1 for roller bearings will be applicable here. The exception will be that no elastohydrodynamic effects will be present because the relative sliding velocity between the two counterrotating rollers will be very high and there will be no rolling component.

3.5 External System Interactions and Constraints

The purpose of modeling the interaction of the bearing with the surrounding external systems is to determine the applied conditions on the bearing. For example, the applied loads, speeds, temperatures, and their time variations should be determined from the operating environment of the complete rotor bearing system. Likewise, geometric constraints which dictate either the initial conditions or the general motion of the bearing elements must be modeled. Although the determination of operating conditions such as loads, speeds, and temperatures is important, these conditions are somewhat trivial compared to certain geometric constraints such as static or quasi-static equilibrium.

Aside from the equilibrium constraints for ball and roller bearings, there are certain system interactions which influence both the geometry of the bearing and the applied transient loads on the bearing elements. Shrink fits of the races, centrifugal growth of the rotating race, thermal expansion of the bearing elements due to high operating temperatures, and thermal gradients across the bearing are some of the factors which have significant impact on bearing geometry. A model for these interactions is quite straightforward and is outlined in Appendix II. Factors which influence the applied loads may be either external, or internal to the bearing. The external factors include vibrational loading, sudden accelerations, dynamic unbalance, etc., while lubricant drag forces and churning moments are examples of the internal factors. For a given system, once again, dynamic conditions, such as external vibrations, race accelerations, rotating unbalanced loads, and similar factors can be easily modeled, but a realistic model for lubricant churning and drag can be very difficult. A discussion of these models is the subject of Chapter 5.

3.5.1 Equilibrium Constraint for Ball Bearings

As discussed earlier in Chapter 1, equilibrium solutions generally need only consider the normal contact forces, and use kinematic hypotheses for determining the velocities, such as "race control." The ball centrifugal forces are added to the applied contact forces and therefore the equilibrium is of a quasi-static nature. Such a solution does not always completely represent the steady-state behavior of a ball bearing. In fact, it may provide no more

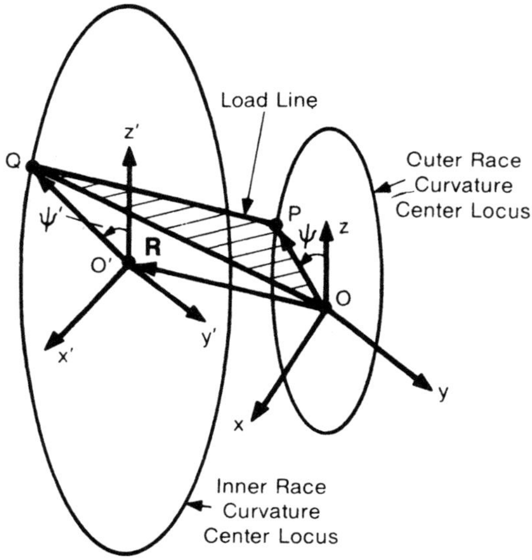

Figure 3-24 Geometry of interaction under static equilibrium.

than a point on the complete steady-state response, which can only be defined by a time cycle solution. However, the equilibrium solution does provide a fairly accurate load distribution on the rolling elements and it can be very effectively used for computing the fatigue life and stiffness of the bearing. The formulas generally used for computing the fatigue life are well known and they are briefly described in Appendix III. Bearing stiffness can be easily derived from the load deflection relationship discussed earlier for the rolling element/race interactions. In addition to these practical parameters the equilibrium solution is often helpful in determining the initial conditions for integrating the equations of motion and determining the true steady-state behavior. Before presenting the equilibrium equations some geometric and kinematic features should be briefly discussed.

Geometrical Considerations. A ball bearing is commonly subjected to a thrust load, a radial load, and a misalignment. Figure 3-24 shows the relative position of the races for a combined axial and radial load; for simplicity in the diagram misalignment is not shown. In the event that the bearing is subjected to a much more sophisticated loading, the initial conditions may still be set by a simplified loading, and dynamic equations may be integrated to obtain the ultimate steady-state or equilibrium solution. The static equilibrium formulation may, therefore, be considered in terms of the following three degrees of freedom:

- Relative axial deflection of the races, $\mathscr{X}$, along the x axis.
- Relative radial deflection of the races, $\mathscr{Z}$, along the z axis.
- Relative misalignment Θ about the y axis.

It may be sufficient to consider the misalignment Θ as fixed, thereby eliminating the moment equilibrium equation. Thus equilibrium has to be performed only in the axial and the radial directions. The geometric interactions entering the equilibrium equations are very similar to the ball/race interactions discussed earlier. Quasi-static equilibrium, however, will demand that, in addition to determining the relative position of the races under equilibrium conditions, the position vectors of the balls be determined such that ball equilibrium is satisfied and the centrifugal forces are duly accounted for. Thus, for a specified azimuth location of the balls, the axial and radial coordinates of the geometric center of every ball form additional unknowns. Let the race geometric center and the geometric center of the ball at azimuth location ψ be given by the following vectors:

$$\text{Race geometric center: } \mathbf{r}_{gr} = \begin{Bmatrix} \mathscr{X} \\ 0 \\ \mathscr{Z} \end{Bmatrix} \tag{3.102}$$

$$\text{Ball geometric center: } \mathbf{r}_{gb} = \begin{Bmatrix} x \\ -r\sin\psi \\ r\cos\psi \end{Bmatrix} \tag{3.103}$$

If the race is misaligned by an angle Θ, the ball center relative to the race center in the race frame may be given by

$$\mathbf{r}_{brc} = \begin{Bmatrix} x - \mathscr{X} - \Theta(r\cos\psi - \mathscr{Z}) \\ -r\sin\psi \\ \Theta(x - \mathscr{X}) + r\cos\psi - \mathscr{Z} \end{Bmatrix} \tag{3.104}$$

and the race azimuth angle, ψ', is

$$\psi' = \arctan\left(\frac{r\sin\psi}{\Theta(x - \mathscr{X}) + r\cos\psi - \mathscr{Z}}\right) \tag{3.105}$$

The vector $\mathbf{r}_{br}$, locating the ball center relative to the race curvature center, is now defined in the inertial frame as

$$\mathbf{r}_{br}^{i} = \begin{Bmatrix} x - \mathscr{X} - \Theta\mathscr{R}\cos\psi' \\ -r\sin\psi + \mathscr{R}\sin\psi' \\ r\cos\psi - \mathscr{Z} - \mathscr{R}\cos\psi' \end{Bmatrix} \tag{3.106}$$

where $\mathscr{R}$ is the radius of the locus of the race curvature centers.

For the purpose of computing contact angles, the above vector may be transformed in the ball azimuth frame and the components may be denoted as follows:

$$\mathbf{r}_{br}^{a} = \begin{Bmatrix} a_1 \\ a_2 \\ a_3 \end{Bmatrix} = \begin{Bmatrix} x - \mathscr{X} - \Theta\mathscr{R}\cos\psi' \\ -\mathscr{R}\sin(\psi - \psi') - \mathscr{Z}\sin\psi \\ -\mathscr{R}\cos(\psi - \psi') + r - \mathscr{Z}\cos\psi \end{Bmatrix} \tag{3.107}$$

Similar to the definitions shown in equations (3.7) the contact angles are expressed as

$$\tan\alpha_1 = \frac{a_1}{a_3} \quad \text{and} \quad \tan\alpha_2 = \frac{-a_2}{\sqrt{a_1^2 + a_3^2}} \tag{3.108}$$

Also, the contact deflection δ is determined in terms of the argument of $\mathbf{r}_{br}$, the race curvature factor f, and the ball diameter d:

$$\delta = \sqrt{a_1^2 + a_2^2 + a_3^2} - (f - 0.5)d \tag{3.109}$$

Clearly, no contact will exist if δ is negative. For positive δ, the contact load is determined by the Hertzian solutions described in Appendix I. It may suffice here to denote the contact load Q by the expression

$$Q = \Upsilon\delta^{3/2} \tag{3.110}$$

where Υ is a constant that is dependent on the geometric parameters and material properties of the ball and race.

It may be noted that although the above formulation is applicable to any ball/race contact, one of the races may be assumed as a reference race for which

$$\mathscr{X} = \mathscr{Z} = \Theta = 0 \quad \text{and} \quad \psi' = \psi$$

Kinematic Considerations. For a given ball position, the conventional race control-type kinematic hypothesis may be used in order to determine the ball angular velocities. The hypothesis has been expressed by Jones [60] in terms of the inequality

$$Q_m a_m \mathscr{E}_m \cos(\alpha_i - \alpha_0) > Q_n a_n \mathscr{E}_n \tag{3.111}$$

where α_i and α_0 are contact angles [corresponding to the nominal angle α_1 in equation (3.108)] for the inner and outer races, respectively; Q is the normal contact load; a is the semimajor axis of the contact ellipse; and $\mathscr{E}$ is the complete elliptic integral of second kind of modulus (a/b), where b is the semiminor axis of the contact ellipse:

$$\mathscr{E} = \int_0^{\pi/2} [1 - (1 - b^2/a^2)\sin^2\phi]^{1/2}\, d\phi \tag{3.112}$$

If $m = 0$ and $n = i$ for the inequality (3.111) then outer race control exists, i.e., the ball is constrained from spinning on the outer race. Similarly, if $m = i$ and $n = 0$, inner race control exists and the ball may spin only on the outer race.

The physical justification for the above constraint lies in the fact that the ball may generally spin such that the spin moment produced by friction is minimal. Thus, the race that produces the large spin moment is the race that restricts the ball from spinning. In high-speed bearings, therefore, a large centrifugal force generally establishes outer race control; in other words, pure rolling exists at the outer race contact. However, since the contact surface is curved, as shown earlier in Figure 3-1, pure rolling can exist either at the center of the contact ellipse or, possibly, at two points on the major axis of the contact ellipse. For a constant traction coefficient when there is no net tractive force in the rolling direction, and when pure rolling exists at point P (see Figure 3-4), then the abscissa $\bar{x}_p$ can be determined by the equation

$$\int_0^{\bar{x}_{p/a}} (1 - \bar{x}^2)\, d\bar{x} = \int_{\bar{x}_{p/a}}^{1} (1 - \bar{x}^2)\, d\bar{x}$$

or

$$(\bar{x}_{p/a}) = 0.34729636 \tag{3.113}$$

The ball angular velocity can now be determined from the constraint on relative ball spin, the prescribed points of pure rolling and the angular velocity of the race. For the purpose of simplicity, all mass and geometric centers may be assumed to be coincident in the following analysis. The justification for this simplicity lies in the fact that the analysis presented here will be employed only for determining the initial conditions to a more complete and sophisticated dynamic simulation.

As shown in Figure 3-25, let point P be located by a vector $\mathbf{c}$ with respect to the center of the contact zone, and let vectors $\mathbf{r}_p$ and $\mathbf{R}_p$ denote the center of the contact zone with respect to the ball and race centers, respectively, as described in Section 3.1. Also, it may be noted that Figure 3-25 is drawn in the x–z plane of the ball azimuth frame. The ball angular velocity has x and z components only; the orbital velocity is $\dot{\theta}$, and the race angular velocity has a component only along the x axis.

In terms of the above simplifications and assumptions, the various velocities are written in the contact frame as

$$\text{(1) Race angular velocity:}\quad \boldsymbol{\omega}_{ri} = \begin{Bmatrix} \Omega_i \cos\alpha_i \\ 0 \\ \Omega_i \sin\alpha_i \end{Bmatrix}$$

where $i = 1, 2$ denotes the outer and inner raceways respectively, and Ω is the nonzero angular velocity component about the x axis.

$$\text{(2) Ball angular velocity:}\quad \boldsymbol{\omega}_{bi} = \begin{Bmatrix} \omega_x \cos\alpha_i - \omega_z \sin\alpha_i \\ 0 \\ \omega_x \sin\alpha_i + \omega_z \cos\alpha_i \end{Bmatrix}$$

where ω_x and ω_z are the components about the x and z axes.

$$\text{(3) Ball orbital velocity:}\quad \boldsymbol{\omega}_{oi} = \begin{Bmatrix} \dot{\theta} \cos\alpha_i \\ 0 \\ \dot{\theta} \sin\alpha_i \end{Bmatrix}$$

$$\text{(4) Vector:}\quad \mathbf{c}_i = \begin{Bmatrix} x_0 \\ 0 \\ \sqrt{\rho^2 - x_0^2} - \rho \end{Bmatrix}$$

where ρ is the radius of curvature of the deformed pressure surface.

The relative slip velocity at the point P is given by

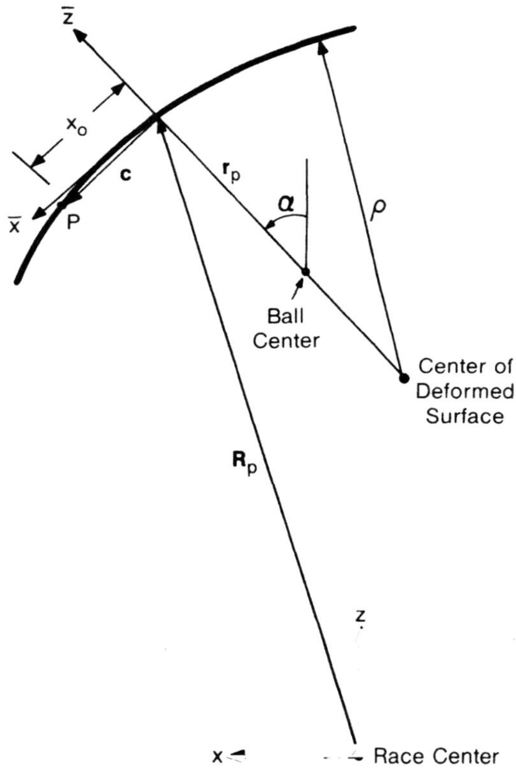

Figure 3-25 Coordinate frames for ball/race kinematics.

$$\mathbf{u}_i = (\boldsymbol{\omega}_{ri} - \boldsymbol{\omega}_{oi}) \times (\mathbf{R}_{pi} + \mathbf{c}_i) - \boldsymbol{\omega}_{bi} \times \mathbf{r}_{pi} \tag{3.114}$$

The slip vector $\mathbf{U}_i$ in the above equations will result only in a y component, and the slip velocity will therefore be in the rolling direction. For the controlling race this slip will vanish for any specified x_o; and for the race on which spinning is allowed, the slip will vanish for $x_o = 0$. These two conditions will provide two algebraic equations for the three unknowns, ω_x, ω_z, and $\dot{\theta}$. The third equation will be derived from the fact that the relative spin angular velocity on the controlling race must vanish. If the controlling race is denoted by the subscript I and the other race is denoted by the subscript J, straightforward algebra will show that the three simultaneous equations are written as

$$\begin{bmatrix} \sin\alpha_I & \cos\alpha_I & \sin\alpha_I \\ (r_{p_3} + c_3)\cos\alpha_I & -(r_{p_3} + c_3)\sin\alpha_I & \bar{X} \\ -r_{p_1}\sin\alpha_J + r_{p_3}\cos\alpha_J & -r_{p_1}\cos\alpha_J - r_{p_3}\sin\alpha_J & \bar{Y} \end{bmatrix} \begin{Bmatrix} \omega_x \\ \omega_z \\ \dot{\theta} \end{Bmatrix}$$

$$= \begin{Bmatrix} \Omega_1 \sin\alpha_I \\ \Omega_I \bar{X} \\ \Omega_J \bar{Y} \end{Bmatrix} \tag{3.115}$$

where

$$\bar{X} = r_{p_1} \sin\alpha_I - R_{p_1} \sin\alpha_I + (R_{p_3} + c_3) \cos\alpha_I$$
$$\bar{Y} = -R_{p_1} \sin\alpha_J + R_{p_3} \cos\alpha_J$$

A solution to equations (3.115) will give the required ball velocities. Now, if m denotes ball mass and r_b denotes its radial position, then the centrifugal force is simply

$$F_c = m r_b \dot{\theta}^2 \tag{3.116}$$

Equilibrium Equations. Equilibrium equations for the ball in the axial and radial directions are written as

$$\begin{aligned} \sum_{i=1}^{2} Q_i \sin\alpha_{1_i} \cos\alpha_{2_i} &= 0 \\ \sum_{i=1}^{2} Q_i \cos\alpha_{1_i} \cos\alpha_{2_i} &= F_c \end{aligned} \tag{3.117}$$

where the index i denotes the outer and inner race contacts respectively for the values 1 and 2.

The race equilibrium equations are

$$\begin{aligned} \sum_{i=1}^{N} Q_i \sin\alpha_{1_i} \cos\alpha_{2_i} + F_x &= 0 \\ \sum_{i=1}^{N} Q_i \cos\alpha_{1_i} \cos\alpha_{2_i} \cos\psi_i + F_r &= 0 \end{aligned} \tag{3.118}$$

where F_x and F_r are the axial and radial components of the applied forces, and N is the number of balls.

As discussed earlier, one of the races may be assumed to be a reference and the relative displacement for the other race has to be determined from the equilibrium equations. Also, for the reference race, α_2 will vanish.

Solutions to Equilibrium Equations. Both the ball and race equilibrium equations are nonlinear algebraic equations and an iterative method is required for their solutions. The procedure will involve satisfying ball equilibrium for a given relative position of the races and then checking for race equilibrium. The conventional Newton-Raphson method may be used for determining the successive iterates for both the ball and race positions. Assuming that the outer race is the reference race, the equilibrium equations may be written in the form

$$\begin{aligned} Q_1 \sin\alpha_{1_1} + Q_2 \sin\alpha_{2_1} \cos\alpha_{2_2} &= g_1 \\ Q_1 \cos\alpha_{1_1} + Q_2 \cos\alpha_{2_1} \cos\alpha_{2_2} - F_c &= g_2 \end{aligned} \tag{3.119}$$

and

$$\sum_{i=1}^{N} Q_{2i} \sin\alpha_{1i} \cos\alpha_{2i} + F_x = \bar{g}_1$$
$$\sum_{i=1}^{N} Q_{2i} \cos\alpha_{1i} \cos\alpha_{2i} \cos\psi_i + F_r = \bar{g}_2 \tag{3.120}$$

The unknown variables are the positions for each ball (x_i, r_i), $1 \leq i \leq N$, and the position of the inner race $(\mathscr{X}, \mathscr{Z})$ relative to the outer.

Since both Q and α are expressed in terms of the vector $\mathbf{r}^a_{br}$ given by equation (3.107), the various derivatives appearing in the Jacobian matrices can be expressed in terms of the derivatives of the components $\mathbf{r}^a_{br}$. The required derivatives may be written as

$$\frac{\partial Q}{\partial b} = \frac{3}{2}\Upsilon\delta^{1/2}\frac{\partial \delta}{\partial b} = \frac{3}{2}\Upsilon\delta^{1/2}\left[\frac{a_1\dfrac{\partial a_1}{\partial b} + a_2\dfrac{\partial a_2}{\partial b} + a_3\dfrac{\partial a_3}{\partial b}}{\sqrt{a_1^2 + a_2^2 + a_3^2}}\right] \tag{3.121a}$$

$$\frac{\partial \sin\alpha_1}{\partial b} = -\frac{a_3(a_3 - a_1)\dfrac{\partial a_1}{\partial b} - a_1(a_1 + a_3)\dfrac{\partial a_3}{\partial b}}{(a_1^2 + a_3^2)^{3/2}} \tag{3.121b}$$

$$\frac{\partial \cos\alpha_1}{\partial b} = -\frac{a_1(a_1 - a_3)\dfrac{\partial a_3}{\partial b} - a_3(a_1 + a_3)\dfrac{\partial a_1}{\partial b}}{(a_1^2 + a_3^2)^{3/2}} \tag{3.121c}$$

$$\begin{aligned}\frac{\partial \cos\alpha_2}{\partial b} = &\left[\{(a_1 + a_3)a_2^2 - a_2(a_1^2 + a_3^2)\}\left(\frac{\partial a_1}{\partial b} + \frac{\partial a_3}{\partial b}\right)\right.\\ &\left. - (a_1^2 + a_2^2)(a_1 + a_2 + a_3)\frac{\partial a_2}{\partial b}\right]\\ &\div \left[(a_1^2 + a_2^2 + a_3^2)^{3/2}\sqrt{a_1^2 + a_3^2}\right]\end{aligned} \tag{3.121d}$$

where b is any one of the unknown variables. The various derivatives of the components a_1, a_2, and a_3 are determined by straightforward differentiations of the expressions described in equation (3.107).

In summary the computational process for the quasi-static solution for ball bearings may be described by the following steps:

1. Make an initial estimate for the ball and race positions. This may represent a very lightly loaded configuration.
2. Compute all ball loads, contact angles, Hertzian half-widths, etc., at each ball/race contact, using the analysis described in section 3.1.
3. Use equation (3.114) to determine type of control for each ball.
4. Compute ball angular and orbital velocities, taking into account the kinematic considerations. Also, determine the centrifugal force F_c for each ball.
5. Solve equations (3.119) by the Newton-Raphson method and determine the equilibrium positions (x, r) for each ball.

6. Test for race equilibrium as per equations (3.120). If equilibrium is satisfied, then the current solution will be the final solution. If not, compute the corrections in the race position $(\mathscr{X}, \mathscr{Z})$ using the Newton-Raphson algorithm.
7. Correct the race position $(\mathscr{X}, \mathscr{Z})$ and go to step 2.

Although the above procedure involves an iteration within an iteration, the convergence on the inner iteration has been found to be very fast and the scheme has proven to be satisfactory.

3.5.2 Equilibrium Constraint for Roller Bearings

For post practical applications, a cylindrical roller bearing is generally subjected to a radial load along the z axis and a misalignment about the y axis. The equilibrium formulation may, therefore, be considered in terms of the following two degrees of freedom:

- Relative radial deflection of the races, $\mathscr{Z}$, along the z axis.
- Relative misalignment, Θ, about the y axis.

Once again, assuming a fixed misalignment and solving the equilibrium equation in the radial direction only may be sufficient.

As shown in Figure 3-25, let the relative race deflection be $\mathscr{Z}$. Then, for a prescribed roller azimuth ψ the race and roller geometric centers are written as

$$\text{Race geometric center:} \quad \mathbf{r}_{gr} = \begin{Bmatrix} 0 \\ 0 \\ \mathscr{Z} \end{Bmatrix} \tag{3.122}$$

$$\text{Roller geometric center:} \quad \mathbf{r}_{gb} = \begin{Bmatrix} 0 \\ -r\sin\psi \\ r\cos\psi \end{Bmatrix} \tag{3.123}$$

If the race is misaligned by an angle Θ, the roller center relative to the race center in the race frame may be given by

$$\mathbf{r}_{brc} = \begin{Bmatrix} -\Theta(r\cos\psi - \mathscr{Z}) \\ -r\sin\psi \\ +r\cos\psi - \mathscr{Z} \end{Bmatrix} \tag{3.124}$$

and the race azimuth angle ψ' is

$$\psi' = \arctan\left(\frac{r\sin\psi}{r\cos\psi - \mathscr{Z}}\right) \tag{3.125}$$

The vector $\mathbf{r}_{br}$, locating the roller center relative to the race land center,

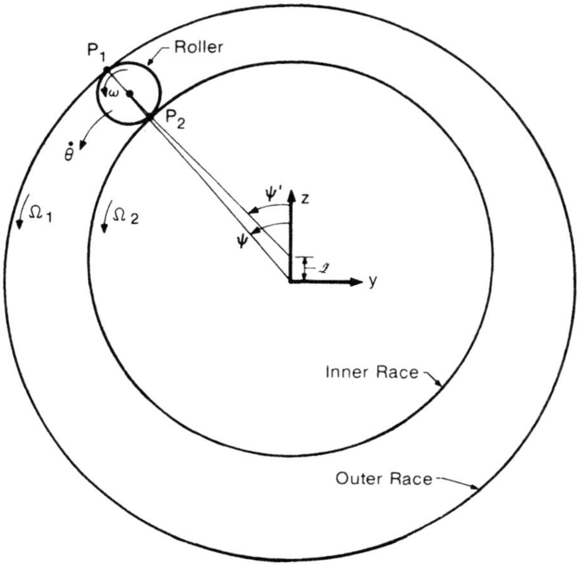

Figure 3-26 Schematic roller and race position in a radially loaded roller bearing.

is now defined in the inertial frame as

$$\mathbf{r}_{br}^{i} = \begin{Bmatrix} -\Theta\mathscr{R}\cos\psi' \\ -r\sin\psi + \mathscr{R}\sin\psi' \\ r\cos\psi - \mathscr{R} - \mathscr{R}\cos\psi' \end{Bmatrix} \tag{3.126}$$

where $\mathscr{R}$ is the radius of the locus of the race land centers.

The elastic deflection at the roller/race contact is readily given by the expression

$$\delta = \tfrac{1}{2}d - |r_{br}| \tag{3.127}$$

where d is the roller diameter.

Contact will exist only for a positive δ, and the contact load is determined by the line contact load–deflection relations described for roller/race interactions.

Kinematic Considerations. The computation of roller angular and orbital velocities is much simpler than that for ball bearings because the orientation of the roller angular velocity vector is well defined by the polar axis of the roller, which is generally parallel to the bearing axis. Thus, it is possible to obtain an epicyclic solution which assumes pure rolling at both the outer and inner races at points P_1 and P_2 (see Figure 3-26). When P_1 and P_2 lie in the y–z plane the equations determining the roller angular velocity ω and the orbital velocity $\dot{\theta}$ can be written in terms of the prescribed angular velocities Ω_1 and Ω_2 for the outer and inner races, respectively, as

$$\begin{bmatrix} c_{b_1} & c_{r_1} \\ c_{b_2} & c_{r_2} \end{bmatrix} \begin{Bmatrix} \omega \\ \dot{\theta} \end{Bmatrix} = \begin{Bmatrix} \Omega_1 \ c_{r_1} \\ \Omega_2 \ c_{r_2} \end{Bmatrix} \tag{3.128}$$

where c_{b_1} and c_{b_2} locate the points P_1 and P_2 relative to the roller center, and c_{r_1} and c_{r_2}, respectively, locate the point P_1 relative to the outer race center and point P_2 relative to the inner race center. The centrifugal force acting on the roller can now be written in terms of its mass m and the radial position r_b as

$$F_c = m r_b \dot{\theta}^2 \tag{3.129}$$

Equilibrium Equations. The equilibrium equation for the roller in the racial direction is written as

$$\sum_{i=1}^{2} Q_i = F_c \tag{3.130}$$

where the index i denotes the outer and inner race contacts for the values 1 and 2, respectively.

The race equilibrium equation is written as

$$\sum_{i=1}^{N} Q_i \cos\psi_i + F_r = 0 \tag{3.131}$$

where F_r is the applied radial load on the bearing and N is the number of rollers.

The relative displacement of the races will be determined from the solution to the above equilibrium equation.

Solutions to Equilibrium Equations. Both the roller and race equilibrium equations will be nonlinear algebraic equations and, using an iterative method, a simultaneous solution will be obtained. For the purpose of a numerical solution, the equilibrium equations (3.130) and (3.131) may be written as

$$Q_1 + Q_2 = F_c = g \tag{3.132}$$

and

$$\sum_{i=1}^{N} Q_i \cos\psi_i + F_r = \bar{g} \tag{3.133}$$

The unknown variables are radial positions for each roller (r_i), $1 \leq i \leq N$, and the position of inner race $\mathscr{X}$ relative to the outer.

The computation of derivatives, which is quite straightforward from equation (3.119), will basically involve differentiation of $|\mathbf{r}_{br}|$ with respect to r and $\mathscr{X}$. Similar to the ball bearing case, the computation process for the quasi-static solution for a roller bearing can be summarized by the following steps:

1. Obtain an initial guess for the relative roller and race positions.

2. Compute all roller loads at each roller/race contact, using the analysis described earlier for roller/race interactions.
3. Compute roller angular and orbital velocities from the kinematic considerations and estimate the centrifugal force F_c for each roller.
4. Solve equation (3.132) by the Newton–Raphson method and determine the equilibrium position (r) for each roller.
5. Test for race equilibrium as per equation (3.133). If equilibrium is satisfied, then the current solution will be the final solution. If not, compute the correction in the race position ($\mathscr{X}$) using the Newton–Raphson algorithm.
6. Correct the race position ($\mathscr{X}$) and go to step (2).

Once again, the solution process consists of a rapidly converging roller equilibrium iteration within the iteration for race equilibrium.

3.6 Summary

Models for the geometric interaction between the various elements of a rolling bearing are considered, and the applied force and moment vectors are derived. Formulation of the model consists of the following:

- geometrical considerations
- constitutive relations for normal and traction forces
- integration of local interactions

First, the vectors locating the position of the interacting elements are used to determine the geometric interaction. The relative slip or sliding speed at the point of interaction is determined from the velocity vectors. Second, appropriate constitutive equations are used to determine the normal and traction forces corresponding to the computed geometric interaction and the relative slip at the point of interaction. Finally, the local forces are integrated over the contact zone to obtain the total interaction between the elements under consideration.

In the first four sections of this chapter the above generic approach is applied to the rolling element/race, rolling element/cage, and cage/race interactions, and the interaction between rolling elements in a cageless bearing. The fifth section is devoted to the quasi-static equilibrium constraints for both ball and roller bearings.

CHAPTER 4

Elastohydrodynamic Lubrication

An elastohydrodynamic model which simulates the tractive behavior of the lubricant in rolling–sliding concentrated contacts is presented. The model consists of two parts: the computation of lubricant film thickness from the prescribed viscosity–temperature–pressure relation in the low-pressure region, and the prediction of traction based on the lubricant constitutive equation in the high-pressure region. For the latter, the coefficients are estimated by a least-squares analysis of the experimental traction data. The coefficients defining the behavior of a number of lubricants are presented.

The significance and theory of elastohydrodynamic lubrication (commonly abbreviated as EHD or EHL) in concentrated rolling–sliding contacts is fairly well understood and a large number of publications have appeared over the last couple of decades. Although a fair review of the literature is beyond the scope of this chapter, some of the noted works are by Grubin [33], Crook [18], Cheng and Sternlicht [14], Dowson and Higginson [23], Cameron [11], Herrebrugh [50], Cheng [15], and more recently, Hamrock and Dowson [46], which also provides an extensive list of the available literature. The problems of line contact configuration in roller bearings and the more complicated elliptical point contacts in ball bearings have been solved. The available works cover various aspects of elastohydrodynamic lubrication, such as lubricant rheology, film thickness computation, traction modeling, surface and asperity interaction, etc. In view of the enormous breadth of the subject, the objective of this chapter is restricted to the discussion of the models and formulations used in the computer program ADORE presented later in Chapter 7.

Aside from the common fact that lubrication in rolling bearings prevents metallic contacts, reduces friction and wear of interacting elements and, in the case of liquids, serves as a coolant for the bearing, the tractive behavior of the lubricant has been a significant consideration in the simulation of the performance of rolling bearings. For prescribed operating conditions, heat generation in the bearing is dependent on the traction coefficients*

*The friction forces in a rolling–sliding contact are generally referred to as traction forces; the effective coefficient of friction is called the traction coefficient, and it represents the ratio of tractive to normal forces at a contact.

at the various rolling–sliding contacts. The tractive forces at the rolling element/race interface greatly influence the orbital accelerations of the rolling elements which, in turn, determine the extent of rolling element/cage collisions and associated cage instabilities. In bearings where excessive sliding or skidding exists at the rolling element/race interface, the traction–slip relationship of the lubricant employed directly impacts the stability of the bearing.

Due to the complexities associated with the behavior of any lubricant under varying pressures and temperature, the development of a generalized model simulating the tractive behavior of a lubricant is extremely difficult. The general approach has been to experimentally test the lubricants in use and then fit an appropriate constitutive relation to the experimental traction data, and thereby obtain a traction–slip relationship for the lubricant under question. Using this general approach, a number of traction models have appeared in the literature [2, 3, 4, 5, 12, 19, 41, 55, 56, 57, 58, 63, 87, 90, 92, 93, 94, 97]. However, only one such model (after Kannel and Walowit [63] and Smith et al. [87]) is discussed in this chapter and incorporated in ADORE since the effectiveness of other models has yet to be investigated.

4.1 General Consideration in Lubrication Traction Modeling

The pressure–temperature–viscosity relationship in the environment of a rolling–sliding contact is the key factor in defining the traction behavior of the lubricant under question. However, the determination of such a relation over the wide range of conditions existing in the contact zone has been very difficult and semi-empirical models have, therefore, been promoted. Some of the general features of a typical contact, lubricant behavior, and the resulting traction are briefly discussed below.

4.1.1 Rolling Element/Race Contact Zone

A typical contact zone representing a rolling element/race interaction is shown in Figure 4-1. The interaction is divided into three regions: the inlet region where the lubricant is "pumped up" to form a film between the interacting surfaces; the high-pressure, or Hertz contact region, where a full lubricant film is established and the shear through the film results in traction; and the exit region where the classical pressure spikes and film constriction are observed. The minimum film thickness is about 70–75 percent of the nominal film thickness. For all practical purposes, the tractive behavior can be modeled by the inlet and high-pressure regions. Since the lubricant is pumped from the inlet region into the high-pressure contact zone, the nominal film thickness is determined by the lubricant characteristics in the inlet region. The net traction is then determined by the lubricant behavior in the high-pressure region. Thus, lubricant properties under

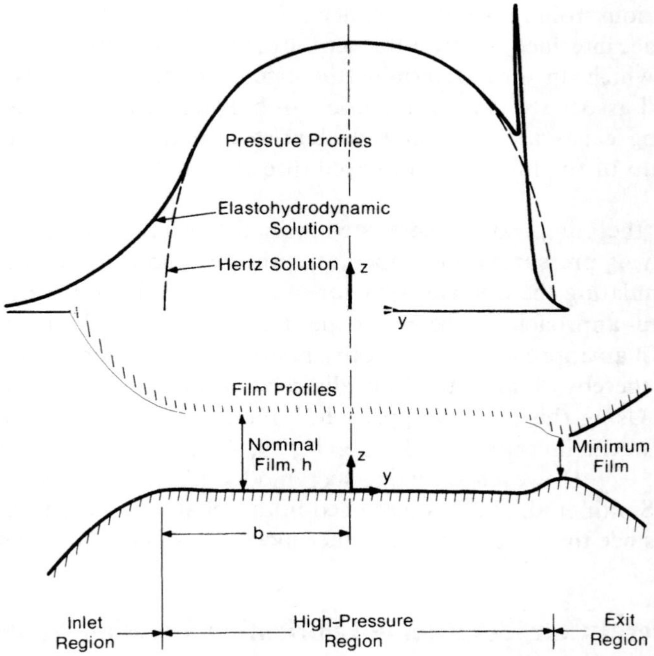

Figure 4-1 Schematic representation of an elastohydrodynamic contact zone.

ambient pressure determine the nominal film thickness, while high-pressure properties determine traction. It is, therefore, necessary to understand lubricant behavior in both high-pressure and ambient pressure environments.

4.1.2 Lubricant Rheology

The variation in lubricant viscosity as a function of temperature and pressure is the required constitutive relation in simulating a lubricated contact. Some of the general characteristics, such as exponential decrease in viscosity with temperature rise and an exponential increase in viscosity with increased pressure, are well established for most lubricating oils. However, the details of such variations are unique to each lubricant, and it may not be possible to fit any one type of constitutive equation to all lubricants in a wide range of operating environments. Nonetheless, over a certain range of pressures and temperatures, variation in viscosity can be approximated by either one of the following equations:

$$\mu = \mu_0 \exp\left[\alpha p + \beta\left(\frac{1}{T} - \frac{1}{T_0}\right)\right] \tag{4.1}$$

$$\mu = \mu_0 \exp\left[\alpha p + \beta(T_0 - T)\right] \tag{4.2}$$

here μ is the viscosity at pressure p (above ambient) and temperature T,

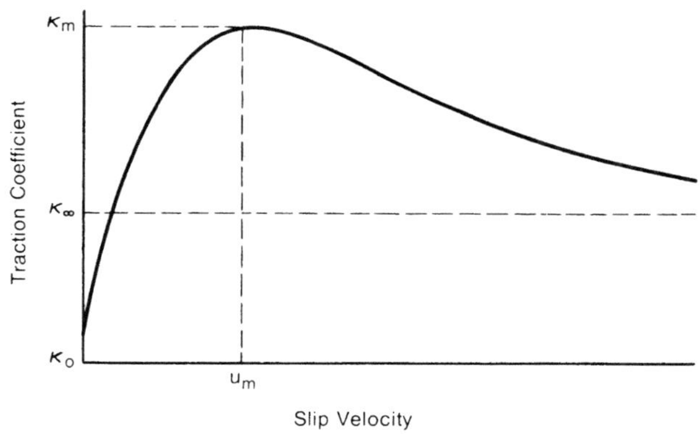

Figure 4-2 Typical traction–slip behavior in an elastohydrodynamic contact.

while μ_0 is the viscosity at a reference temperature T_0 and at ambient pressure; the coefficients α and β are known as the pressure–viscosity and temperature–viscosity coefficients, respectively.

Equations (4.1) and (4.2), although not unique by any means, are convenient in the computation of both film thickness and traction, equation (4.1) being more useful in the low-pressure region, while equation (4.2) is more convenient in simulating behavior in the high-pressure zone. Thus, equation (4.1) is useful for the film thickness computation and equation (4.2) is convenient for the computation of traction. It should be noted that the values of the coefficients α and β may be quite different in the low- and high-pressure regions. Also, the units of β in the two equations must be carefully noted.

4.1.3 Typical Traction–Slip Behavior

As will be demonstrated later, the general behavior of traction with the relative slip velocity between the interacting race and rolling element can be represented by Figure 4-2. The traction coefficient initially increases with increasing slip, but as the shear heating in the lubricant becomes significant at high slip velocities, it peaks to a maximum value and then begins to drop with further increase in slip. This general form of the relationship has also been observed with some solids, where the interaction is simulated by the following equation [67]:

$$\kappa = (A + Bu)\exp(-Cu) + D \tag{4.3}$$

here κ is the traction coefficient at a slip velocity u and A, B, C, and D are coefficients for a particular lubricant.

It may often be very convenient to compute the coefficients A, B, C, and D from actual traction data over the range of operating conditions of interest and then to use equation (4.3) to compute traction as a function of slip velocity. Such an approach, when possible, is certainly more efficient, in terms of the computational effort, than the more detailed traction model presented later in this chapter.

In general, the coefficients A, B, C, and D will be a function of the operating conditions and lubricant properties. If it is postulated that the model is valid for a given lubricant at prescribed operating conditions, then the values of the coefficients may be derived from the various parameters shown in Figure 4-2, which can be expressed as

$$u = 0 \qquad \kappa = \kappa_0 \tag{4.4a}$$

$$u = \infty \qquad \kappa = \kappa_\infty \tag{4.4b}$$

$$u = u_m \qquad \kappa = \kappa_m \qquad \frac{\partial \kappa}{\partial u} = 0 \tag{4.4c}$$

Combining (4.3), (4.4a), and (4.4b) gives

$$A = \kappa_0 - \kappa_\infty \tag{4.5}$$

$$D = \kappa_\infty \tag{4.6}$$

Also, from equation (4.3) and (4.4c),

$$-C(A + Bu_m) + B = 0 \tag{4.7}$$

and

$$(A + Bu_m)\exp(-Cu_m) + D = \kappa_m$$

which form the two simultaneous equations for determining B and C, after A and D are substituted from equations (4.5) and (4.6) respectively. Eliminating C in equations (4.7) will give a nonlinear equation in B,

$$(\kappa_0 - \kappa_\infty + Bu_m)\exp\left\{-\frac{Bu_m}{\kappa_0 - \kappa_\infty + Bu_m}\right\} = \kappa_m - \kappa_\infty \tag{4.8}$$

which is solved by standard bisection methods. Hence, all four coefficients can be computed from the prescribed conditions shown in Figure 4.2.

It must be remembered that although the above model may be realistic for a solid lubricant, it may overly simplify the behavior of any lubricating oil, in which case the parameters A, B, C, and D will be a function of rolling speeds, Hertzian stresses, etc., and such variations may often be quite significant. It is, therefore, necessary to develop a more sophisticated model for liquid lubricants.

4.2 An Elastohydrodynamic Traction Model

An approach to the development of an elastohydrodynamic traction model consists of basically two parts: the computation of film thickness and the computation of shear stress or traction across the film. The actual film thickness is determined by first computing the isothermal film thickness under ideal conditions where the lubricant floods the inlet zone (commonly referred to as the fully flooded condition), and then applying appropriate correction factors for thermal effects and starvation if the inlet zone is not flooded and the lubricant available is limited. The shear stress distribution through the film is obtained by solving the energy equation with appropriate geometric compatibility and constitutive equations, and finally the shear stress may be integrated over the contact zone to compute traction. The details of each of these steps associated with the traction model development are presented below.

4.2.1 Film Thickness Computation

As discussed above the actual film thickness h in any elastohydrodynamic contact is expressed in terms of the fully flooded isothermal film thickness h_{iso}, and the thermal reduction and starvation factors, denoted as ϕ_T and ϕ_S, respectively:

$$h = h_{\text{iso}}\phi_T, \phi_S \tag{4.9}$$

Until very recently, the side leakage factor used by Cheng [15] was widely adopted to modify the line contact film thickness to obtain an equivalent film thickness for elliptical contacts. However, advances in the elastohydrodynamic theory of elliptical contacts credited to Hamrock and Dowson [46] and Evans and Snidle [25, 26] have replaced the use of the side leakage factors with the more realistic formulas for elliptical contacts.

Isothermal Film Thickness. There are a number of formulas available for computing isothermal film thickness:

Grubin Formula [33]. This is perhaps one of the first relations which appeared in the literature for the computation of the isothermal film thickness h in a line contact configuration:

$$\frac{h}{R} = 1.95\frac{(GU)^{8/11}}{W^{1/11}} \tag{4.10}$$

where

$R = \dfrac{R_1 R_2}{R_1 \pm R_2}$, the effective radius in rolling direction

$G = \alpha E'$, the elasticity parameter

$$\frac{1}{E'} = \frac{1}{2}\left\{\frac{1-\sigma_1^2}{E_1} + \frac{1-\sigma_2^2}{E_2}\right\}$$

$U = \dfrac{\mu_0 \bar{U}}{E'R}$, the speed parameter

$\bar{U} = \frac{1}{2}(U_1 + U_2)$, the rolling speed

$W = \dfrac{q}{E'R}$, the load parameter

and σ, E, μ_0, α, q, U_1, U_2, and R_1, R_2 are, respectively, Poisson's ratio, elastic modulus, inlet viscosity, pressure–viscosity coefficient, applied load per unit length, surface velocities, and radii of curvature in the rolling direction of interacting surfaces 1 and 2.

Dowson–Higginson Formula [23]. With the load, speed, and elasticity parameters as defined for Grubin's formula, the Dowson–Higginson relationship is written as

$$\frac{h}{R} = 2\frac{G^{0.60}U^{0.70}}{W^{0.13}} \tag{4.11}$$

Hamrock–Dowson Formula [46]. Whereas both of the above formulas are for line contact configurations, in a more recent investigation, Hamrock and Dowson presented a formula for an elliptical point contact:

$$\frac{h}{R} = 2.69\frac{U^{0.67}G^{0.53}(1 - 0.61e^{-0.73k})}{W^{0.067}} \tag{4.12}$$

Here, $W = \dfrac{Q}{E'R}$, Q being the applied load; and $k = \dfrac{a}{b}$, the ellipticity ratio, where a and b are the semimajor and minor axes of the contact ellipse. The parameters U and G are the same as those described for Grubin's formula.

Thermal Reduction Factor. McGrew et al. [73] tabulated the thermal reduction factor ϕ_T based on the analysis by Cheng and Sternlicht [14] for line contact as a function of the dimensionless Hertzian pressure p_H/E' and the following parameters:

$$Q_m = \frac{\mu_0(U_1 + U_2)^2}{2KT_0} = \frac{2\mu_0\bar{U}^2}{KT_0}$$

$$\alpha^* = \tfrac{1}{2}\pi p^*\alpha$$

$$\beta' = \frac{\beta}{T_0}$$

$$s = \frac{U_2 - U_1}{u_2}$$

K, β, T_0, and p^* represent thermal conductivity, the temperature–viscosity coefficient, the inlet temperature, and a characteristic pressure (10^5 lb_f/in^2 or 6.8947×10^8 N/m^2), respectively; all other parameters are the same as discussed earlier.

Cheng and Sternlicht [14] have shown that the most important parameter in determining the thermal reduction factor is the viscous heating parameter Q_m. Additionally, it is shown that ϕ_T can be fitted to the relationship

$$\phi_T = f_1(1 - 0.10s)\left(1 - f_2\frac{p_H}{E'}\right) \tag{4.13}$$

Table 4-1 Variations of f_1 With Respect to α^*, β'/α^*, and Q_m†

α^*	15.710	15.710	15.710	23.564	23.564	23.564	31.419	31.419	31.419
Q_m \ β'/α^*	0.35	0.50	0.75	0.35	0.50	0.75	0.35	0.50	0.75
0.00	1.00	1.00	1.00	1.00	1.00	1.00	1.00	1.00	1.00
0.01	0.99	0.98	0.99	0.99	0.97	0.97	0.98	0.98	0.98
0.02	0.98	0.96	0.98	0.98	0.95	0.94	0.96	0.95	0.94
0.05	0.96	0.94	0.93	0.95	0.92	0.90	0.93	0.90	0.89
0.10	0.93	0.90	0.89	0.90	0.87	0.84	0.90	0.85	0.82
0.20	0.89	0.86	0.81	0.86	0.81	0.76	0.84	0.78	0.74
0.50	0.80	0.78	0.66	0.77	0.70	0.62	0.75	0.66	0.59
1.00	0.72	0.65	0.56	0.67	0.60	0.50	0.66	0.56	0.46
2.00	0.62	0.53	0.44	0.57	0.47	0.36	0.56	0.45	0.35
5.00	0.47	0.37	0.30	0.44	0.32	0.24	0.43	0.31	0.23
10.00	0.36	0.27	0.22	0.34	0.22	0.17	0.34	0.21	0.15

†Reproduced from Reference [73], courtesy of Aero Propulsion Laboratory, Wright-Patterson Air Force Base, Ohio.

Table 4-2 Variation of f_2 With Respect to α^*, β'/α^*, and Q_m†

α^*	15.710	15.710	15.710	23.564	23.564	23.564	31.419	31.419	31.419
Q_m \ β'/α^*	0.35	0.50	0.75	0.35	0.50	0.75	0.35	0.50	0.75
0.0	0.00	0.00	0.00	0.00	0.00	0.00	0.00	0.00	0.00
0.5	0.00	3.60	3.95	0.00	3.78	4.34	0.00	8.30	4.60
1.0	0.00	12.80	12.40	0.00	11.40	8.20	4.10	12.40	12.00
5.0	47.50	48.80	40.70	26.50	17.60	30.10	12.90	13.75	19.40
10.0	107.80	75.60	51.40	52.90	26.50	25.60	25.60	20.90	19.50

†Reproduced from Reference [73], courtesy of Aero Propulsion Laboratory, Wright-Patterson Air Force Base, Ohio.

where f_1 and f_2 are functions of the parameters Q_m, α^*, and β'. Over the range of parameters analyzed, these functions are tabulated in Tables 4-1 and 4-2 respectively. For any arbitrary set of parameters, a logorithmic interpolation of the data in Tables 4-1 and 4-2 is found to be very satisfactory.

Recently, a thermal analysis for an elliptical contact [10] appeared in the literature. For most ball bearings, however, the contact ellipses are very narrow (major to minor axis ratio of 5 or more), and, therefore, the contact is quite close to a line contact configuration. Consequently, the use of the preceding thermal analysis may be adequate for point contact in ball bearings also.

Starvation Factor. When a limited quantity of oil is present and the inlet zone is not fully flooded, a starvation factor ϕ_S must be applied to reduce the effective film thickness. The general approach in computing such a factor has been to prescribe an inlet boundary condition, in terms of distance, from the Hertzian contact zone to the point of film termination.

For a line-contact configuration, the starvation factor ϕ_S can be expressed, after Wolveridge et al. [98], as

$$\phi_S = 1 - \exp[-1.68005\Phi^{0.8315} + 0.260137\Phi^{1.558} - 0.0161460\Phi^{2.296}] \tag{4.14}$$

with

$$\Phi = \frac{x_0}{b}\left(\frac{b^2}{2h_\infty R}\right)^{2/3}$$

where x_0 is the distance to film termination measured from the edge of the contact zone; h_∞ is the fully flooded film thickness; b is the contact half-width; and R is the equivalent radius in the rolling direction, as defined earlier.

Equation (4.14) is a least-squares fit to the results presented by Wolveridge et al. [98], and the stated coefficients represent the available data with a maximum error of 1 percent.

For an elliptical point-contact configuration, Hamrock and Dowson [46] presented a starvation factor to modify their fully flooded film thickness [equation (4.12)], in terms of two distance parameters:

$$\tilde{m} = \frac{\tilde{x}}{b} \tag{4.15}$$

and

$$m^* = \frac{x^*}{b}$$

where $\tilde{x}$ is the distance measured from the center of the contact zone to the point of film termination, and x^* is the distance which defines a boundary between the fully flooded and starved conditions. In other words, the contact is fully flooded if $\tilde{x} \geq x^*$ and starved if $\tilde{x} < x^*$.

After performing the computation over a certain range of parameters, Hamrock and Dowson [46] obtained the following least-squares fit to their data:

$$\phi_S = \left[\frac{\tilde{m} - 1}{m^* - 1}\right]^{0.29} \tag{4.16}$$

and

$$m^* = 1 + 3.06\left[\left(\frac{R}{b}\right)^2 h_\infty\right]^{0.58} \tag{4.17}$$

The above is applicable to the central film thickness; similar results are also available for the minimum film thickness.

Most of the above formulas have been fairly well validated experimentally. For a more thorough discussion on the elastohydrodynamic film thickness and its interaction with the lubricant properties, the reader is referred to

Cameron [11], Gentle et al. [31], Greenwood and Kauzlarich [32], and many of the other references cited by Hamrock and Dowson [46]. In the area of starvation effects, the works of Wedevan et al. [95], Castle and Dowson [13], Wedevan [96], and Dowson et al. [24] are of great interest.

4.2.2 Computation of Traction

The model presented below for computing traction is after Kannel and Walowit [63] and as used by Gupta et al. [41]. The problem is basically formulated in terms of the following equations:

$$\text{Energy equation: } K\frac{\partial^2 T}{\partial z^2} = -\tau\dot{s} \tag{4.18}$$

$$\text{Geometric compatibility: } \frac{\partial u}{\partial z} = \dot{s}(\tau, p, T) \tag{4.19}$$

$$\text{Constitutive equation: } \dot{s}(\tau, p, T) = \frac{\tau}{\mu(p, T)} \tag{4.20}$$

where

K = thermal conductivity
T = absolute temperature
τ = shear stress
$\dot{s}$ = shear strain rate
u = lubricant velocity
p = pressure
μ = lubricant viscosity
z = coordinate across the film (as shown in Figure 4-1).

With the nominal film thickness h computed by the formulas described in the preceding sections, the applicable boundary conditions in terms of the coordinates shown in Figure 4-1 are

$$\begin{aligned} &\text{at } z = 0: \quad T = T_1 \text{ and } u = U_1 \\ &\text{at } z = h: \quad T = T_2 \text{ and } u = U_2 \end{aligned} \tag{4.21}$$

With T_0 as a certain reference temperature at which the viscosity is μ_0, all the variables may be nondimensionalized as follows:

$$T^* = T/T_0,\ \mu^* = \mu/\mu_0,\ z^* = z/h; \qquad U^* = u/\bar{U} \tag{4.22}$$

where $\bar{U} = \frac{1}{2}(U_1 + U_2)$

Combining equations (4.18) through (4.20) and (4.22), the problem may be expressed in terms of the following dimensionless equations:

$$\frac{\partial^2 T}{\partial z^2} = -\frac{\tau \bar{U} h}{K T_0}\frac{\partial U}{\partial z} \tag{4.23}$$

and

$$\frac{\partial U}{\partial z}=\frac{h\tau}{\mu_0\bar{U}}\frac{1}{\mu} \tag{4.24}$$

where the asterisk has been dropped for typographical brevity.

Note that the viscosity μ is a function of pressure and temperature as specified in equation (4.20).

Integrating equation (4.23) with the first boundary condition in equation (4.21) gives

$$\frac{\partial T}{\partial z}-\left.\frac{\partial T}{\partial z}\right|_{z=0}=-\frac{\tau\bar{U}h}{KT_0}(U-U_1) \tag{4.25}$$

Dividing equation (4.25) by (4.24) results in the governing equation

$$\frac{1}{\mu}\frac{\partial T}{\partial U}=\frac{\mu_0\bar{U}}{h\tau}\left.\frac{\partial T}{\partial z}\right|_{z=0}-\frac{\mu_0\bar{U}^2}{KT_0}(U-U_1) \tag{4.26}$$

which may be integrated through the film with the boundary conditions in equation (4.21) to obtain

$$\left.\frac{\partial T}{\partial z}\right|_{z=0}=\frac{h\tau}{\mu_0\bar{U}(U_2-U_1)}\left[\int_{T_1}^{T_2}\frac{dT}{\mu}+\frac{\mu_0\bar{U}^2}{KT_0}\frac{(U_2-U_1)^2}{2}\right] \tag{4.27}$$

By substituting equation (4.27) in (4.26) and by denoting the viscosity μ as a function of pressure and temperature

$$\mu=f(p,T) \tag{4.28}$$

the governing equation through the film is written as

$$\frac{\partial T}{\partial U}=\left[\frac{I}{U_2-U_1}+\frac{\mu_0\bar{U}^2}{KT_0}(1-U)\right]f(p,T) \tag{4.29}$$

where

$$I=\int_{T_1}^{T_2}\frac{dT}{f(p,T)}$$

Also, the shear stress is written by combining equations (4.19), (4.20), and (4.22) as

$$\tau=\frac{\mu_0\bar{U}}{h}\int_{U_1}^{U_2}f(p,T)\,dU \tag{4.30}$$

Equations (4.29) and (4.30) may be solved simultaneously to obtain the shear stress. The solution process will consist of first integrating (4.29) to establish a temperature–velocity relation, which when combined with (4.28) will give the viscosity–velocity relation required in (4.30). Also, for the purpose of computing the velocity and temperature profiles, the coordinate z, corresponding to temperature T and velocity U, determined by integrating (4.29), is obtained by equations (4.19), (4.20), and (4.22):

$$z = \frac{\mu_0 \bar{U}}{\tau h} \int_{U_1}^{U} f(p, T)\, dU \tag{4.31}$$

It should be noted that the boundary condition $T = T_2$ at $U = U_2$ is implicit in equation (4.29), since this condition has been used in equation (4.27).

For any arbitrary viscosity–temperature–pressure relation assumed in equations (4.28), (4.29), and (4.30) can, in general, be solved numerically to obtain the shear stress at any point within the contact zone. The shear stress can then be integrated over the contact zone to determine the total tractive force. A traction coefficient may be defined either for a local point in the contact zone or in terms of the integrated traction:

$$\begin{aligned} &\text{Local traction coefficient,} \qquad &\kappa_{\text{local}} = \frac{\tau}{p} \\ &\text{Global traction coefficient,} \qquad &\kappa_{\text{global}} = \frac{F}{Q} \end{aligned} \tag{4.32}$$

where F is the integrated traction, and Q is the applied normal load at the contact.

Since in any bearing the slip velocity varies greatly within the contact zone, the local traction coefficient is generally the required input in computing the applied forces on the rolling elements and the races. As will be seen in the next section, however, the global traction coefficient is significant in terms of estimating a constitutive relation (4.28) from experimental traction data obtained on a single rolling contact.

Equation (4.28), which defines the pressure–viscosity–temperature relation, is the primary input in the above model. The complexities associated with the computation of shear stress, which is directly related to the general form of equation (4.28), have created considerable interest in extensive investigations of lubricant rheology. Some of the recent work in this area has been done by Kannel and Dow [64], Houpert et al. [54], and Bair and Winer [6].

Although the formulation presented above will treat any form of equation (4.28), the two most common forms will be considered here. These forms are written in terms of the dimensionless temperature defined in equation (4.22) as

$$\text{Type I relation:} \quad \frac{\mu}{\mu_0^*} = f(p, T) = \exp\{\alpha^* p + \beta^* T_0(1 - T)\} \tag{4.33}$$

$$\text{Type II relation:} \quad \frac{\mu}{\mu_0^{**}} = f(p, T) = \exp\left\{\alpha^* p + \frac{\beta^{**}}{T_0}\left(\frac{1}{T} - 1\right)\right\} \tag{4.34}$$

Smith et al. [87] and Walowit and Smith [94] used the Type I relation and have shown that with such a relation the shear stress can be expressed in closed form. In the case of a Type II relation, however, a closed form integral is not straightforward and a numerical approach is preferred.

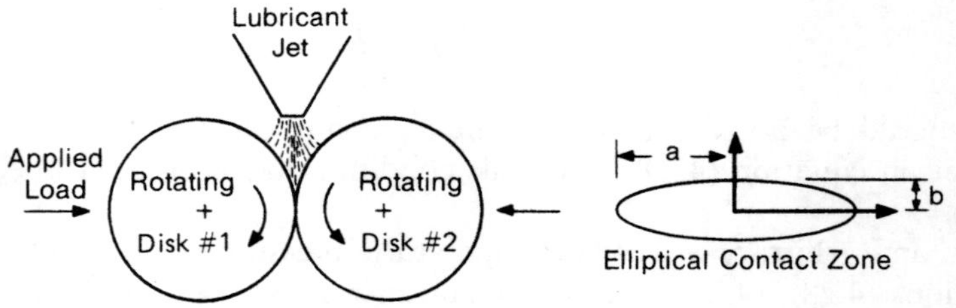

Figure 4-3 Schematic representation of a rolling disk machine for lubricant traction measurement.

The basic difference between the two relations lies in their physical characteristics. As the sliding speed increases, the temperature goes up due to increased shear heating, and the traction coefficient with Type I relations continues to diminish. However, with the Type II relation, the thermal effect and traction coefficient tend to level off. The difference between the two relations, in terms of predicting the traction coefficient, is significant, therefore, only at higher slip velocities.

4.2.3 Estimation of Lubricant Constitutive Equation

As discussed above, equation (4.28), which defines the constitutive equation of the lubricant in the high-pressure region of the contact zone, is the key input in the traction model discussed above. When either the Type I or Type II relation is used, the estimation of the parameters μ_0^*, α^*, β^*, etc., is an important task. The most direct method of determining these parameters is to measure the viscosity, both as a function of pressure and temperature, and then fit the data to an appropriate relation. However, high-pressure viscometry has been a very difficult subject and the most common approach for determining the required parameters is to run a single-contact experiment where the tractive behavior of the lubricant is measured as a function of contact pressures, temperatures, and rolling–sliding velocities, and the observed behavior is fitted to the above model in order to estimate the relevant parameters. Although such an approach may not provide significant insight into the physics of lubricant behavior, it does provide a realistic means of simulating lubricant traction in a rolling bearing.

As shown schematically in Figure 4-3, the apparatus used for measuring lubricant traction basically consists of two independently rotating disks loaded against each other and an oil jet which supplies the lubricant to the contact zone between the disks. These disks are generally crowned so that the contact zone is elliptical and the size and shape of the contact zone may closely simulate the actual condition in a bearing. The apparatus is commonly known as a "rolling disk machine" and it has been used and discussed by a number of investigators [62, 86, 41]. The experimental details are, therefore, omitted here and the reader is directed to the existing literature.

Data analysis consists of first dividing the available data into subsets representing the traction–slip behavior as a function of contact pressure at fixed rolling speed and lubricant temperature. It may also be assumed that the temperature at the surface of the disk is equal to the inlet lubricant jet temperature, since a good measurement of the disc surface temperature is generally not available. The traction behavior may now be fitted to either the Type I or Type II relation and the required parameters may be estimated by standard least-squares procedures.

In order to illustrate some of the analytical details, consider the elliptical contact zone, shown in Figure 4-3, and assume that the slip velocity is constant over the entire contact area. Using the Type I relation, equations (4.28) through (4.30) may now be written as

$$\mu = \mu_0^* \exp\{\alpha^* p + \beta^* T_0(1 - T)\} \tag{4.35a}$$

$$\frac{\partial T}{\partial U} = \frac{\mu_0^* \bar{U}^2}{K T_0}(1 - U)\exp\{\alpha^* p + \beta^* T_0(1 - T)\} \tag{4.35b}$$

$$\tau = \frac{\mu_0^* \bar{U}}{h}\int_{U_1}^{U_2} \exp\{\alpha^* p + \beta^* T_0(1 - T)\}\, dU \tag{4.35c}$$

In writing equation (4.35c) from (4.29), it may be noted that the integral I has been set to zero since $T_1 = T_2 = 1$.

Integration of equations (4.35b) and (4.35c) will result in the following expression for the shear stress:

$$\tau = \frac{\mu_0^* \bar{U}}{h\Lambda}(U_2 - U_1)e^{\alpha^* p}\frac{\sinh^{-1}\Lambda}{\sqrt{1 + \Lambda^2}} \tag{4.36}$$

where

$$\Lambda^2 = \frac{\mu_0^* p e^{\alpha^* p}\bar{U}^2(U_2 - U_1)^2}{8K}$$

Since the slip velocity is assumed to be constant over the contact ellipse, the shear stress can be readily integrated over the contact ellipse to obtain the tractive force F:

$$F = \int_0^1\int_0^{2\pi} ab\tau\, d\theta\, r\, dr = 2\pi ab\int_0^1 \tau r\, dr \tag{4.37}$$

The global traction coefficient κ_{global} can now be written as

$$\kappa_{\text{global}} = \frac{F}{Q} = \frac{2\pi ab}{Q}\int_0^1 \tau r\, dr \tag{4.38}$$

where Q is the normal contact load and a and b are the semimajor and minor axes of the contact ellipse.

Realizing that the pressure p in equation (4.36) may be written as

$$p = p_H\sqrt{1 - r^2} \tag{4.39}$$

where p_H is the maximum Hertz pressure, the traction coefficient may be rendered as

$$\kappa_{\text{global}} = \frac{F}{Q} = \frac{3}{\alpha^* p_H^2 h \Psi} \sqrt{\frac{8K\mu_0^*}{\beta^*}} \left\{ [\sinh^{-1}(\Psi e^{\alpha^* p_H/2})]^2 - \frac{2}{\alpha^* p_H} [f(\Psi e^{\alpha^* p_H/2}) - f(\Psi)] \right\}$$

where

$$\Psi = \bar{U}(U_2 - U_1)\sqrt{\frac{\mu_0^* \beta}{8K}} \quad \text{and} \quad f(\Psi) = \int_0^{\Psi} \frac{[\sinh^{-1}\psi]^2}{\psi'} d\psi' \tag{4.40}$$

The integral $\phi(\psi)$ will have to be determined numerically and it may be readily tabulated, as shown by Smith et al. [87].

The subscript "global" is hereafter dropped for brevity and the traction coefficient is defined by the notation

$$\kappa = \lambda \bar{Y} \bar{Z} \tag{4.41}$$

where

$$\lambda = \frac{3}{2p_H h} \sqrt{\frac{8K\mu_0^*}{\beta}}$$

$$\bar{Y} = \frac{1}{p\Psi}, \; p = \frac{\alpha^* p_H}{2}$$

and

$$\bar{Z} = [\sinh^{-1}(\Psi e^p)]^2 - \frac{1}{p}[f(\Psi e^p) - f(\Psi)]$$

If the measured traction coefficient for any data point i is κ_{mi}, then the squared deviation between the measured and the predicted data is

$$\hat{L} = \sum_{i=1}^{N} (\kappa_i - \kappa_{mi})^2$$

The three variables with respect to which $\hat{L}$ will be minimized are λ, p, and Ψ. It is clear that variation of $\hat{L}$ with λ is linear and for prescribed p and Ψ, therefore, λ can be determined by conventional linear regression [45].

Variations with p and Ψ is nonlinear and for computing these parameters; the problem may be set up as

$$\sum_{i=1}^{N} (\kappa_i - \kappa_{mi}) \frac{\partial \kappa_i}{\partial x_j} \equiv G_j = 0 \qquad j = 1, 2, \; x_1 = \Psi, \; x_2 = p \tag{4.42}$$

and the Jacobian

$$\frac{\partial G_j}{\partial x_k} = \sum_{i=1}^{N} \left\{ \frac{\partial \kappa_i}{\partial x_k} \cdot \frac{\partial \kappa_i}{\partial x_j} + (\kappa_i - \kappa_{mi}) \frac{\partial^2 \kappa_i}{\partial x_j \partial x_k} \right\} \tag{4.43}$$

Once the various derivatives have been expressed in terms of derivatives of Y and Z defined in equation (4.41), the set of two simultaneous equations can then be solved by the conventional Newton–Raphson method. It may be noted that by carrying out such a least-squares analysis, the errors associated with certain assumptions, such as surface temperatures, lubricant behavior, etc., are to some extent already corrected by the final estimates of the coefficients. The model, therefore, provides an acceptable representation of the observed traction behavior.

A similar procedure can be used for the Type II relation, but the complexity of the task increases greatly because the entire process must be done numerically. However, the coefficients of the Type I relation can be used to obtain realistic first estimates for the coefficients of the Type II relation. Furthermore, since α primarily determines the traction behavior of low sliding speeds where thermal effects are absent (and hence the two types of relations will be equivalent), it may be quite reasonable to assume $\alpha^{**} = \alpha^{*}$. The other two coefficients, μ_0^{**} and β^{**}, can now be obtained numerically for a precribed α^{**}.

Once the set of parameters is estimated for each subset of data corresponding to the prescribed lubricant temperature and rolling speed, the data can be tabulated and simple regression analysis can be used to define the effects of rolling speed and temperature. The generalized constitutive equation may then be substituted in the numerical traction model in order to simulate traction in a bearing. Alternatively, if the rolling speed does not vary greatly within the bearing and the temperatures at each rolling-element-to-race interaction are assumed to be constant, then the coefficients may be determined by interpolating the set of coefficients obtained for the various temperature and rolling velocity data subsets.

In reality, the above process is less tedious than it appears, both in terms of data analysis and in terms of the subsequent use of the model. The analysis of the data has to be carried out only once and all of the coefficients may be tabulated for future use. The use of the model is quite straightforward and the computer program ADORE allows for either the closed-form Type I model or any arbitrary relation for which the shear stress is computed numerically. For any particular bearing, if it is found that the traction–slip behavior as predicted by the most realistic traction model does not vary significantly from one rolling element to the other, then it may be reasonable to approximate it with a single curve of the type shown in Figure 4-2. This curve can then be used for all rolling element/race interactions. Such a simplification may generally save one-half of the computational effort.

4.3 Traction Behavior of Some Lubricants

The above model has been used to simulate the traction behavior of a number of lubricants. The procedure consists of two steps: obtaining experimental traction data using the conventional rolling disk-type apparatus, and performing a least-squares fit to the above model and computing the coefficients α^{*}, μ_0^{*}, β^{*}, μ_0^{**}, β^{**}, etc.

Gupta et al. [41] have tabulated such coefficients for a number of lubricants that fall into three basic categories:

1. U.S. Specification MIL-L-23699 type oils
2. U.S. Specification MIL-L-7808 type oils
3. traction fluids

It is shown that both Type I and Type II models fit equally well to most of the experimental data, except in the few cases where the Type II model offers a better fit at high slip velocities, perhaps because of the improved modeling of thermal effects offered by this model. For actual comparison of the data the reader is referred to Gupta et al. [41].

Since the Type I model is attractive from a numerical standpoint and in most cases it is as good as the Type II model, it becomes a logical choice for most lubricants.

The following paragraphs summarize the behavior of several lubricants within each of the above categories. Also, some available data for polyphenyl ether and an SAE-30-type mineral oil is presented.

4.3.1 U.S. Specification MIL-L-23699

The traction behavior of a number of lubricating oils within this specification, the data representing the ambient pressure–viscosity–temperature behavior, and the appropriate coefficients for the traction models are reproduced here from Gupta et al. [41]. Within this category, the five different lubricants studied are

1. Mobil Jet II oil, base stock
2. Mobil Jet II oil, fully formulated MIL-L-23699
3. MCS-1780 or MIL-L-27502, high-temperature version of MIL-L-23699
4. British version of U.S. Specification MIL-L-23699
5. A used sample of the British lubricant listed above.

The viscosity–temperature relations in the inlet zone (at low pressure) required for the film thickness computation are very similar for all of the above lubricants. For lubricants 1, 2, 4, and 5, viscosity as a function of temperature can be written as

$$\mu = \mu_0 \exp\left\{\beta_1\left(\frac{1}{T} - \frac{1}{T_0}\right) + \beta_2\left(\frac{1}{T_2} - \frac{1}{T_0}\right)^2\right\} \text{ Pa s} \qquad (4.44)$$

where $\mu_0 = 27.6 \times 10^{-3}$ Pa s is the reference viscosity at a temperature $T_0 = 311.11$ K, and the coefficients β_1 and β_2 are 3.60958×10^3 K and 7.76568×10^5 K^2.

For the high-temperature oil (#3 in the above list), the relation (4.44) is applicable with the coefficients $\mu_0 = 39.40 \times 10^{-3}$ Pa s, $T_0 = 311.11$ K, $\beta_1 = 3.71079 \times 10^3$ K and $\beta_2 = 7.39040 \times 10^5$ K^2.

The effective temperature viscosity coefficient β in equation (4.1) can

Table 4-3 Traction Coefficients for the MIL-L-23699 Type Lubricants ($\alpha^* = 5.8015 \times 10^{-9}$ 1/Pa)

Temperature (°C)	Rolling Velocity (m/s)	MIL-L-23699 μ_0^* (Pa s $\times 10^3$)	MIL-L-23699 β^* (1/K $\times 10^2$)	MIL-L-27502 μ_0^* (Pa s $\times 10^3$)	MIL-L-27502 β^* (1/K $\times 10^2$)
24	21.17	663.33	3.3768	519.82	3.1077
24	30.47	401.41	2.6833	280.55	2.5972
24	45.53	251.06	2.6951	543.49	3.2764
24	60.95	228.26	2.6629	142.16	4.6312
49	21.17	268.10	3.7145	184.36	3.2782
49	30.47	149.15	2.8552	133.39	2.9407
49	45.53	107.86	2.5423	82.323	2.3845
49	60.95	89.811	2.7670	47.719	2.4043
71	21.17	109.50	4.5290	76.504	3.7607
71	30.47	74.091	3.4250	57.767	3.0245
71	45.53	58.063	2.7610	40.667	2.4190
71	60.95	45.320	2.5810	26.120	2.3870
105	21.17	31.886	5.3618	17.038	4.5868
105	30.47	22.624	3.4432	14.973	3.6680
105	45.53	27.059	2.5294	13.981	2.7742
105	60.95	21.516	2.2864	10.126	2.5634
149	21.17	3.4441	5.2058	3.2150	5.7974
149	30.47	3.9347	4.5812	2.9589	4.8818
149	45.53	3.0121	4.5000	2.9545	4.5000
149	60.95	3.0282	4.5000	4.9675	3.6000

be obtained by differentiating (4.44) in the vicinity of actual operating temperatures.

$$\beta = \frac{\partial(\ln(\mu/\mu_0))}{\partial(1/T - 1/T_0)} = \beta_1 + 2\beta_2\left(\frac{1}{T} - \frac{1}{T_0}\right) \tag{4.45}$$

Since data on viscosity variation as a function of pressure are very limited, even for low-pressure conditions, the estimation of pressure–viscosity coefficients is quite difficult. Reference [74] is suggested to the reader as probably the best source for the procedure of computing α in the vicinity of the operating temperature.

The only other property required for the computation of film thickness is thermal conductivity, K, which can be expressed [73] as

$$K = 0.14867 - 2.22753 \times 10^{-4}(T - 366.67)\ \text{N/s}\,^\circ\text{C} \tag{4.46}$$

where T is the absolute temperature in K.

Based on the above inlet properties, a lubricant film thickness is computed for actual operating conditions over which the traction data were obtained. A least-squares fit is then performed to obtain the optimum coefficients for the Type I and Type II models [41]. The coefficients for the Type I model for the MIL-L-23699 and MIL-L-27502 are summarized in Table 4-3.

4.3.2 U.S. Specification MIL-L-7808

Also from Gupta et al. [41], the behavior of the following three lubricants is presented here.

1. ETD-2389, base stock
2. ETD-2389, fully formulated
3. Turbo Nicoil 13B, a French version of MIL-L-7808

The viscosity–temperature relationship in the inlet zone for all of the above oils may be approximated by the following relation after Walowit and Smith [93]:

$$\mu_0 = 10^{-3}[\exp\{\exp(-3.7048 \ln T_0 + 22.216)\} - 0.87] \text{ Pa s} \quad (4.47)$$

The effective temperature–viscosity coefficient β is obtained by differentiating the above equation:

$$\beta = \frac{\partial \ln \mu_0}{\partial(1/T_0)} \quad (4.48)$$

and the pressure viscosity coefficient and thermal conductivity are given by

$$\alpha = 7.5808 \times 10^{-10} + \frac{1.0742 \times 10^{-6}}{T_0 - 215.55} \text{ 1/Pa} \quad (4.49)$$

and

$$K = 9.65788 \times 10^{-2} \text{ N/s °C} \quad (4.50)$$

where T_0 is the reference temperature in K.

Once again, after computing the lubricant film thickness using the above properties, the traction data are fitted to the Type I traction model and the coefficients are summarized in Table 4-4.

4.3.3 Traction Fluid Santotrac 30

The behavior of Santotrac 30 with the following ambient properties is also reported by Gupta et al. [41]:

$$\log_{10} \log_{10}(\nu + 0.80) = -4.10436 \log_{10} T + 10.3047 \quad (4.51)$$

$$\rho = 0.817 + 6.66 \times 10^{-4}(422.22 - T) \text{ g/cc} \quad (4.52)$$

$$\alpha = 2.90 \times 10^{-8} \exp[-8.31776 \times 10^{-3}(T - 277.78)] \text{ 1/Pa} \quad (4.53)$$

$$K = 0.109041 - 1.70913 \times 10^{-5}(T - 311.11)^{1.322} \text{ N/s °C} \quad (4.54)$$

where

ν = kinematic viscosity (cs)
T = absolute temperature (K)

Table 4-4 Traction Coefficients for the MIL-L-7808 Type Oil ($\alpha^* = 5.2214 \times 10^{-9}$ 1/Pa)

Temperature (°C)	Rolling Velocity (m/s)	μ_0^* (Pa S $\times 10^3$)	β^* (1/K $\times 10^2$)
24	21.17	393.72	4.4500
24	30.47	197.48	2.9360
24	45.53	125.75	2.9156
24	60.95	59.374	2.7421
49	21.17	111.19	5.0504
49	30.47	63.029	3.1572
49	45.53	39.186	2.6644
49	60.95	32.020	2.9632
71	21.17	35.695	5.7654
71	30.47	34.290	4.3429
71	45.53	22.360	3.3048
71	60.95	17.341	2.7974
105	21.17	8.9577	8.5127
105	30.47	14.658	4.2739
105	45.53	6.8372	4.2329
105	60.95	7.4436	3.3487
149	21.17	3.0473	3.6000
149	30.47	2.1709	3.6000
149	45.53	1.9182	3.6000
149	60.95	1.6239	3.6000

ρ = density (g/cc)
α = pressure–viscosity coefficient (1/Pa)
K = thermal conductivity (N/s °C).

In the neighborhood of any reference temperature, an effective value of the temperature–viscosity coefficient β (K) is obtained numerically from equations (4.53) and (4.54).

The coefficients for the Type I model are summarized in Table 4-5. Limited data on a heavier traction fluid, Santotrac 50, has also been obtained by Nagaraj and Gupta [45]. However, it is shown that the above traction models fits the data well only at low slip velocities.

4.3.4 Polyphenyl Ether

In the study of 5P4E-type polyphenyl ether, Smith et al. [87] have shown that the Type I traction model fits the data reasonably well. The lubricant behavior in the inlet zone is given by the relation

$$\mu = \mu_0 \exp\left[\alpha p + (\beta + \gamma p)\left(\frac{1}{T} - \frac{1}{T_0}\right)\right] \tag{4.55}$$

Table 4-5 Traction Coefficients for Santotrac 30-Type Traction Fluid ($\alpha^* = 9.4275 \times 10^{-9}$ 1/Pa)

Temperature (°C)	Rolling Velocity (m/s)	μ_0^* (Pa s $\times 10^3$)	β^* (1/K $\times 10^2$)
24	21.17	783.93	2.7371
24	30.47	865.02	2.9704
24	45.53	519.34	2.7230
24	60.95	414.29	2.7293
49	21.17	299.38	3.8680
49	30.47	269.73	3.1559
49	45.53	206.13	2.7448
49	60.95	182.17	3.0161
71	21.17	99.429	5.1977
71	30.47	62.611	3.6934
71	45.53	74.353	3.0240
71	60.95	52.430	2.8903
105	21.17	18.047	6.1213
105	30.47	12.344	4.2136
105	45.53	9.8126	3.1579
105	60.95	9.7133	3.4547
149	21.17	2.4556	3.8358
149	30.47	0.97561	3.8358
149	45.53	1.0690	3.8358
149	60.95	0.93121	4.5302

where

$$T_0 = 350\ \text{K},\ \mu_0 = 32.75 \times 10^{-3}\ \text{Pa s},\ \alpha = 1.6820 \times 10^{-8}\ \text{1/Pa},$$
$$\beta = 3.6134 \times 10^3\ \text{K},\ \gamma = 1.4749 \times 10^{-5}\ \text{K/Pa}$$

The coefficients for the Type I traction model over the range of available traction data are shown to be

$$\mu_0^* = 9.6527\ \text{Pa s} \qquad \text{for rolling velocity, } U \leq 22.86\ \text{m/s}$$
$$= 9.6527 \exp\{-0.3011(U/22.86 - 1)\}\ \text{Pa s} \qquad \text{for } U > 22.86\ \text{m/s} \tag{4.56}$$

$$\beta^* = 0.0828\ \text{1/K} \tag{4.57}$$

$$\alpha^* = 5.4679 \times 10^{-9}\ \text{1/Pa} \tag{4.58}$$

4.3.5 SAE-30-Type Oil

Based on very limited data, Nagaraj and Gupta [75] presented the behavior of Mobil DTE heavy oil, which conforms to SAE-30 specification. The ambient properties used for film thickness are given by the equation

Table 4-6 Traction Behavior of Mobil DTE Heavy Oil ($\alpha^* = 7.2519 \times 10^{-9}$ 1/Pa); Hertz Pressure Range = 0.70–1.0 GPa

Lub. Temp. (°C)	Rolling Velocity (m/s)	μ^* (Pa s) × 10	β^* (1/K) × 10^2
66	12.70	3.3233	4.9464
66	25.40	1.6637	3.0600
66	30.48	1.0556	3.0510
66	38.10	0.7180	3.0744
93	12.70	1.1494	6.1596
93	25.40	0.7674	3.8790
93	30.48	0.5967	3.3840

$$\mu = 7.1660 \times 10^{-2} \exp\left[1.90 \times 10^{-8} p + 4.1043 \times 10^3 \left(\frac{1}{T} - \frac{1}{311.11}\right)\right] \tag{4.59}$$

and

$$K = 0.09608 \text{ N/s °C} \tag{4.60}$$

The traction data is fitted to the Type I model and the coefficients are summarized in Table 4-6.

It is seen that for most of the lubricants discussed above the traction model coefficients μ_0^* and β^* depend on the operating temperature and the rolling speed. With the objective of describing this dependence in a somewhat concise form the data is fitted to the equations of the following form:

$$\mu_0^* = V_1 \exp[V_2(313 - T) \exp[V_3(25 - U)] \tag{4.61}$$

and

$$\beta^* = B_1 \left(\frac{T_1}{313}\right)^{B_2} \left(\frac{U}{25}\right)^{B_3} \tag{4.62}$$

where T is the temperature (K) and U is the rolling velocity (m/s).

Conventional linear regression analysis is used to estimate the coefficients V_1, V_2, V_3, B_1, B_2, and B_3. The results are summarized in Table 4-7 for the various lubricants discussed above.

4.4 Summary

For a lubricated rolling element/race contact, an elastohydrodynamic model is presented to simulate the tractive behavior of the lubricant in the contact zone. After discussing some general behavior of concentrated elastohydrodynamic contacts, the traction model is discussed in two parts: the computation of film thickness and the determination of shear stress through the film which is integrated to compute the tractive force. Various expressions for computing the isothermal film thickness and the corrective factors for

Table 4-7 Regression Coefficients for the Type I Traction Model

Lubricant	α^* (1/Pa × 10^9)	V_1 (Pa s × 10)	V_2 (1/K × 10^2)	V_3 (s/m × 10^2)	B_1 (1/K × 10^2)	B_2	B_3 (× 10)
MIL-L-23699	5.8015	2.6529	3.6358	1.7054	3.3398	1.3075	−3.9353
MIL-L-27502	5.8015	2.1655	3.9221	0.6441	3.2577	1.0723	−2.7538
MIL-L-7808	5.2214	1.1431	3.3723	2.4079	4.1745	0.60873	−4.7828
Santotrac 30	9.4275	4.0063	5.0459	1.5590	3.4104	1.0930	−2.5647
Mobil DTE	7.2519	2.8407	2.3743	4.5016	2.8939	2.1402	−5.2909

thermal effects and starvation are presented. Shear stress is determined by integrating the energy equation across the lubricant film with appropriate geometric compatibility and constitutive relations. Since the constitutive equation defining the viscosity–pressure–temperature relationship for the high-pressure contact zone is generally not known *a priori*, the coefficients of a preassumed form of constitutive equation are estimated by carrying out a least-squares analysis of the actual traction data. Therefore, the coefficients so obtained closely simulate the behavior of the lubricant. Finally, the data for a number of commonly used lubricants are presented.

CHAPTER 5

Churning and Drag Losses

Based on the classical theories of fluid flow around a moving sphere and a cylinder, simple formulas for obtaining a first approximation to the lubricant drag forces and churning moments in a rolling bearing are presented.

In high-speed turbine engine applications, where a substantial quantity of oil is present in the bearing cavity, churning and drag losses may constitute a relatively large portion of the total power consumed by the bearing. Generally, the force exerted on the bearing element as it translates through the lubricant is called drag, and the moment imposed due to rotation of the element in the lubricant is called churning. Both the internal bearing geometry and the operating environment introduce complications that hinder realistic modeling of churning and drag. The classical laminar and turbulent flow theories are still the best indices, but these allow only a rough estimate of the effects of churning and drag phenomena. Moreover, simulation of any fluid damping associated with any oscillatory motion of bearing elements submerged in lubricant is beyond the existing technology.

The objective of this chapter is to briefly summarize the churning and drag relationships used in the computer code ADORE. Whether using either the computer code or the relations presented herein, the reader should recognize the limitations of the current technology, which lead to an oversimplification of the problem, and should use the calculated results only for the purpose of establishing qualitative trends.

5.1 Estimation of Drag Forces

The drag force exerted on the balls or rollers as they translate through a fluid can be estimated in terms of a drag coefficient C_D defined for spherical and cylindrical bodies (see Schlichtig [83]) as

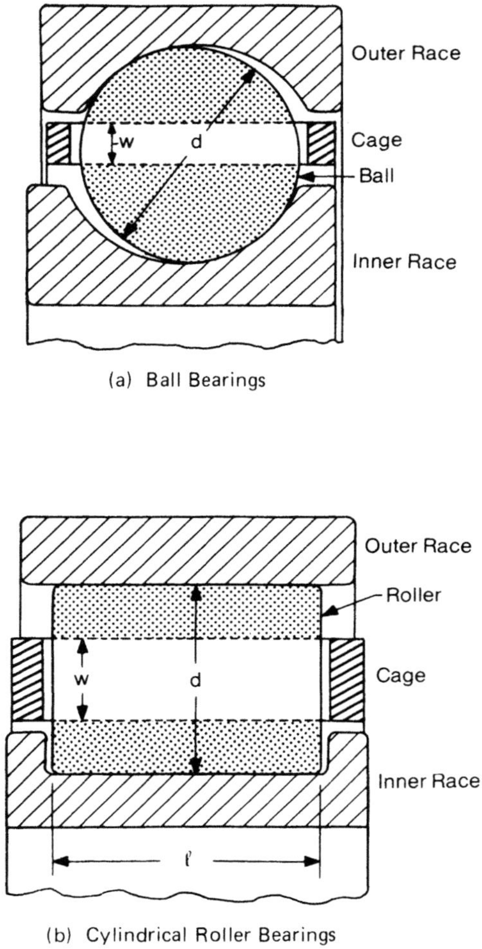

Figure 5-1 Effective frontal areas for drag force estimation. (a) Ball bearings; (b) cylindrical roller bearings.

$$C_D = \frac{D}{\frac{1}{2}\rho V^2 \mathscr{A}} \tag{5.1}$$

where D is the drag force; ρ is the mass density of the fluid; V is the velocity at which the body is traveling; and $\mathscr{A}$ is the frontal area expressed by the body to the flow direction.

The effective frontal areas for a ball and cylindrical roller bearing with cage are shaded in Figure 5-1. With the various dimensions indicated in the figure, the required area may be expressed as

$$\mathscr{A} = \tfrac{1}{4}\pi d^2 - wd \tag{5.2}$$

for ball bearings, and

Table 5-1 Drag Coefficients For Spherical and Cylindrical Bodies*

Reynolds Number (Re)	Drag Coefficient for Spheres (C_D)	Drag Coefficient for Cylinders (C_D)
10^{-1}	275.00	60.00
1	30.00	10.00
10	4.20	3.00
10^2	1.20	1.80
10^3	0.48	1.00
10^4	0.40	1.20
10^5	0.45	1.20
2×10^5	0.40	1.20
3×10^5	0.10	0.90
4×10^5	0.09	0.65
5×10^5	0.09	0.30
10^6	0.09	0.30

* After Schlichtig [83].

$$\mathscr{A} = \ell(d - w) \tag{5.3}$$

for cylindrical roller bearings.

For a cageless bearing, the required frontal area will be

$$\mathscr{A} = \tfrac{1}{4}\pi d^2 \tag{5.4}$$

for ball bearings, and

$$\mathscr{A} = \ell d \tag{5.5}$$

for cylindrical roller bearings.

Similar expressions may also be written for spherical and tapered roller bearings.

With μ being the viscosity of the fluid, the relevant Reynolds number is defined as

$$\mathrm{Re} = \frac{\rho V d}{\mu} \tag{5.6}$$

The corresponding drag coefficient C_D can now be determined from the data presented in Table 5-1. For any prescribed Reynolds number, a logarithmic interpolation of these data can be used to compute the relevant drag coefficients. In the absence of any data for tapered and spherical rollers, the drag coefficient can only be estimated by the available data for cylinders.

5.2 Estimation of Churning Moments

Realistic estimates for churning moments on the various elements in a rolling bearing is an extremely difficult task and the various approaches used only provide a rather gross approximation. The formulas and relationships for cylindrical bodies presented by Rumbarger, Filetti, and Gubernick [82] provide an estimate of churning moments in cylindrical roller bearings, and until any improved models become available these relationships will be used to approximate the churning moments on roller and cage surfaces in roller bearings.

Churning losses on cylindrical bodies occur on both the cylindrical and end surfaces.

5.2.1 Loss on the Cylindrical Surface

The moment M_c on the cylindrical surface is expressed as

$$M_c = \tau \mathscr{A} r \tag{5.7}$$

where $\mathscr{A}$ is the surface area; r is the reference radius from the center of rotation; and the wall shear stress, τ, is given by

$$\tau = \tfrac{1}{2}\hat{f}\rho U^2 \tag{5.8}$$

where ρ is the mass density; U is the mass average velocity of the fluid; and $\hat{f}$ is the friction factor which is defined for two types of flow conditions:

Vortex turbulent flow: $$\frac{\hat{f}}{\hat{f}_L} = 1.3\left(\frac{\text{Ta}}{41}\right)^{0.539474}_{T>41} \tag{5.9}$$

Couette turbulent flow: $$\frac{\hat{f}}{\hat{f}_L} = 3.0\left(\frac{\text{Re}}{2500}\right)^{0.85596}_{\text{Re}>2500} \tag{5.10}$$

where

$\text{Re} = \dfrac{\rho r \omega c}{\mu}$ is the Reynolds number

$\text{Ta} = \dfrac{\rho r \omega c}{\mu}\sqrt{\dfrac{c}{r}}$ is the Taylor number

and

$\hat{f}_L = 16/\text{Re}$ is the laminar friction factor for $\text{Re} < 25000$ or $\text{Ta} < 41$.

Also, r is the radius of the cylindrical surface under consideration and c is the effective clearance between the cylindrical surface rotating with an angular velocity ω in the stationary housing. Since in the case of a roller the estimation of effective clearance is difficult, the pocket clearance can be used but only as a first approximation.

For typical high-speed roller bearings, the flow on the cage surface is usually approximated as Couette turbulent, while that on the roller surfaces is generally best assumed as vortex turbulent. The computations for the inner and outer cylindrical surfaces of the cage are performed separately, using proper clearance and area values for each surfaces.

5.2.2 Loss on the End Surface

The following expressions are used to compute the churning moment on the sides of the cage in a roller bearing as well as on the ends of rollers. On the end faces, the moment is given by the expression

$$M_e = \tfrac{1}{2}\rho\omega r^5 C_n \tag{5.11}$$

where

$$C_n = \begin{cases} 3.87/\mathrm{Re}^{0.50} & \text{for laminar flow, Re} < 300{,}000 \\ 0.146/\mathrm{Re}^{0.20} & \text{for turbulent flow, Re} > 300{,}000 \end{cases}$$

and the Reynolds number

$$\mathrm{Re} = \frac{\rho r^2 \omega}{\mu}. \tag{5.12}$$

The effective radius r for a solid roller is the roller radius; for a hollow roller or the cage with an inner and outer radius of r_{in} and r_{out}, respectively, r can be expressed [82] as

$$\begin{aligned} r^5 &= r_{\mathrm{out}}(r_{\mathrm{out}}^4 - r_{\mathrm{in}}^4) && \text{for laminar flow} \\ r^5 &= r_{\mathrm{out}}^{0.40}(r_{\mathrm{out}}^{4.60} - r_{\mathrm{in}}^{4.60}) && \text{for turbulent flow} \end{aligned} \tag{5.13}$$

and the Reynolds number will be based on the outer radius r_{out}.

No straightforward expression is available for computing the churning moments on a ball. However, if the loss is approximated as that occurring on a projected area on a plane normal to the ball angular velocity vector, then the ball can be represented by a thin disk. Equation (5.11) can now be used to compute the moment on the projected area or the end surface of the disk. The disk, being infinitesimally thin, will have no cylindrical surface and the moment computed on the end faces will constitute the total churning moment.

5.3 Effective Lubricant Viscosity and Density

All of the above formulas assume that the element on which the churning and drag losses are being computed is completely submerged in a fluid having a viscosity μ and a density ρ. For most bearing applications, however, only a small fraction of the bearing cavity is filled with the lubricating oil; therefore, the effective contribution to churning and drag effects is made by an oil–air mixture. Since the shear force through such a mixture is determined primarily by the component having a higher viscosity, it is reasonable to assume the viscosity of the oil as the effective viscosity for the churning and drag computations. The effective density, on the other hand, depends on the actual mass of the fluid to be moved and it may therefore be expressed as

$$\rho = \frac{\rho_{oil} \times \text{Oil Volume} + \rho_{air} \times \text{Air Volume}}{\text{Total Volume of Bearing Cavity}} \tag{5.14}$$

This type of volume average density approximation has been traditionally used in rolling bearing analyses [82].

5.4 Summary

In view of the complicated internal geometry of the bearing and the operating environment, it is emphasized that, as yet, a realistic simulation of churning and drag effects is not possible. The formulas presented for computing drag forces and churning moments are based on classical theories of fluid flow, and they provide only a first approximation to the actual losses in a bearing. The values predicted by the formulas are intended to be used only to establish general trends in a parametric design study.

CHAPTER 6

Numerical Integration of the Equations of Motion

The equations of motion of each bearing element are nondimensionalized and reduced to a set of simultaneous first-order differential equations. Both explicit and implicit integrating algorithms are presented for integrating the set of differential equations. All algorithms provide an estimate of the local truncation error at each time step, which is used to optimize the step size for prescribed limits on the local truncation error.

Since the real-time simulation of any dynamic system consists of the integration of a large set of differential equations in real time, the computational efficiency of the integrating algorithm has substantial practical significance. Although the advent of modern high-speed computers has greatly facilitated the use of sophisticated computer codes, the effectiveness of most advanced computer codes is still dependent on computer resource-related factors. For example, the improper selection of the integrating algorithm, or the failure to optimize the size of the time step, may result in the use of prohibitively large computer time.

An integrating algorithm may be either explicit or implicit. The explicit algorithms make use of the solutions only from the previous step to determine the solution at the next step. The implicit algorithms generally constitute a predictor–corrector process, where previously computed solutions at a number of steps are first extrapolated to the next step (predictor process), after which the differential equation is used to apply a correction to the predicted solution (corrector process) in an iterative manner. In the explicit methods, depending on the order of the method, several derivative computations may be required over one time step, while relatively few derivative evaluations over a time step will be necessary in the implicit methods. Also, estimation of the local truncation error is quite straightforward in an implicit representation and, therefore, the step size optimization is relatively easy. However, for certain prescribed initial conditions and depending on the order of the method, the implicit formula is not self-starting, since solutions at a number of previous steps are required. In this case, an explicit formula may be required to generate a sufficient number of solutions to start the predictor–corrector process.

The implicit predictor–corrector methods often permit a relatively large step size, and for most systems two or three corrector iterations are generally adequate. Therefore, the predictor–corrector methods may considerably speed up the integration process over the prescribed time domain. In the case of rolling element bearings, however, there are two general characteristics which must be carefully studied before selecting an implicit method. First, the general motion of the bearing elements contains cyclic components of greatly different frequencies; for example, the rolling element/race elastic contact vibration frequency is several orders of magnitude larger than the shaft rotational frequency [38]. Second, the solution may contain a large number of discontinuities resulting from rolling element/cage and cage/race collisions. Since the extrapolation process would be quite inefficient under both of these conditions, the explicit methods may be generally preferred for rolling bearing simulation. However, if the bearing is cageless, and certain constraints eliminate high-frequency vibration, the implicit methods are useful once the integration process is initiated. It is for this reason that both methods have been incorporated in ADORE.

The objective of this chapter is to discuss the various integrating algorithms incorporated in the ADORE. The equations of motion formulated in Chapter 2 are first nondimensionalized and then several explicit formulas are discussed. Based on the Adams-type methods, a predictor–corrector algorithm is developed in divided difference form so that the change in step size can be relatively easy. Finally, the use of certain constraints in speeding up the integration over the prescribed time domain is discussed.

6.1 Dimensional Organization

Since estimation of the local truncation error is an important part of any integrating algorithm, it is necessary to properly nondimensionalize the equations of motion and determine the various scales. In a rolling bearing, the highest frequency present in the general motion of the rolling element generally corresponds to the rolling element/race elastic contact vibrations [38]; this frequency is readily seen in the axial and radial motion of the rolling element. The axial equation of motion is simply

$$m\ddot{x} = F_1 \tag{6.1}$$

where F_1 is the applied force in the axial direction, m is the mass of the rolling element, and $\ddot{x}$ is the axial acceleration.

Due to a large number of parameters entering into the computation of the geometric interaction and the determination of the applied forces, the selection of realistic force and length scales is extremely difficult. However, it is found that the applied load on the bearing Q_0 and the nominal radius of the rolling element r_0 can be satisfactorily used to nondimensionalize the equation of motion. Thus, equation (6.1) is written as

$$\frac{md^2(x^* r_0)}{dt^2} = F_1^* Q_0 \tag{6.2}$$

where $x^* = x/r_0$ and $F_1^* = F_1/Q_0$ are the dimensionless position and applied force in the axial direction, respectively.

Equation (6.2) can be further reduced to determine a dimensionless time t^* as

$$\frac{d^2 x^*}{dt^{*2}} = F_1^* \tag{6.3}$$

where

$$t^* = t/\sqrt{mr_0/Q_0} \tag{6.4}$$

Thus, the characteristic force, length, and time scales ($\mathscr{Q}$, $\mathscr{L}$, $\mathscr{T}$ respectively) may be summarized as

$$\begin{aligned} \mathscr{Q} &= Q_0 \\ \mathscr{L} &= r_0 \end{aligned} \tag{6.5}$$

and

$$\mathscr{T} = \sqrt{\frac{mr_0}{Q_0}}$$

These scales may be used to nondimensionalize all of the equations of motion; then the generalized solution and derivative vectors, **X** and **Y** respectively, defined earlier in equations (2.15) and (2.24) can be written in a dimensionless form. For convenience, the current time is included in the solution vector **X** as a first element and the corresponding element in the time derivative vector **Y** is set to unity. The generalized solution and derivative vectors may therefore be written as

$$\mathbf{X} = \begin{matrix} \{t \\ x_1, r_1, \theta_1, \dot{x}_1, \dot{r}_1, \dot{\theta}_1, \eta_1, \xi_1, \lambda_1, \dot{\eta}_1, \dot{\xi}_1, \dot{\lambda}_1, \\ \cdots\cdots\cdots\cdots\cdots\cdots\cdots\cdots\cdots, \\ x_n, r_n, \theta_n, \dot{x}_n, \dot{r}_n, \dot{\theta}_n, \eta_n, \xi_n, \lambda_n, \dot{\eta}_n, \dot{\xi}_n, \dot{\lambda}_n\}^T \end{matrix} \tag{6.6}$$

and

$$\mathbf{Y} = \dot{\mathbf{X}} = \begin{matrix} \{1, \\ \dot{x}_1, \dot{r}_1, \dot{\theta}_1, \ddot{x}_1, \ddot{r}_1, \ddot{\theta}_1, \dot{\eta}_1, \dot{\xi}_1, \dot{\lambda}_1, \ddot{\eta}_1, \ddot{\xi}_1, \ddot{\lambda}_1, \\ \cdots\cdots\cdots\cdots\cdots\cdots\cdots\cdots\cdots, \\ \dot{x}_n, \dot{r}_n, \dot{\theta}_n, \ddot{x}_n, \ddot{r}_n, \ddot{\theta}_n, \dot{\eta}_n, \dot{\xi}_n, \dot{\lambda}_n, \ddot{\eta}_n, \ddot{\xi}_n, \ddot{\lambda}_n\}^T \end{matrix} \tag{6.7}$$

Note that all variables are dimensionless and that the asterisks have been dropped for brevity.

The problem to be solved may now be written in terms of the solution and derivative vectors as

$$\mathbf{X} = \mathbf{X}_0 \quad \text{at} \quad t = t_0 \tag{6.8}$$

and

$$\mathbf{Y} = \dot{\mathbf{X}}$$

Using any numerical integration algorithm, the integration of the above will give the required simulation of the general motion of various bearing elements.

6.2 Explicit Algorithms

Explicit algorithms for integrating a set of differential equations are perhaps the easiest to use because they require only one set of solutions at the preceding time step. They are well suited to a system with discontinuities and any arbitrary cyclic variations. Also, explicit algorithms are useful in starting or restarting the predictor–corrector process of any implicit method. Although a few explicit algorithms provide an estimate of the local truncation error, a realistic estimation of the truncation error is the most common limitation of an explicit method, and the general procedure for estimating the error consists of repeating the calculations with the step size doubled. This, obviously, results in a large number of derivative evaluations and the computational effort becomes fairly extensive. The permissible step size, for a prescribed limit on the truncation error, may also be quite small compared to that allowable by a predictor–corrector algorithm; therefore, a large number of steps may be required for the integration over a prescribed time domain. In spite of these limitations, the explicit methods are well suited to integrating the equations of motion of the elements of a rolling bearing because of both the discontinuities resulting from rolling element/cage and cage/race collisions and the high-frequency rolling element/race vibrations.

The classical Runge–Kutta-type formulas are the most widely noted explicit methods. Some of these formulas also provide an estimate of the local truncation error. Although the estimate of error is not as accurate as the one provided by the implicit method, it can be effectively used to optimize the step size and it therefore adds significantly to the capabilities of the method.

The Runge–Kutta–Merson method is a fourth-order process, requiring an additional function evaluation for estimating the truncation error. There have been some questions about the accuracy of the error estimate provided by this method, particularly when the equations are nonlinear. The Scranton method [68] also uses five-function evaluations but the error estimate is only realistic for a single equation and its validity to a system of equations is questionable. The England method [68], which attempts to resolve this problem by an additional function evaluation per step, is therefore a six-stage process. Fehlberg [27, 28] has developed explicit formulas, of order 1 to 8,

Table 6-1 Coefficients of the First-Order Runge–Kutta–Fehlberg Formula

i \ j	$\mathscr{B}_{ij}$ 0	1	$\mathscr{G}_i$	$\mathscr{G}_i^*$
0	0		$\frac{1}{512}$	$\frac{1}{512}$
1	$\frac{1}{2}$		$\frac{255}{256}$	0
2	$\frac{1}{256}$	$\frac{255}{256}$	$\frac{1}{512}$	$-\frac{1}{512}$

Table 6-2 Coefficients of the Second-Order Runge–Kutta–Fehlberg Formula ($n = 3$)

i \ j	$\mathscr{B}_{ij}$ 0	1	2	$\mathscr{G}_i$	$\mathscr{G}_i^*$
0	0			$\frac{533}{2106}$	$-\frac{23}{1782}$
1	$\frac{1}{4}$			0	$\frac{1}{33}$
2	$-\frac{189}{800}$	$\frac{729}{800}$		$\frac{800}{1053}$	$-\frac{350}{11583}$
3	$\frac{214}{891}$	$\frac{1}{33}$	$\frac{650}{891}$	$-\frac{1}{78}$	$\frac{1}{78}$

Table 6-3 Coefficients of the Third-Order Runge–Kutta–Fehlberg Formula ($n = 4$)

i \ j	$\mathscr{B}_{ij}$ 0	1	2	3	$\mathscr{G}_i$	$\mathscr{G}_i^*$
0	0				$\frac{229}{1470}$	$-\frac{4}{735}$
1	$\frac{2}{7}$				0	0
2	$\frac{77}{900}$	$\frac{343}{900}$			$\frac{1125}{1813}$	$-\frac{75}{3626}$
3	$\frac{805}{1444}$	$-\frac{77175}{54872}$	$-\frac{97125}{54872}$		$\frac{13718}{81585}$	$\frac{5776}{81585}$
4	$\frac{79}{490}$	0	$\frac{2175}{3626}$	$\frac{2166}{9065}$	$\frac{1}{18}$	$\frac{1}{18}$

Table 6-4 Coefficients of the Fourth-Order Runge–Kutta–Fehlberg Formula ($n = 5$)

i \ j	$\mathscr{B}_{ij}$ 0	1	2	3	4	$\mathscr{G}_i$	$\mathscr{G}_i^*$
0						$\frac{16}{135}$	$-\frac{1}{360}$
1	$\frac{1}{4}$					0	0
2	$\frac{3}{32}$	$\frac{9}{32}$				$\frac{6656}{12825}$	$\frac{128}{4275}$
3	$\frac{1932}{2197}$	$-\frac{7200}{2197}$	$\frac{7296}{2197}$			$\frac{28561}{56430}$	$\frac{2197}{75240}$
4	$\frac{439}{216}$	-8	$\frac{3680}{513}$	$-\frac{845}{4104}$		$-\frac{9}{50}$	$-\frac{1}{50}$
5	$-\frac{8}{27}$	2	$-\frac{3544}{2565}$	$\frac{1859}{4104}$	$-\frac{11}{40}$	$\frac{2}{55}$	$-\frac{2}{55}$

Table 6-5 Coefficients of the Fifth-Order Runge–Kutta–Fehlberg Formula ($n = 7$)

i \ j	$\mathscr{B}_{ij}$ 0	1	2	3	4	5	6	$\mathscr{G}_i$	$\mathscr{G}_i^*$
0	0							$\frac{7}{1408}$	$\frac{5}{66}$
1	$\frac{1}{6}$							0	0
2	$\frac{4}{75}$	$\frac{16}{75}$						$\frac{1125}{2816}$	0
3	$\frac{5}{6}$	$-\frac{8}{3}$	$\frac{5}{2}$					$\frac{9}{32}$	0
4	$-\frac{8}{5}$	$\frac{144}{25}$	-4	$\frac{16}{25}$				$\frac{125}{768}$	0
5	$\frac{361}{320}$	$-\frac{18}{5}$	$\frac{407}{128}$	$-\frac{11}{80}$	$\frac{55}{128}$			0	$\frac{5}{66}$
6	$-\frac{11}{640}$	0	$\frac{11}{256}$	$-\frac{11}{160}$	$\frac{11}{256}$	0		$\frac{5}{66}$	$-\frac{55}{66}$
7	$\frac{93}{640}$	$-\frac{18}{5}$	$\frac{803}{256}$	$-\frac{11}{160}$	$\frac{99}{256}$	0	1	$\frac{5}{66}$	$-\frac{55}{66}$

with a greatly improved estimate of the local truncation error. Primarily due to an accurate error estimate these formulas are found to be very efficient for the present problem. However, based on some preliminary calculations with these formulas it is concluded that the order of the most efficient method really depends on the particular problem under consideration. The computer program ADORE, therefore, offers Runge–Kutta–Fehlberg formulas of orders 1 to 7. It is found that an order greater than 7 greatly increases the computing costs without an appreciable increase in accuracy.

All of the explicit formulas can be best summarized in terms of tabular notation. Thus, with the given solution y_n at the nth step, the step size Δ_n, and the derivative $f(y)$, the solution y_{n+1} at the $(n + 1)^{\text{th}}$ step is given by

$$y_{n+1} = y_n + \sum_{i=0}^{n} \mathscr{G}_i k_i \tag{6.9}$$

where

$$k_i = \Delta_n f\left(y_n + \sum_{j=0}^{i-1} \mathscr{B}_{ij} k_j\right), \qquad i = 1, \ldots, n \tag{6.10}$$

and the estimate of local truncation error E_{n+1} is given by

$$E_{n+1} = \sum_{i=0}^{n} \mathscr{G}_i^* k_i \tag{6.11}$$

The values of $\mathscr{B}$, $\mathscr{G}$ and $\mathscr{G}^*$ for the various order formulas are presented in Tables 6-1 through 6-7 after Fehlberg [27, 28].

6.2.1 Step-Changing Criterion

The procedure for computing the truncation error in the Runge–Kutta–Fehlberg method of order m essentially consists of generating an additional formula of order $(m + 1)$. The two solutions obtained with these formulas are then subtracted to obtain an estimate of the local truncation error of

Table 6-6 Coefficients of the Sixth-Order Runge–Kutta–Fehlberg Formula ($n = 9$)

i \ j	$\mathscr{B}_{ij}$ 0	1	2	3	4	5	6	7	8	$\mathscr{G}_i$	$\mathscr{G}_i^*$
0	0									$\frac{11}{864}$	$\frac{11}{270}$
1	$\frac{2}{33}$									0	0
2	0	$\frac{4}{33}$								0	0
3	$\frac{1}{22}$	0	$\frac{3}{22}$							$\frac{1771561}{6289920}$	0
4	$\frac{43}{64}$	0	$\frac{165}{64}$	$\frac{77}{32}$						$\frac{32}{105}$	0
5	$-\frac{2383}{486}$	0	$\frac{1067}{54}$	$-\frac{26312}{1701}$	$\frac{2176}{1701}$					$\frac{243}{2560}$	0
6	$\frac{10077}{4802}$	0	$\frac{5643}{686}$	$\frac{116259}{16807}$	$-\frac{6240}{16807}$	$\frac{1053}{2401}$				$\frac{16807}{74880}$	0
7	$-\frac{733}{176}$	0	$\frac{141}{8}$	$-\frac{335763}{23296}$	$\frac{216}{77}$	$-\frac{4617}{2816}$	$\frac{7203}{9152}$			0	$\frac{11}{270}$
8	$\frac{15}{352}$	0	0	$-\frac{5445}{46592}$	$\frac{18}{77}$	$-\frac{1215}{5632}$	$\frac{1029}{18304}$	0		$\frac{11}{270}$	$-\frac{11}{270}$
9	$-\frac{1833}{352}$	0	$\frac{141}{8}$	$-\frac{51237}{3584}$	$\frac{18}{7}$	$-\frac{729}{512}$	$\frac{1029}{1408}$	0	1	$\frac{11}{270}$	$-\frac{11}{270}$

Table 6-7 Coefficients of the Seventh-Order Runge–Kutta–Fehlberg Formula ($n = 12$)

$i \backslash j$	$\mathscr{B}_{ij}$: 0	1	2	3	4	5	6	7	8	9	10	11	$\mathscr{G}_i$	$\mathscr{G}_i^*$
0	0												0	$\frac{41}{840}$
1	$\frac{2}{27}$												0	0
2	$\frac{1}{36}$	$\frac{1}{12}$											0	0
3	$\frac{1}{24}$	0	$\frac{1}{8}$										0	0
4	$\frac{5}{12}$	0	$-\frac{25}{16}$	$\frac{25}{16}$									0	0
5	$\frac{1}{20}$	0	0	$\frac{1}{4}$	$\frac{1}{5}$								$\frac{34}{105}$	0
6	$-\frac{25}{108}$	0	0	$\frac{125}{108}$	$-\frac{65}{27}$	$\frac{125}{54}$							$\frac{9}{35}$	0
7	$\frac{31}{300}$	0	0	0	$\frac{61}{225}$	$-\frac{2}{9}$	$\frac{13}{900}$						$\frac{9}{35}$	0
8	2	0	0	$-\frac{53}{6}$	$\frac{704}{45}$	$-\frac{107}{9}$	$\frac{67}{90}$	3					$\frac{9}{280}$	0
9	$-\frac{91}{108}$	0	0	$\frac{23}{108}$	$-\frac{976}{135}$	$\frac{311}{54}$	$-\frac{19}{60}$	$\frac{17}{6}$	$-\frac{1}{12}$				$\frac{9}{280}$	0
10	$\frac{2383}{4100}$	0	0	$-\frac{341}{164}$	$\frac{4496}{1025}$	$-\frac{301}{82}$	$\frac{2133}{4100}$	$\frac{45}{82}$	$\frac{45}{164}$	$\frac{18}{41}$			0	$\frac{41}{840}$
11	$\frac{3}{205}$	0	0	0	0	$-\frac{6}{41}$	$-\frac{3}{205}$	$-\frac{3}{41}$	$\frac{3}{41}$	$\frac{6}{41}$	0		$\frac{41}{840}$	$-\frac{41}{840}$
12	$-\frac{1777}{4100}$	0	0	$-\frac{341}{164}$	$\frac{4496}{1025}$	$-\frac{289}{82}$	$\frac{2193}{4100}$	$\frac{51}{82}$	$\frac{33}{164}$	$\frac{12}{41}$	0	1	$\frac{41}{840}$	$-\frac{41}{840}$

the mth-order method. It is often desirable to use the solution obtained by the $(m + 1)^{th}$-order method as the final solution, because in such a case the computed truncation error, which is actually for an mth-order method, becomes a fairly conservative estimate and the step-changing criterion can be very efficient [29, 84]. A close examination of the formulas will show that generating the final solution of order $(m + 1)$ often requires one additional function evaluation compared to the case when a solution of order m and difference in the solutions of order m and $(m + 1)$ are computed. However, this additional effort greatly outweighs the efficiency gained in the step-changing algorithm. Thus the coefficient $\mathscr{G}_i$ summaried in Tables 6-1 through 6-7 are actually for the method of order one greater than that stated in each table.

For a method of order m, the local truncation error E_{n+1} at the $(n + 1)^{th}$ step varies as $(m + 1)^{th}$ power of the step size:

$$E_{n+1} \sim (\Delta_n)^{m+1} \tag{6.12}$$

Although this error may be compared to a prescribed limit on the permissible error and Δ_{n+1} can be estimated, it is often desirable to limit the error per unit size of the step and thereby obtain some bounds on the global truncation error to a certain extent. Such an error will clearly vary as the step size to the mth power.

$$\bar{E}_{n+1} = \frac{En + 1}{m} \sim (\Delta n)^m \tag{6.13}$$

An estimate of the step size for advancing the solution from step $(n + 1)$ to step $(n + 2)$ is now given by

$$\Delta_{n+1} = \Delta n \left(\frac{\varepsilon}{\bar{E}_{n+1}} \right)^{1/m} \tag{6.14}$$

where ε is the prescribed limit on the local truncation error per unit size of step.

In spite of the conservative estimate of the error it is found that for the present problem, the criterion stated in the above equation sometimes fails, particularly when severe rolling element/cage collisions introduce a distinct discontinuity in the solution. For this reason a multiplier of 0.80 is used as follows:

$$\Delta_{n+1} = 0.80\Delta_n \left(\frac{\varepsilon}{E_{n+1}} \right)^{1/m} \tag{6.15}$$

Also, as suggested by Shampine [84], an increase in the step size by more than a factor of 5.0 at any step is not permitted.

In addition to the above step-changing criterion, ADORE allows for prescribed upper and lower limits on the size of the time step. Such a feature can completely eliminate the automatic step-changing scheme if so desired for a particular case. Also, if the truncation limit criterion is not met with the minimum permissible step size for five consecutive steps further integration is terminated in ADORE.

6.3 Implicit Algorithms

A substantial savings of computer time is realized for problems where an implicit algorithm is workable. Of the numerous investigations dealing with the different aspects of the predictor–corrector process in an implicit method that have appeared in the literature, perhaps the most noted works are by Gear [30], Shampine and Allen [84], Lambert [68], Hindmarsh [52], Hindmarsh and Byrne [53], and Forsythe et al. [29]. Since the predictor formula requires an extrapolation of previously computed solutions, the variation in step size over the available solutions is an important consideration. For an arbitrary variation in step size, an interpolation is generally performed to generate equally spaced solutions and then the classical Newton formula in backward differences [51] is used to extrapolate the solutions. This is probably the most cumbersome step in the implicit method.

The algorithm employed in ADORE is essentially an Adams-type predictor–corrector representation in divided differences developed by Gupta [40]. The divided difference representation eliminates the need for interpolation in the predictor stage, whereby an arbitrary variation in step size can be very effectively treated. The method is presented below for a single differential equation and the extension to a system of equations will be straightforward.

Let the governing differential equation be written as

$$\frac{\partial y}{\partial x} = f(x, y) \tag{6.16}$$

with the initial condition, $y_0 = y(x_0)$.

It may be further assumed that solutions in the domain $0 \leq x \leq x_n$ have been obtained by some explicit algorithm, and the predictor–corrector formula, to be developed here, will advance the solution from x_n to x_{n+1}. The process can be written as

$$y_{n+1} = y_n + \int_{x_n}^{x_{n+1}} f(x, y)\, dx \tag{6.17}$$

With certain polynomial approximations to $f(x, y)$, the required algorithm can be developed. Similar to the well-known Adams–Bashford–Moulton formula [30], the predictor and corrector formulas may now be developed in divided differences.

6.3.1 Predictor Formula

Let the solutions $(y_n, y_{n-1}, \ldots y_{n-N})$ and the derivatives $(f_n, f_{n-1}, \ldots f_{n-N})$ be available at the $(N+1)$ points $(x_n, x_{n-1}, \ldots x_{n-N})$. For any x, let the derivative $f(x)$† be an Nth-order polynomial passing through the points

† Note the change in notation: the derivative is actually a function of both x and y and the notation used is $f(x) \equiv f(x, y)$.

$(x_n, x_{n-1}, \ldots x_{n-N})$, i.e.,

$$\begin{aligned} f(x) = f(x_n) &+ \sum_{i=1}^{N} (x - x_n)(x - x_{n-1}) \cdots (x - x_{n-i+1}) f[x_n, x_{n-1} \cdots x_{n-i}] \\ &+ (x - x_n)(x - x_{n-1}) \cdots (x - x_{n-N}) f[x_n, x_{n-1} \cdots x_{n-N}, \xi] \end{aligned} \tag{6.18}$$

where $x_{n-N} \leq \xi \leq x_{n+1}$ and the conventional divided difference notation [51] is used. The divided difference notation will be convenient for step-size changes, and will eliminate the need for interpolation to points spaced uniformly with a constant grid spacing.

The predictor formula is obtained by substituting the polynomial approximation, equation (6.18), in equation (6.17). Thus,

$$\begin{aligned} y_{n+1} = y_n &+ \sum_{i=0}^{N} f[x_n, x_{n-1} \cdots x_{n-i}] \sum_{j=0}^{i} \frac{x_{n+1}^{j+1} - x_n^{j+1}}{(j+1)} A_j \\ &+ f[x_n, x_{n-1} \cdots x_{n-N}, \xi] \sum_{j=0}^{N+1} \frac{x_{n+1}^{j+1} - x_n^{j+1}}{(j+1)} A_j, \end{aligned} \tag{6.19}$$

where A_j are the coefficients of the polynomial

$$(x - x_n)(x - x_{n-1}) \cdots (x - x_{n-i}) \equiv A_0 + A_1 x + A_2 x^2 + \cdots A_{i+1} x^{i+1}$$

In approximating ξ in order to estimate the truncation error terms of equation (6.19), it is reasonable to assume ξ equal to x_{n+1} and to accept the corresponding solution as the predicted value.

From a computational standpoint, it may be noted that as the order N increases, a round-off problem may be experienced with the terms multiplying the polynomial coefficients A_j. This problem can be eliminated by scaling the independent variable as

$$S \equiv \frac{x - x_p}{x_{p+1} - x_p} = \frac{x - x_p}{\Delta_p} \tag{6.20}$$

so that the polynomial

$$(x - x_n)(x - x_{n-1}) \cdots (x - x_{n-j}) \equiv \Delta_p^{j+1} (S - x_n^*)(S - x_{n-1}^*) \cdots (S - x_{n-j}^*)$$

where $x_j^* = (x_j - x_p)/\Delta_p$.

Also, if $p = n$, the integral of equation (6.17) takes the convenient form

$$\begin{aligned} &\int_{x_n}^{x_{n+1}} \{(x - x_n) \cdots (x - x_{n-j})\}\, dx \\ &= \Delta_n^{j+2} \int_0^1 \{(S - x_n^*)(S - x_{n-1}^*) \cdots (S - x_{n-j}^*)\}\, dS \\ &= \Delta_n^{j+2} \sum_{k=0}^{j+1} \frac{A_k^*}{k+1} \end{aligned} \tag{6.21}$$

where A_k^* are coefficients of the polynomial

$$(S - x_n^*) \cdots (S - x_{n-j}^*) \equiv A_0^* + A_1^* S + \cdots A_{n-j+1}^* S^{n-j+1}$$

Substituting (6.20) and (6.21) in equation (6.19) gives the predictor formula

$$\begin{aligned} y_{n+1} = y_n &+ \sum_{i=0}^{N} f[x_n, x_{n-1}, \ldots, x_{n-i}]\Delta_n^{i+1} \sum_{j=0}^{i} \frac{A_j^*}{j+1} \\ &+ f[x_n, x_{n-1}, \ldots, x_{n-N}, x_{n+1}]\Delta_n^{N+2} \sum_{j=0}^{N+1} \frac{A_j^*}{j+1} \end{aligned} \tag{6.22}$$

It can easily be seen from equation (6.20) that the coefficient A_j^* will only have to be computed if the grid in x is nonuniform.

6.3.2 Corrector Formula

For the general corrector formula, let $f(x)$ be a polynomial of degree M, passing through the $(M + 1)$ points $(x_{n+1}, x_n, x_{n-1}, \ldots, x_{n-M+1})$. Again, using the divided difference notation,

$$\begin{aligned} f(x) &= f(x_{n+1}) \\ &+ \sum_{i=0}^{M-1} (x - x_{n+1})(x - x_n) \cdots (x - x_{n-i+1}) f[x_{n+1}, x_n, \ldots, x_{n-i}] \\ &+ (x - x_{n+1})(x - x_n) \cdots (x - x_{n-M+1}) f[x_{n+1}, x_n, \ldots, x_{n-M+1}, \xi] \end{aligned} \tag{6.23}$$

Substitution of (6.23) in equation (6.17) gives the required corrector formula:

$$\begin{aligned} y_{n+1} = y_n &+ \sum_{i=-1}^{M-1} f[x_{n+1}, x_n, \ldots, x_{n-i}] \sum_{j=0}^{i=1} \frac{x_{n+1}^{j+1} - x_n^{j+1}}{(j+1)} B_j \\ &+ f[x_{n+1}, x_n, \ldots, x_{n-M+1}, \xi] \sum_{j=0}^{M+1} \frac{x_{n+1}^{j+1} - x_n^{j+1}}{(j+1)} B_j \end{aligned} \tag{6.24}$$

where B_j are coefficients of the polynomial

$$(x - x_{n+1})(x - x_n)(x_{n-1}) \cdots (x - x_{n-i}) \equiv B_0 + B_1 x + \cdots B_{i+2} x^{i+2}.$$

The value of ξ may be assumed as x_{n-M} in order to estimate the truncation error. The resulting demand for one additional data point is easily met if $M \leq N$, which is normally the case.

Scaling the independent variable, as for the predictor formula, will give the final corrector algorithm:

$$\begin{aligned} y_{n+1} = y_n &+ \sum_{i=-1}^{M-1} f[x_{n+1}, x_n, \ldots, x_{n-i}]\Delta_n^{i+2} \sum_{j=0}^{i+1} \frac{B_j^*}{j+1} \\ &+ f[x_{n+1}, x_n, \ldots, x_{n-M+1}, x_{n-M}]\Delta_n^{M+2} \sum_{j=0}^{M+1} \frac{B_j^*}{j+1} \end{aligned} \tag{6.25}$$

Again, the coefficients B_j^* will only have to be computed for a nonuniform grid. Also, it is readily discernible that the truncation error will employ the last term in the divided difference table and it will be proportional to Δ_n^{M+2}.

In order to implement equations (6.22) and (6.25), a formula will be required for computing the coefficients of the polynomial:

$$P \equiv (x - x_n)(x - x_{n-1}) \cdots (x - x_{n-p})$$

By expanding the polynomial for $p = 1, 2, 3, 4$ and by noting the algebraic similarities in the expansions, the polynomial can be written in the following general form:

$$\begin{aligned} P \equiv x^{p+1} + \sum_{k=1}^{p+1} x^{p-k+1}(-1)^k \sum_{i_1=1}^{p-k+2} x_{n-i_1+1} \sum_{i_2=i_1+1}^{p-k+3} x_{n-i_2+1} \\ \cdots \sum_{i_j=i_{j-1}+1}^{p-k+j+1} x_{n-i_j+1} \cdots \sum_{i_k=i_{k-1}+1}^{p+1} x_{n-i_k+1} \end{aligned} \tag{6.26}$$

The above formula can be used to compute the required coefficients, once the maximum value of p is assigned.

6.3.3 Step-Changing Criterion

The step-size in (6.26) can be changed at any time if the truncation error at the preceding step and the allowable limit are known. Let the order of the corrector be equal to that of the predictor ($M = N$); then the above algorithm becomes an $(N + 1)$ step process. The truncation error of the predictor–corrector step will now be given by the truncation error of the corrector step only [30]. Thus, the truncation error committed in arriving at x_{n+1} from x_n is given by

$$\frac{E_{n+1}}{\Delta_n^{N+2}} = f[x_{n+1}, x_n, \ldots, x_{n-N}] \sum_{j=0}^{N+1} \frac{B_j^*}{j+1} \tag{6.27}$$

As discussed earlier for the explicit methods, it may be desirable to limit the truncation error per unit time with the objective of obtaining certain bounds on the global error. Thus an estimate of step Δ_{n+1} in going from x_{n+1} to x_{n+2} can be obtained from the relation

$$\Delta_{n+1} = 0.80\Delta_n \left(\frac{\varepsilon}{\bar{E}_{n+1}} \right)^{1/(N+1)} \tag{6.28}$$

where ε is the allowable truncation limit and the constant 0.80 is used, with the expectation that the tolerance criterion will be met at x_{n+2}. Equation (6.28) can be used for both increasing and decreasing the step size.

Unlike the explicit method, anytime the step size is changed, all coefficients for the predictor–corrector formulas will have to be computed for $N + 1$ steps, assuming that no step change occurs within these steps. Since this added work may outweigh the savings in increased step size, it is recommended that the step size be changed only if the following condition is met:

$$\frac{\Delta_{n+1}}{\Delta_n} > 1.25 \tag{6.29}$$

In addition to the above criterion, ADORE does not permit a step-size change by more than a factor of 5.0 at any step, as done in the case of explicit methods. Also, the step size is varied within the prescribed upper and lower limits and in the event the truncation limit criterion is not met for five consecutive steps, further integration is terminated.

6.3.4 Change of Order

In lieu of a criterion for change of order, a trial-and-error approach is used. At a given time, solutions with the order reduced and increased by 1 are computed and the order giving the minimum truncation is selected. Here, too, maximum and minimum bounds on the order should be assigned. The maximum order can be determined by examining the divided differences which will show a general decrease with an increase in order. Due to machine round-off errors, however, the computed values tend to increase at very high orders. Therefore, the maximum permissible order is limited by machine round-off. Selection of the minimum order will be accomplished primarily by experience with the problem under investigation.

6.3.5 Computational Considerations

The following guidelines are helpful in the development of any computer code for implementing the above formulation:

1. Set a maximum limit on the order and hence the number of solutions to be stored.
2. Compute divided differences of derivatives at the first call and upgrade table at subsequent calls by adding the latest solution and deleting the earliest solution.
3. Check grid uniformity, and in the case of nonuniformity resulting from step-size change, compute coefficients (in subsequent calls).
4. Ensure that the current order is less than the maximum allowable order shown in the divided difference table.
5. Store the entire divided difference table in disk file at the end of the run for subsequent restarts of the program.

6.4 Selection of a Method

Generally, due to the discontinuities introduced by the rolling element/cage and cage/race collisions, an explicit formula will be most useful in obtaining a solution to the equations of motion of bearing elements. Also, for roller bearing where significant roller skew and the interaction of roller and race flange is very dynamic, the explicit formulas may be better than the predictor–corrector techniques. However, while using ADORE for certain applications it is found that a third-order predictor–corrector algorithm resulted in noticeable reduction in the computing costs over the fourth-order explicit formula.

For the prescribed problem, the selection of a particular method should, therefore, be done on a trial basis and the fourth-order explicit method may be used as a starting point.

For cageless bearings without significant collisions between rolling elements, the implicit predictor–corrector algorithm rapidly simulates the bearing performance over very large time intervals. In particular, if external constraints are used to suppress all rolling element/race high-frequency vibrations, then the predictor–corrector technique is extremely efficient. Also, when the solutions obtained by explicit methods demonstrate a quasi-steady-state and no significant transients or discontinuities are present, the predictor–corrector formula very conveniently generates the steady-state behavior over several shaft revolutions.

6.5 External Constraints

Since rolling element/race vibration generally constitutes the highest frequency in the motion of the rolling elements, careful suppression of this vibratory motion can result in a significant increase in the permissible size of the time step in both the explicit and implicit methods of integrating the equations of motion. The low-frequency behavior of the bearing can therefore be very effectively investigated with such constrained motion. Gupta [40] has investigated two types of external constraints: equilibrium constraints, where the rolling element/race interaction is primarily determined by the equilibrium of normal contact forces, and fictitious damping, where high-frequency vibration is selectively damped.

6.5.1 Equilibrium Constraints

Equilibrium constraints primarily consist of replacing the axial and radial equations of motion of the rolling element by the equilibrium of the normal forces in the respective directions, as discussed in Chapter 3. Thus, in the generalized acceleration vectors, the axial and radial velocities are set to zero, and the positions of the mass centers of the rolling elements are determined from the solution to the equilibrium equations. Since the equilibrium equations are nonlinear, this involves an iterative solution at each time step. However, since the change in equilibrium solutions over a time step is rather small, the convergence of the iteration is quite rapid when the preceding solution is used as a starting point. In any event, the additional effort involved in solving the equilibrium equations is merited in view of the computer time saved by the large step size allowable by the equilibrium constraint.

Although the equilibrium constraints can be very effectively used in ball bearings, some care must be exercised while working with roller bearings. Since roller skew, when significant, greatly alters the roller/race normal contact force, a solution to the radial and/or axial equilibrium of forces, as is provided in ADORE, may not be justifiable. Also, since the roller/race-flange vibrations may be of the same order as the nominal roller/race contact, suppression of the vibratory motion resulting from the nominal roller/race contact may not be adequate. The use of the equilibrium constraints for roller bearings is therefore suggested only for conditions where roller skew is absent.

6.5.2 Fictitious Damping

External damping is often effective in selectively suppressing certain high frequencies. Since the actual damping at any contact interface in the bearing is quite small, fictitious damping is introduced in order to eliminate the high-frequency response without significantly altering any low-frequency behavior.

Any contact interface can be modeled in terms of a spring, with stiffness k (N/m), in parallel with a dashpot, with damping C (N s/m). The contact stiffness will be known from the Hertzian contact parameters at the particular contacts. The required input of the damping coefficient may best be described in terms of a damping ratio C/C_c, where C_c is the critical damping [21] that corresponds to the undamped natural frequency of the system.

$$\text{Damping ratio } \zeta = C/C_c$$

where

$$C_c = 2m\omega_n;$$

and

$$\omega_n = \sqrt{\frac{k}{m}}$$

Thus ζ can be specified for the ball/race, ball/cage, and cage/race interfaces. For the purpose of "real" damping, ζ will be very small for most steels, but for certain nonmetals used for fabricating the cages, it may be appreciable. In any event, any nonzero value of ζ will only affect the general response in the neighborhood of the undamped natural frequency ω_n [21], and the response at very low frequencies will be practically unchanged.

Gupta [40] has shown that for ball bearings, although damping constraints suppress vibratory motion, the permissible increase in step size is not always as large as that permitted by the equilibrium constraints. In terms of computer time, any real benefit provided by the damping constraints is therefore not fully established.

6.6 Summary

The equations of motion of the bearing elements are properly nondimensionalized with preselected force, length, and time scales. This facilitates the estimation of the local truncation error encountered in any numerical method of integrating the differential equations of motion. Both explicit and implicit methods of integration are discussed. Among the explicit algorithms, the Runge–Kutta–Fehlberg formulas of orders one to seven are presented. Each of these formulas provides an estimate of the local truncation error. The conventional step-doubling procedure commonly used to estimate the truncation error in most explicit formulas is, therefore, not necessary. In the implicit category, an Adams-type predictor–corrector algorithm is presented in divided difference form, which greatly eases the prediction process when the size of the time step varies arbitrarily. An estimation of the local truncation error at each step is used to optimize the step size in both the explicit and implicit methods. Finally, some considerations in selecting a method of integration and the use of external constraints in speeding up the integration over large time intervals are discussed. Equilibrium constraints, which replace the axial and radial equations of motion of a rolling element with the equilibrium of normal forces in the respective directions, are shown to be very useful in simulating low-frequency behavior of ball bearings.

CHAPTER 7

The Computer Program ADORE

The operation, structure, and capabilities of a general-purpose computer program ADORE (Advanced Dynamics Of Rolling Elements) are discussed. The general capabilities of the program range from a simple equilibrium analysis to highly sophisticated dynamic simulations, where the influence of geometric imperfections of the bearing elements and the transient operating conditions on the overall dynamic performance of the bearing may be investigated. Also, the highly modular structure of ADORE makes it amiable to program modifications and simulation of special effects.

It is evident from the analyses presented in the preceding chapters that the simulation of the overall performance of a rolling bearing consists of a rather sophisticated ensemble of fairly simple to extremely complicated interactions. The application of the analytical models, simulating the interactions between the bearing elements, therefore lies in the development of a general-purpose computer program, which allows bearing designers, manufacturers, and users to model actual bearing configurations with varying levels of sophistication. The computer program ADORE discussed in this chapter is aimed at fulfilling such a need. It is expected that the extremely modular nature of ADORE will not only help simulate a wide range of variations in bearing geometry and operating conditions, but it will encourage the reader to develop his own version of the program for any particular application.

Over the past couple of decades, a number of computer programs dealing with the various aspects of rolling bearing analysis and performance simulation have appeared in the literature. As discussed earlier, the work may be divided into two categories: static equilibrium analysis and real-time dynamic analysis. Perhaps the first computer program solving the static equilibrium problem can be attributed to Jones [59, 60], which has been subsequently extended by Harris [48, 49] and Mauriello et al. [72]. Among some of the more recent programs aimed at solving the static-equilibrium problem are SHABERTH, credited to Crecelius and Pirvics [17], TRANSIM, by Ragen [80], CYBEAN, by Kleckner et al. [65], and SPHERBEAN, by Kleckner and Pirvics [65]. SHABERTH, a system program that solves the equilibrium problem for a system of bearings, takes into consideration both mechanical and thermal interactions. TRANSIM further extends the scope of SHABERTH by including gears in the system, thereby allowing the analysis of transmission systems consisting of gears and bearings. The more recent programs, CYBEAN and SPHERBEAN, are more advanced versions for cylindrical roller and spherical roller bearing analysis, respectively. Although these programs primarily carry out a static equilibrium, the analysis may be called quasi-static, since centrifugal forces and gyroscopic moments are included in the equilibrium of forces and moments.

Walters [95] probably produced the first computer program for solving the differential equations of motion of the elements in a ball bearing. The computer program known as BASDAP simulates the motion of balls and cage in the ball bearing. Although cage motion is in a complete six-degree-of-freedom system, ball motion is constrained in the axial and radial directions, and the ball/cage and race/cage interactions are constrained such that, upon contact, both interacting elements have the same velocity normal to the contact plane. Gupta [34] produced a simulation of ball motion in an angular-contact thrust ball bearing with all six degrees of freedom for ball motion. Gupta's computer program DREB [36, 37] allows for all six degrees of freedom for each element in the bearing and, in addition, it relaxes the cage interaction constraints of BASDAP. DREB analyzes both ball and cylindrical roller bearings.

With some constraints placed on the roller motion, Brown et al. [9] presented a computer code, TRIBO1, to analyze a cylindrical roller bearing for high-speed turbine engine applications. More recently, Gupta's RAPIDREB [40] has shown that certain constraints on ball motion, similar to the ones used by Walters, can really speed up the integration process when all constraints on cage interactions are relaxed. Similar to the models of Gupta [39], Conry [16] has presented a transient analysis and a computer program for cylindrical roller bearings. Based on all of this available work, it is clear that the computer simulation of rolling bearings is continually receiving increased attention. The review papers by Gupta [35] and Sibley and Pirvics [85] provide a more detailed discussion of some of the available computer programs.

The computer program presented in this text is a generalization of the DREB program, and in addition to ball and cylindrical roller bearings, it also considers tapered, spherical tapered, and to some extent spherical roller bearings. Also, any arbitrary variation in bearing internal geometry and the influence of manufacturing tolerances are treated. ADORE, therefore, provides the most advanced and general computer code for simulating the performance of a rolling bearing. A more in-depth discussion of ADORE's capabilities, including details on program input/output, will be discussed in this chapter following a general overview of the program and some comments on its structure.

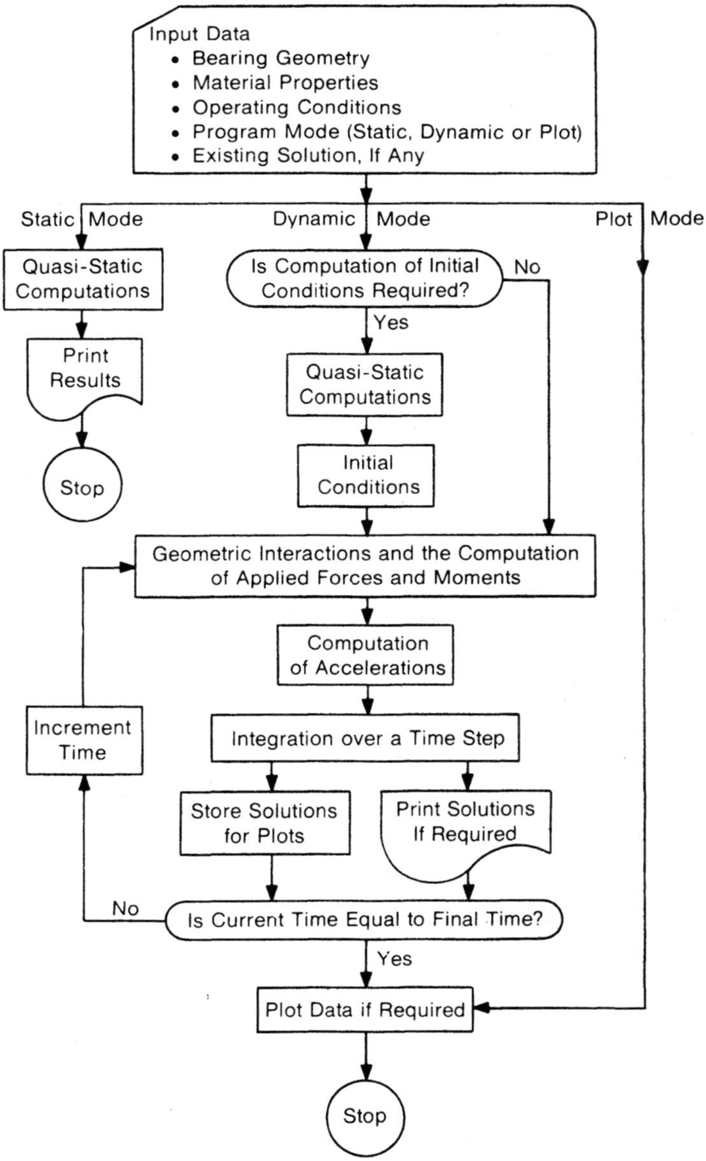

Figure 7-1 Simplified overview of the computer program ADORE.

7.1 Program Overview

A schematic overview of ADORE is presented in Figure 7-1. The program essentially has three modes of operation: the quasi-static mode, the dynamic mode, and a postprocessing or plot mode. Each mode may be executed independently. An equilibrium analysis is performed in the quasi-static mode, and characteristics such as fatigue life, stiffness, general load distribution, etc., are computed. The dynamic mode is really the prime mode of operation. From the given bearing geometry and operating conditions, a quasi-static analysis may first be performed to set the initial conditions for starting the integration of the differential equations of motion. If the initial conditions are known from previously computed solutions, this step may be bypassed. Once the initial conditions are known, geometric interaction is analyzed to calculate all of the applied forces and moments acting on the various elements. Also, forces due to lubricant drag and churning moments are computed according to the method of analysis presented in Chapter 5.

Next, the equations of motion developed in Chapter 2 are used to compute the various accelerations on each element in the bearing. An integration over a time step is then performed, using a preselected integrating algorithm. The solution so obtained is stored in a data file and plotted later. The integration continues until a predefined final time is reached; then the plot routines for postprocessing of the data and plotting may be called, if so desired. While running the program in plot mode, the control immediately goes to the plotting package and the already computed data may be plotted.

An important aspect, not shown in Figure 7-1, is the step size optimization and check on local truncation error. After any preselected number of steps, the step size is increased within a permissible range. Also, whenever the truncation error exceeds the prescribed limit, the current solution is deleted and a smaller time step is attempted. If the truncation error criterion is not satisfied even with the smallest possible step, then execution is terminated after a predefined number of steps and a diagnostic message is printed.

Figure 7-1 is by no means a flowchart of ADORE, and it presents a grossly oversimplified picture of the program. In fact, a detailed flowchart would be impossible to present due to both the size of ADORE and the complicated nature of the code.

7.2 Structure of ADORE

The structure of ADORE is highly modular and the entire code is divided into a large number of subprograms. As shown schematically in Figure 7-2, the nine basic modules of ADORE are

1. ADRAn: Input/output and quasi-static computation
2. ADRBn: Computation of derivatives or accelerations
3. ADRCn: Rolling element/race geometric interaction and normal contact forces

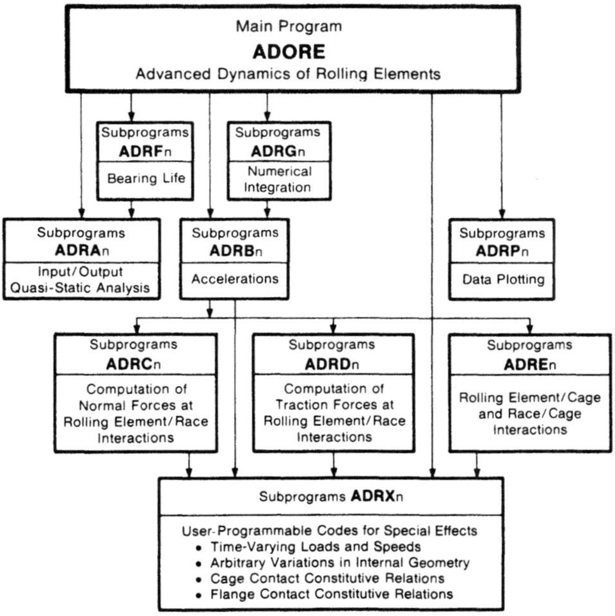

Figure 7-2 Subsections of the computer program ADORE.

4. ADRDn: Rolling element/race traction and lubricant effects
5. ADREn: Rolling element/cage and race/cage interaction
6. ADRFn: Computation of fatigue life
7. ADRGn: Numerical integration algorithms
8. ADRPn: Postprocessing and plotting
9. ADRXn: User-programmable routines for special effects

The first three letters, ADR, in the module name represent an abbreviation of ADORE; the fourth letter denotes the module name; and the last letter, n, may assume any numeric value depending on the number of subprograms in the module.

ADORE calls the module ADRAn for input/output and the computation of the quasi-static solution. Bearing life is computed by calling ADRFn. For the dynamic analysis, the two primary modules called by ADORE are ADRBn and ADRGn for computing the accelerations and integrating the differential equations of motion, respectively. Since the integrating algorithms used are generally of order greater than one, ADRBn is also called by the integrator module ADRGn. The module ADRPn is called by ADORE for plotting purposes and a few initial calls to ADRXn are simply for the purpose of special input/output which may be required by the user-programmable subprograms ADRXn.

The heart of ADORE is in the module ADRBn, which calls the three basic modules ADRCn, ADRDn, and ADREn for the computation of rolling element/race normal forces, traction forces, and the cage interactions, respectively. All of the user-programmable subprograms may be called by any or all of these three modules and the derivative module ADRBn.

The listing of the mainprogram and the user-programmable is subroutines, ADRXn is presented in Appendix IV, and a test case for an angular contact ball bearing is included in Appendix V.

7.3 ADORE Capabilities

The modular nature of ADORE provides extensive capabilities for analyzing rolling bearings to obtain real-time performance simulation under any prescribed operating conditions. Aside from providing a full dynamic analysis, ADORE also provides options for a quick quasi-static computation for estimating bearing fatigue life and stiffness.

7.3.1 Bearing Types

There are five bearing types considered by ADORE:

1. Ball bearing (angular contact and radial)
2. Cylindrical roller bearing
3. Spherical roller bearing
4. Tapered roller bearing
5. Spherical tapered roller bearing

Although each of the bearings may be with or without cage, only one row of rolling elements is permitted; therefore, the treatment of spherical roller bearings is somewhat limited.

7.3.2 Types of Cages

The cages in any of the bearing types may be either outer- or inner-race guided. There may be a maximum of two guiding lands, one on each side of the rolling elements, and each guiding land may be located on either the inner or the outer race. In other words, the cage may be guided on the outer race on one side and guided on the inner race on the other side. Cage pocket geometry is cylindrical for ball bearings, rectangular for cylindrical and spherical roller bearings, and trapezoidal for tapered roller bearings.

For ball and cylindrical roller bearings, the cage may also be rolling-element guided. In this case, the cage pocket is spherical for ball bearings and cylindrical for roller bearings, so that the pocket surface conforms to the rolling element.

The cage may be either a one piece element or segmented into several pieces. Segmentation may take place either between or across the pockets.

All of the segments are outer-race guided and no interaction between segments is assumed.

7.3.3 Operating Conditions

Any arbitrary set of constant or time-varying loads and speeds may be imposed on the bearing. While a rotating radial load is relatively easy to simulate by simply selecting an input option, more sophisticated variations in load and speed require the use of subroutine ADRX1. This subject is discussed in more detail later in this chapter.

It is also possible to specify arbitrary displacements, either constant or time varying, in lieu of the applied forces. The races may be misaligned relative to each other and either the applied moments or angular displacements may be prescribed as constant or time varying. In the quasi-static portion of ADORE, however, only one angular displacement, rotation about the shaft's bending axis, is considered. The static equilibrium configuration for any prescribed moment may be easily determined by carrying out a parametric study over varying misalignments.

The races may have prescribed shrink fits with the housing and shaft. Also, the temperature of races, cages, and rolling elements may all be independently prescribed. All thermal and mechanical distortions of the bearing elements due to any prescribed temperatures, shrink fits, and centrifugal growth are considered. It should be remembered, however, that ADORE does not perform any thermal interaction analysis, because such an analysis often requires the integration of the bearing with the entire rotor system in which it is used.

7.3.4 External Constraints

For the purpose of eliminating very-high-frequency motion, ADORE provides the option of equilibrium constraint. Under such a constraint, the radial and axial positions of the mass centers of rolling elements and races are determined by the equilibrium of normal forces at each time step. This constraint greatly speeds up the integration process, and the performance simulation can be obtained over several shaft revolutions within a reasonable computational effort.

7.3.5 Radial Preloads

In radially loaded bearings the outer race is often fitted into a noncircular hole in the housing, thereby producing an out-of-round race and changing bearing internal clearance. This effect can be easily simulated by specifying the resulting out-of-roundness as a function of the circumferential position on the race. Similarly, as discussed later, any arbitrary out-of-roundness of the races may be modeled.

7.3.6 Material Properties

The material behavior of rolling elements, races, and cages may be prescribed arbitrarily by any set of conventional elastic properties. For standard cases, ADORE defaults to conventional bearing steels and all of the default properties may be printed out by executing the program once with the default option.

All materials are considered to be isotropic, so that the classical theory of elasticity is applicable. Considerable caution should be exercised in simulating nonisotropic materials for any of the bearing elements.

7.3.7 Lubricant Traction

ADORE contains traction models for a number of lubricants. The user may select any of the models by exercising a simple input option. The models range from a simple traction–slip curve, which is ideal for solid lubricants and can be used for liquids under some operating conditions, to sophisticated models where lubricant rheology in high-pressure environments is considered.

7.3.8 Churning and Drag

Lubricant churning and drag are known to be extremely complex phenomena and a realistic modeling of these effects is extremely difficult. Similar to the models used in the past, ADORE uses simple relationships for estimating churning and drag losses.

7.3.9 Roller Skew

Since ADORE uses a generalized six-degree-of-freedom system for the motion of a roller, the simulation of roller skew is fairly well treated. Although the roller/race-flange interaction defaults to simplified Hertzian contact analysis, the user-programmable routine may be used to greatly extend the contact analysis and simulate any effects known from available experimental data. Thus, roller skew predictions can be very realistic.

7.3.10 Rolling Element Skid

Skidding of rolling elements can be simulated both as a function of the lubricant tractive behavior and the radial preload discussed above. Skidding is basically determined by examining the variations in the orbital velocity of the rolling element mass center.

7.3.11 Cage Instability

By evaluating the variations in cage angular velocity, the mass center whirl velocity and the radius of the whirl orbit, a very realistic simulation of cage instability can be obtained. The six-degree-of-freedom system used for simulating the cage motion is the key to the prediction of cage instability.

7.3.12 Bearing Power Loss

The frictional loss at each interaction is computed from the known frictional forces and the sliding velocities. These losses are added to the churning and drag losses to compute the total power loss in the bearing. In the event the churning and drag losses are a large fraction of the total losses, the predicted power loss will greatly depend on the churning and drag coefficients, which are inputs to ADORE. In such a case, the predicted power loss should be viewed with some caution.

7.3.13 Wear

Based on the prescribed wear coefficients, the wear rates of the various elements are computed as a function of time. The effect of changes in the bearing internal geometry due to wear may be approximated by first estimating the geometrical changes from the predicted wear rates and then by rerunning ADORE with the predetermined geometrical imperfections.

7.3.14 Geometric Imperfections

Although ADORE defaults to the ideal geometry of the bearing, certain geometrical imperfections can be easily simulated.

Variations of Rolling Element Size. The variation in ball and roller diameters is very easily treated because nowhere in the program are the rolling element diameters assumed to be equal. Any arbitrary size variation may be specified as an input.

Race Coning and Out-of-Roundness. For cylindrical or tapered roller bearings, the radius of the interacting surface of the roller may be specified as a function of the axial and circumferential position coordinates, through one of the user-programmable subroutines. Thus, any arbitrary out-of-roundness and/or coning can be simulated. The out-of-roundness simulation also applies to race waviness.

Race Curvature Variation. The curvature across the race groove in ball and spherical roller bearings may also be varied as a function of the circumferential position on the race.

Bearing Element Imbalance. Since the mass and geometric centers of each element may be different, the effect of imbalance introduced by any geometric imperfection is fully treated. Also, the inertial axes may be different from the geometric axes and all rotational effects due to imperfect geometry can be simulated.

Cage Geometry Variations. In addition to the treatment of cage imbalance and biased rotation as discussed above, the variations in cage pocket clearance can be treated. Also, the position of the cage pocket center and the orientation of the pocket can be arbitrarily specified. Thus, most geometric irregularities associated with the cage may be treated.

7.3.15 Bearing Noise

Although the noise generated by the bearing may depend heavily on its interaction with the system external to the bearing, the mass center acceleration patterns of the various bearing elements do give some estimate of bearing noise. Since these accelerations are computed in real time, a fast Fourier analysis of the data provides a useful spectrum of noise associated with the bearing. Also, since the entire noise analysis can be performed as a function of bearing imperfections, ADORE may be a powerful tool in the noise evaluation of a rolling bearing.

7.3.16 Bearing Life for Arbitrary Load and Speed Cycles

By exercising the equilibrium constraint ADORE may be effectively executed over the entire load and/or speed cycle and the variation in computed fatigue life may be computed. This will provide a realistic estimate of average life over the cycle.

7.3.17 Flexibility in Units

Since most of the analysis in ADORE is dimensionless, the units of the different variables are encountered only in the input/output subroutines. It has therefore been possible to treat both the English (inch–pound force–second) and the widely used SI system (meter–Newton–second) of units.

7.3.18 Graphic Output

The graphic output significantly adds to the practical utility of ADORE. Although all of the results may be printed at any time step, a graphic representation over the entire time interval is very helpful in understanding the overall dynamic behavior of the bearing.

7.3.19 Integrating Algorithms

Both the explicit Runge–Kutta-type formulas and the implicit predictor–corrector algorithms are incorporated in ADORE. Proper selection of the integrating algorithm may appreciably reduce the computational effort required for a particular problem.

7.3.20 Restart Capabilities

The solution vectors are stored in data files at preselected time steps. Any of these solutions may be read in as initial conditions for any subsequent runs; therefore, the performance simulation may be continued further in time from the point where an earlier run was terminated. Also, the last solution vector may be stored for each run. This solution vector may be used as an initial condition, under the arbitrary initial conditions option, and the simulation may be continued even when all data files have been deleted. The arbitrary initial condition option greatly facilitates a perturbation analysis where the steady-state solution under ideal conditions is known and the effect of a small change in either the bearing design or the operating conditions is to be investigated.

7.4 Input/Output Data

Most of the input data to ADORE are fairly self-explanatory, as described in the beginning of the program listing in Appendix IV. Also, both the print and plot outputs, presented in Appendix V, do not need extensive elaboration. However, a brief discussion of some of the input/output variables, the

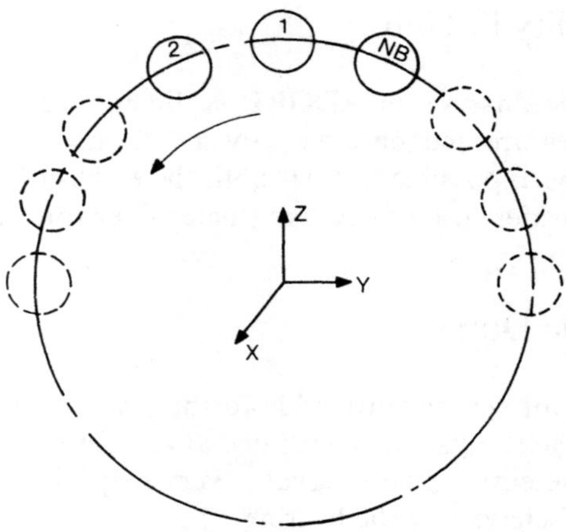

Figure 7-3 The numbering of the rolling element in a bearing.

user-programmable subroutines, and the data management system employed in ADORE will greatly benefit the reader.

7.4.1 Input Data

Aside from the description of the various input variables presented in Appendix IV, a general discussion of some of the variables may be helpful to the user.

An (X, Y, Z) coordinate frame and the negative and positive X axes are mentioned frequently in the input. A description of this coordinate frame is presented in Figure 1-1, where both the space-fixed and rolling-element-fixed coordinate frames are described. The numbering of rolling elements is simply done in the sense of a positive rotation about the X axis, as shown in Figure 7-3, where NB is the total number of rolling elements in the bearing. All of the rolling elements, cages or cage segments, and the outer and inner races are referred to as bearing elements and the numbering sequence of the various elements is as follows.

1 to NB	NB rolling elements
NB + 1 to NB + KS	KS cage segments
NB + KS + 1	Outer race
NB + KS + 2	Inner race

For a cageless bearing the outer and inner races are referred to as element numbers (NB + 1) and (NB + 2), respectively.

In terms of bearing geometry, the dimensions of a cylindrical roller are described in Figure 3-9. For tapered rollers the reference roller diameter is the diameter at the large end which will be located on the negative X axis of the roller-fixed coordinate frame (see Figure 1-1). The orientation of the

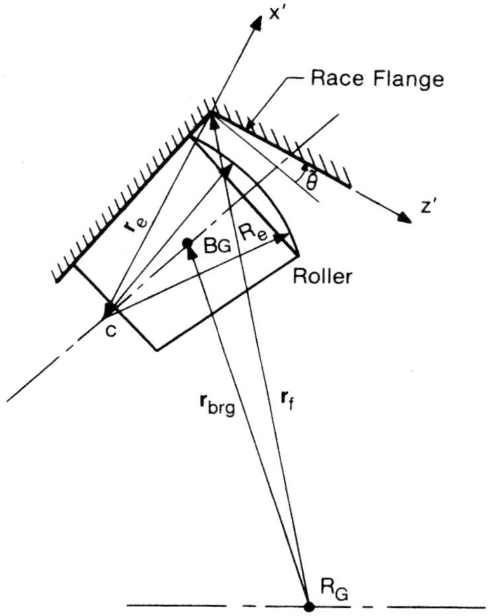

Figure 7-4 Race flange layback angle in a roller bearing.

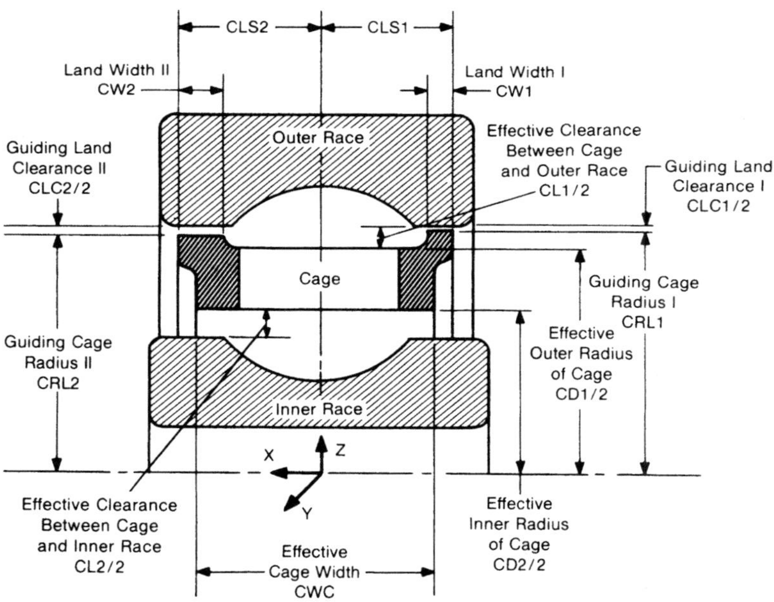

Figure 7-5 Generalized cage geometry.

race flange angle in roller bearings is shown in Figure 7-4. The flange layback angle $\tilde{\theta}$ is an angle between the normal to the nominal race surface and the flange surface. When the net angle extented between the flange and the raceway surface is greater than 90°, $\tilde{\theta}$ is positive; it is negative if the net extended angle is less than 90°. For most of the bearing the layback angle will be a small positive quantity.

The general features of the cage geometry are described in Figure 7-5. Although for most bearings the cage may be essentially cylindrical the geometric input parameters allow for specialized designs. The clearance between the nominal surface of the cage and the race may in general be different from that at the guide lands. In fact, the two guide lands may not be on the same raceway. This is often the race in cage designs for tapered roller bearings. When the cage is segmented, each pocket in a segment is numbered along the positive rotation about the X axis as shown in Figure 7-6, where a full one-piece cage is also shown for reference. Each rolling element/cage interaction is now referenced by the rolling element number, cage segment number, and pocket number in the segment in which the rolling element resides.

At both the rolling element/cage and cage/race interactions, when the minimum clearance or geometric interactions between the mating surfaces is less than a certain critical value a direct contact is established. This critical value is called the critical film thickness for metal contact in the input data and it is generally of the order of the composite surface roughness of the mating surfaces.

In terms of the operating conditions and constraints on the bearing races, either a force or a displacement may be prescribed along any of the (X, Y, Z) axes. The constraint is defined by setting a switch to either a 0 (for prescribing a force) or a 1 (for prescribing a displacement or an acceleration). A similar definition applies to defining the rotational motion of the race by either specifying a moment or an angular acceleration. The constraint switches in the input data are denoted as JF1 and JF2 on record 3.3 and IRF1, IRF2, IRM1, and IRM2 on records 6.2.1 and 6.2.2.

7.4.2 Print Output

Typical print output from ADORE for a ball bearing is summarized in Appendix V. The first few pages of the output consists of a listing of all of the input data records, bearing geometry, material properties, inertial parameters, lubrication parameters, initial operating conditions, scale factors, and output controls. Most of this data is essentially input to ADORE. The translational and rotational constraints listed under initial operating conditions correspond to the specification of either a force or an acceleration, as discussed above. The six components listed under the translational constraints represent the outer and inner race constraints along the (X, Y, Z) axes. The first three components are for the outer race while the latter three

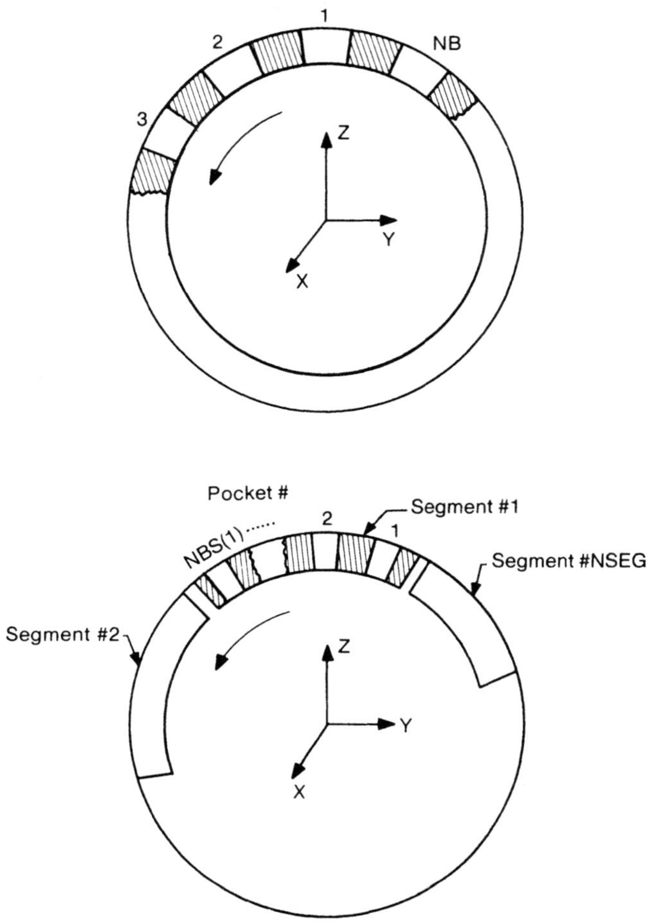

Figure 7-6 Cage pocket numbering for a full one-piece and a segmented cage.

are for the inner race. Rotational constraints are specified only along the X axis and the two components printed correspond to the outer and inner races. Along the Y and Z axes the constraint switch is always set to 1, meaning that only angular accelerations resulting in rotating or time-varying misalignments can be prescribed along these axes. The data control parameters listed under output control just denote the variables JPT and KSTEP specified on input record 1, while the autoplot code denotes the bearing elements (array IBW on input record 3.3) for which the plot data is stored.

The output at each time step is divided into four sections which contain rolling element parameters, race and cage parameters, applied parameters, and the time step summary. A few output variables which need some elaboration follow.

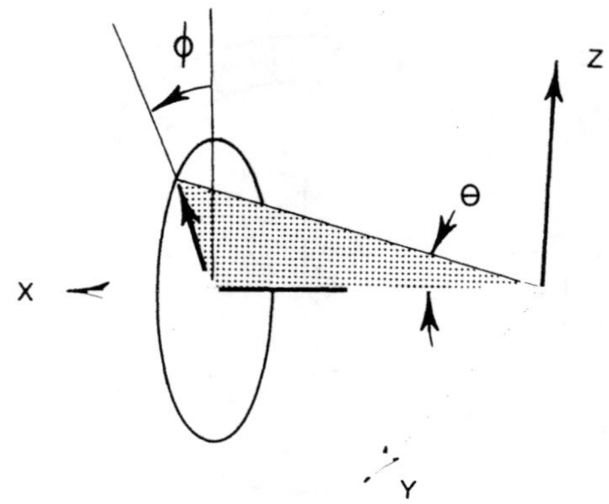

Figure 7-7 Orientation of angular velocity vectors.

1. Angular Velocities. All angular velocity vectors are printed in terms of an amplitude and the orientation THETA and PHI. The two angles define the orientation of the angular velocity vectors as shown in Figure 7-7. The (X, Y, Z) coordinate frame shown in this figure corresponds to the rolling element azimuth frame in the case of rolling element velocities and it represents the inertial frame for the cage and the race angular velocities.

2. Angular Position. The angular position of any bearing element is defined as the orientation of the principal axis of inertia (X) in a certain coordinate frame. The coordinate frame used is the azimuth frame for rolling elements and the inertial frame for the cage and races.

3. Time Average Wear Rates. The wear rate at any interaction is computed by the well-known Archard's wear equation

$$W_r = K_a \frac{QV}{H}$$

where W_r is the volume wear rate, K_a is the wear coefficient, Q is the normal contact load, V is the sliding velocity, and H is the hardness of the material.

Since for rolling element/race contacts the sliding velocities and loads may greatly vary over the contact zone, the product QV is replaced by an integral of the load–slip product over the contact zone. Also for these interactions the wear is divided between the races and the rolling elements according to the prescribed wear coefficients.

For the rolling element/cage and race/cage interactions all of the wear is assumed to occur on the cage. This is quite reasonable since in most cases the material of the cage will be softer than that of the rolling element or the races. The instantaneous wear rate is time-averaged from zero to the current time and the data in the print output represents this average wear rate.

4. Rolling Element/Cage Contact Angle. This output variable denotes the angular position of rolling element/cage interaction in a cage pocket coordi-

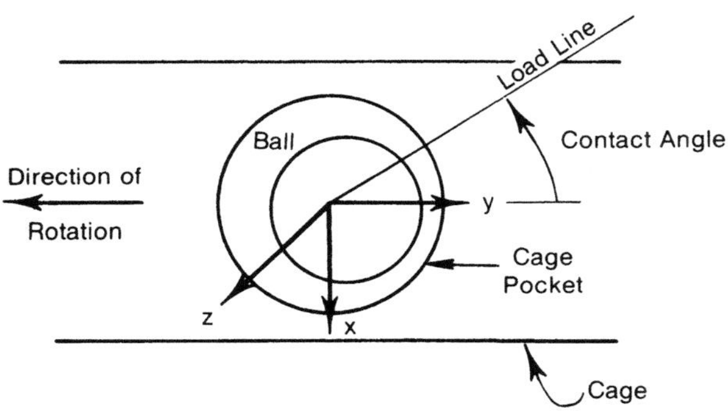

Figure 7-8 Definition of rolling element/cage contact angle.

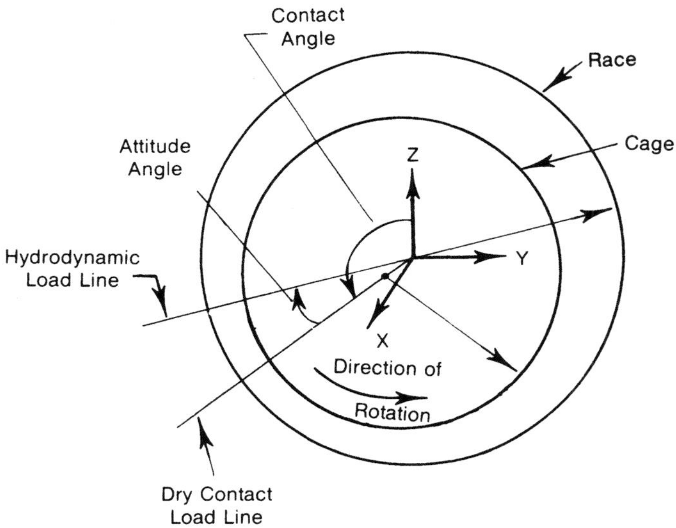

Figure 7-9 Definition of cage/race contact and attitude angles.

nate frame, as shown in Figure 7-8. The rolling element drives the cage when the contact angle is 180° and the cage drives the rolling element if the contact angle is zero. Clearly, the contact angle can be anywhere from 0 to 360° for a ball bearing, but for a roller bearing it will only be either 0° or 180°.

5. Cage/Race Contact and Attitude Angles. The cage/race contact angle defines the angular position of the cage/race interaction in an inertial coordinate frame as shown in Figure 7-9. The attitude angle is only relevant when the hydrodynamics at the cage/race interface is considered. It essentially denotes the angle between the line of minimum clearance and the hydrodynamic load.

6. Power Loss. The frictional dissipation at each interaction is computed and printed in the print output. A sum of all of these losses and the loss due to lubricant churning and drag is printed out as the total power loss. The fraction of this loss due to churning and drag effects is indicated as the churning loss fraction.

7. Internal Clearance and Operating Fits. The internal bearing clearance, outer and race fits, and the cage/race effective diametral play denote actual operating values after allowing for thermal and centrifugal growths.

8. Rolling Element Orbital Velocity Ratio. This variable is essentially a ratio of the rolling element orbital velocity to the angular velocity of the inner race relative to the outer race. The value printed in section 4 of the print output denotes an average over all rolling elements and the time over which the performance simulation is obtained.

9. Cage Angular Velocity Ratio. This parameter is the ratio of the angular velocity of the cage to the angular velocity of the inner race relative to the outer race. In the case of a segmented cage the value printed in the step summary represents an average over all segments and over the time of integration.

10. Cage Whirl Ratio. The ratio of the mass center whirl velocity to the angular velocity of the inner race relative to the outer race is denoted by this output variable. Again an average is computed over the time of integration and all of the cage segments if the cage is segmented.

At the end of the run some statistics of the time-step size are printed and if the plots are generated during the run, a message indicating a successful completion of the plot is printed.

7.4.3 Plot Output

In view of the large amount of output generated by ADORE, the plot output is essential in determining the general dynamic behavior of the bearing. All of the plots are essentially divided into four sets.

1. Power Dissipation and Life. This set contains three plots and the data necessary for generating the plots is always stored. The first plot contains the variation in power loss, the churning loss fraction, and the bearing fatigue life. The moments exerted on the outer and inner races are plotted in the second plot. The x component essentially denotes bearing torque. The time-average wear rates for rolling element #1, the outer and inner races, and the cage are plotted in the third plot in this set. In the case of a segmented cage, the plotted cage wear rate corresponds to cage segment #1.

2. Rolling Element Motion. The first two plots in this set contain the mass center acceleration and position of the rolling element in a cylindrical coordinate frame. The contact load, contact angle, and orbital velocities are plotted in the third plot as a function of the orbital position of the rolling element. Plots #4 and #5 contain the angular velocity of the rolling element, spin-to-roll ratios, rolling element/race slip, power loss, and the variation in lubricant film if the bearing is oil lubricated. Rolling element/cage interaction is the subject of plot #6. In the case of a roller bearing plots #7 and #8 in this set will contain the roller/race flange interactions.

3. Cage Motion. The cage or cage segment angular and whirl velocities are plotted as a function of time in the first plot of this set. Plots #2 and #3 contain the cage/race interaction at the two lands. Plot #4 shows the mass center accelerations of the cage or the cage segment, and the mass center orbit is plotted in Plot #5. The subsequent plots in this section contain the variation in the rolling element/cage contact load and the geometric interaction or clearance in the cage pocket.

4. Race Motion. In the case of any applied load variation or acceleration of the races, these plots may be useful in understanding the overall motion of the races. Similar to the cage motion plots, the first plot in this set contains the variation in the mass center whirl and the race angular velocity. Plots #2 and #3 contain the applied forces and moments on the race. The mass center accelerations of the race are contained in Plot #4, while the mass center orbit is plotted in Plot #5. The final plot (#6) contains the variation in the mass center position of the race as a function of time.

Some details of the various data files used to store the plot data will be discussed later in this chapter and a list of all files and the relevant device codes used in ADORE is included in the program documentation in Appendix IV.

7.4.4 User-Programmable Subroutines

In addition to the flexibility in the input data, several user-programmable subroutines in the ADRX module allow a number of special effects to be very easily programmed. Although the purpose of each subroutine and the programming instructions are well decumented in the program listing in Appendix IV, the scope of these subroutines may be briefly recalled here.

Subroutine ADRX1. Any time variations in the applied loads and race speeds can be programmed in this subroutine to any degree of complexity. Often experimental data available from laboratory tests of the system can be incorporated to obtain bearing performance simulations under actual laboratory conditions.

Subroutine ADRX2. The roller/race-flange contact behavior can be incorporated here in terms of a load–deflection relation. If any such data are available, and it is programmed in this subroutine, then the roller skew predictions may be very realistic.

Subroutine ADRX3. The purpose of this subroutine is to prescribe any force-deflection relation for rolling-element-to-cage contact in the cage pocket. Such a relation is often obtained experimentally and, if available, it should replace the simplified Hertz contact analysis used in ADORE. Appropriate user-specified code in this subroutine will help accomplish this update.

Subroutine ADRX4. This subroutine is similar to ADRX3, but it applies to cage/race interactions. Since the load–deflection relation for line contact is often determined experimentally, this subroutine will help implement any available semi-empirical constitutive equation for the cage/race contact.

Subroutine ADRX5. Variation in roller radius as a function of the axial and circumferential position on the roller surface can be programmed in this subroutine. Thus, roller out-of-roundness, roller coning, and similar effects can be very easily programmed.

Subroutine ADRX6. This subroutine is identical in scope to ADRX5 except that it provides the variation in the radius of the interacting surface of the race. Also, for ball and spherical roller bearings, the variation in curvature across the groove may be programmed in this subroutine.

As may be apparent from the above discussion, the primary objective of all of the above subroutines is to help simulate special effects without any modifications to ADORE. Depending on the problem under consideration, the code in any of the above routines may assume any degree of sophistication. However, it must be remembered that all of the above subroutines are called in some of the innermost loops; therefore, a very complicated code in these routines may substantially impact the total computational effort required. It is therefore suggested that very simple algebraic expressions be used and that any interpolation of extensive data be avoided. If some computations are required to be performed only once, they may be placed along with the input/output statements, which are executed only once in all of the above subroutines.

7.4.5 Data Management in ADORE

Depending on the number of rolling elements in a given bearing, the number of output variables can be quite large. Since ADORE provides a time-transient analysis, the net output of any typical run containing a few hundred time steps may be prohibitively large. It is for this reason that some type of data management is necessary.

ADORE produces both print and plot output and it uses a number of data files to monitor the data. As shown in the list of data files included with the program listing in Appendix IV, there are 11 different data films employed by ADORE. Aside from the normal INPUT and OUTPUT files, the master data file, MASTER, contains the dimensionless solution vectors at the final time step. The file, FINAL, also contains the last solution vector of the run. Using the restart capabilities of ADORE, the computation may be restarted from any of the steps at which the solutions are available in master data file MASTER. The solution vector written in file FINAL may be used for any subsequent runs when the arbitrary initial condition option is used. This is helpful when the influence of a small change in a certain bearing parameter on the steady-state bearing performance is to be investigated.

In order to limit the print output, the various solutions are printed only at certain preselected time steps, with a maximum of six bearing element solutions stored at each time step in files SOL1 to SOL6. These files are used later to generate the required plots. The overall bearing parameters are stored in file SOL7.

If at any time the computation is restarted using any of the solutions contained in file MASTER, it is necessary that all of the plot data files (SOL1 to SOL7) be attached so that the results of each subsequent run can be combined with the preceding runs to produce a comprehensive plot to that point. In the input data, a 36-character string, called the bearing specification code, is specified when first starting any run. This specification code identifies each of the data files corresponding to a run. At subsequent runs ADORE checks the specification code from each file against the one in the master data file. In the case of any mismatch, a diagnostic message is printed and execution is terminated. The specification code therefore provides a safeguard against runs made with incorrect data files.

7.5 Computer Resource Requirement

ADORE has been developed on a CDC (Control Data Corporation) Cyber system. It uses FORTRAN 77 language and it conforms to standard ANSI (American National Standards Institute) specifications. The required central memory is about 250,000-octal 60-bit words, including the memory required by the graphic package. All of the external subrountines, with the exception of the plot package, are contained in the program listing in Appendix IV. The well-known plot package DISSPLA (Display Integrated Software System and Plotting Language) is used to generate all of the plots.

7.6 Summary

The computer program ADORE which implements the rolling bearing dynamics analysis to actual applications is presented. A simplified schematic diagram is presented to demonstrate the general operation of the computer program. It is shown that the highly modular structure of ADORE leads to the simulation of some of the most complex effects in a very amiable fashion. The overall capabilities of the program are shown to range from the simplest quasi-static equilibrium analysis to highly sophisticated dynamic simulations. The latter include cage instability, roller skew, skidding, bearing noise, and similar dynamic factors as a function of operating conditions, bearing design, and geometric imperfections encountered in the manufacturing process and as a result of wear of the interacting bearing elements. Some details of the data management process and the computer resources required are also discussed in order to provide a more in-depth view of the operation and structure of ADORE, and encourage the reader to modify and develop customized versions of the code for specific applications.

CHAPTER 8

Some Dynamic Performance Simulations

Dynamic performance simulations as obtained by ADORE are presented to demonstrate the design significance of computer modeling of bearing performance. The capabilities of ADORE in modeling vibrational characteristics of rolling bearings, ball skid in angular-contact ball bearings, cage instabilities, roller misalignment, and skew in cylindrical roller bearings are discussed.

In view of the large number of geometric and operational parameters it is extremely difficult to present the results of dynamic performance simulations, as obtained by ADORE, in a form that may be useful to a large number of practical applications. Generally, each application is analyzed individually, and the optimum bearing design unique to the application under consideration is developed by carrying out a systematic parametric evaluation of bearing performance. Thus, unlike most classical problems where the solutions are available in the form of tables and graphs in dimensionless variables useable in a large number of applications, rolling bearing computer programs such as ADORE essentially serve as a design and performance diagnosis tool. The objective of this chapter is, therefore, to present a few selected examples to illustrate some of the capabilities of the analytical models and the computer program presented in this text.

8.1 Numerical Considerations

Similar to any time-transient analysis, the numerical considerations in determining the method of integration and the limit on truncation error are very important to the effective use of the computer programs. As discussed in Chapter 6, ADORE employs a number of different integration formulas; therefore, the selection of the optimum method for a given application can only be determined by trial. To illustrate the significance of this problem, consider a radially loaded cylindrical roller bearing, where the races are fixed relative to each other (the relative position is determined by the applied load) and the roller is free to move arbitrarily between the races. Using a fifth-order Runge–Kutta–Fehlberg method, the variation in roller/race contact load is shown in Figure 8-1* with a local truncation error limit of 10^{-7} per unit dimensionless time. The cyclic pattern indicates vibration of the roller between the inner and outer races. When instead of the fifth-order formula a fourth-order algorithm is used with the same truncation error, the solution obtained is shown in Figure 8-2. Clearly the two solutions are identical; however, for an identical number of time steps (100 in this case), the fifth-order formula simulated almost twice as much roller travel in comparison to that obtained by the fourth-order scheme. Since the number of derivative evaluations per time step are 6 and 8 respectively for the fourth- and fifth-order

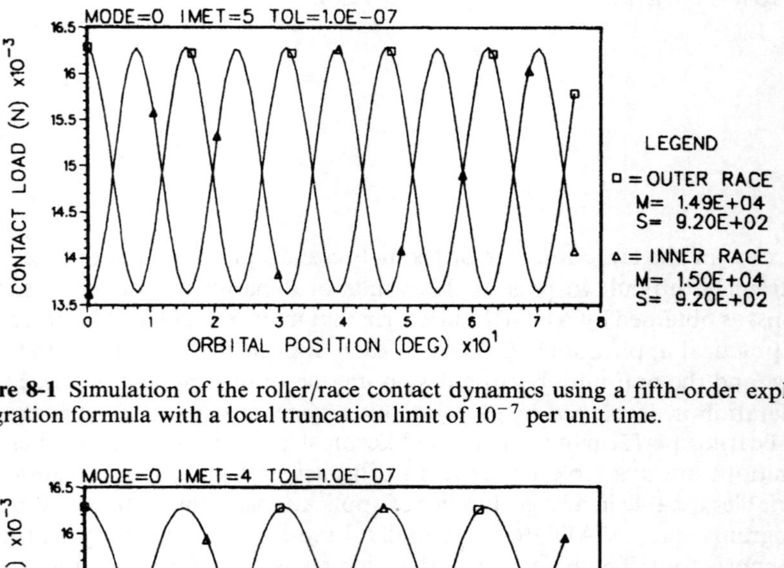

Figure 8-1 Simulation of the roller/race contact dynamics using a fifth-order explicit integration formula with a local truncation limit of 10^{-7} per unit time.

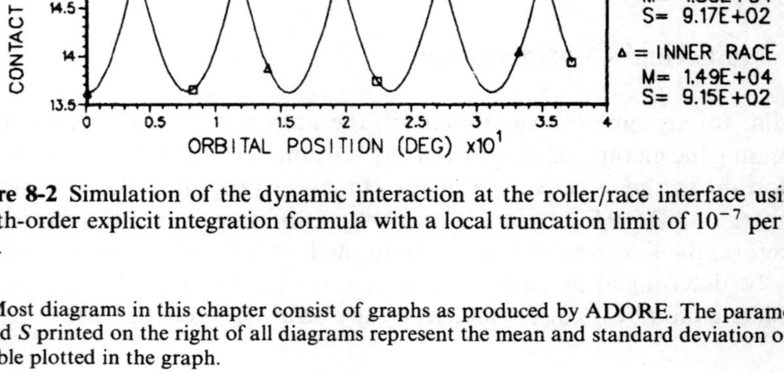

Figure 8-2 Simulation of the dynamic interaction at the roller/race interface using a fourth-order explicit integration formula with a local truncation limit of 10^{-7} per unit time.

* Most diagrams in this chapter consist of graphs as produced by ADORE. The parameters *M* and *S* printed on the right of all diagrams represent the mean and standard deviation of the variable plotted in the graph.

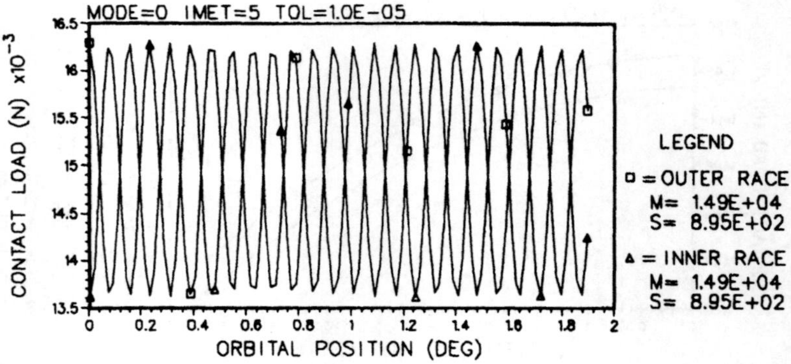

Figure 8-3 Roller/race contact dynamics as simulated by a fifth-order explicit algorithm with a local truncation limit of 10^{-5} per unit time.

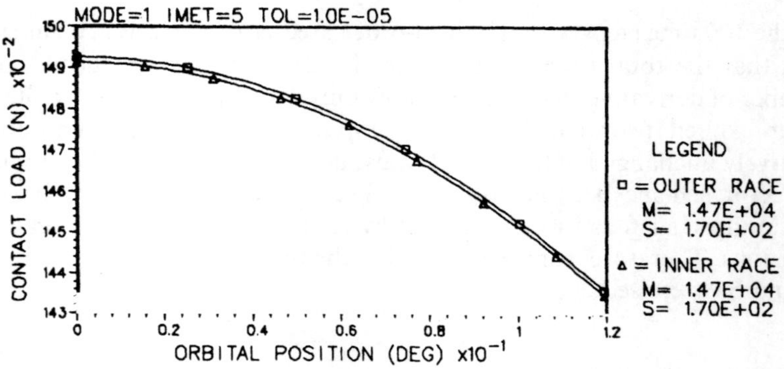

Figure 8-4 Roller/race contact as simulated by a radial equilibrium constraint using a fifth-order integration algorithm and a local trunction limit of 10^{-5} per unit time.

formulas, the computational effort per step is somewhat increased with the higher-order formula, but the savings in terms of the increase in the permissible step size outweighs the increased effort per step. Thus, the fifth-order formula is more effective for this example.

Now, if the limit on truncation error is increased to 10^{-5}, the fifth-order formula results in a further increase in time step, but the solutions now lose some smoothness, as shown in Figure 8-3.

When the interest is essentially in the steady-state solutions, the high-frequency motions can be suppressed by an equilibrium constraint which forces the roller to be in radial equilibrium at any orbital position. The solution for the roller/race load with such a constraint is shown in Figure 8-4 for 100 time steps, using the fifth-order formula with a truncation limit of 10^{-5}. Clearly, the absence of the cyclic variation results in a further increase in the permissible time step. When no cyclic variations are present, a predictor–corrector method may be effective. Figure 8-5 shows the solution obtained

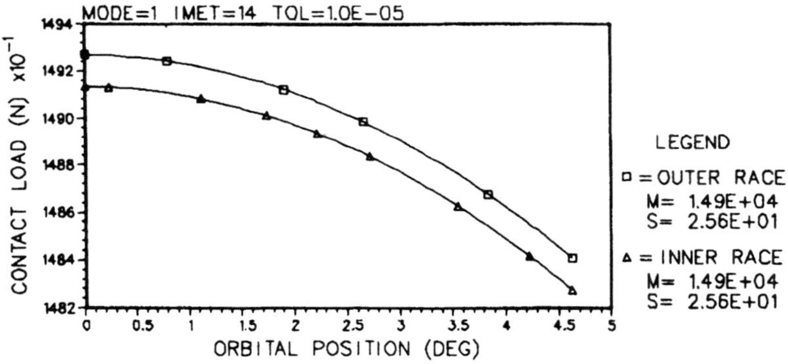

Figure 8-5 Roller/race contact simulation with a radial equilibrium constraint and a predictor–corrector formula of third order with a local truncation limit of 10^{-5} per unit time.

for the 100 time steps with the third-order predictor–corrector formula. It is seen that the total travel of the roller is now somewhat reduced, but the number of derivative evaluations is now only 3, compared to 8 with the fifth-order explicit formula. Thus, the computational effort per unit time is relatively unchanged in this case. It must be remembered that the predictor–corrector scheme becomes quite inefficient when the solutions have discontinuities, such as those produced by rolling element/cage and cage/race collisions. Therefore, some care in using the predictor–corrector techniques should be exercised.

8.2 Vibrational Characteristics

Since the rolling elements are elastically supported between the inner and outer races, some characteristic vibrations of the rolling elements are expected. The general motion simulated by ADORE indicates this; typical results for the roller and ball bearings are discussed below.

8.2.1 Cylindrical Roller Bearings

In a radially loaded cylindrical roller bearing, the roller is free to vibrate between the races. The roller/race loads shown earlier in Figure 8-1 indicate this. The motion can be further elaborated by plotting the mass center accelerations of the roller in Figure 8-6. Initially a small disturbance is introduced to perturb the solution obtained by quasi-static equilibrium. The radial acceleration shows a cyclic pattern which will be a characteristic of the bearing. The frequency of this vibratory motion can be easily determined from the load–deflection relation at the roller/race interaction.

Let the load, Q, be related to the deflection, δ, by the equation

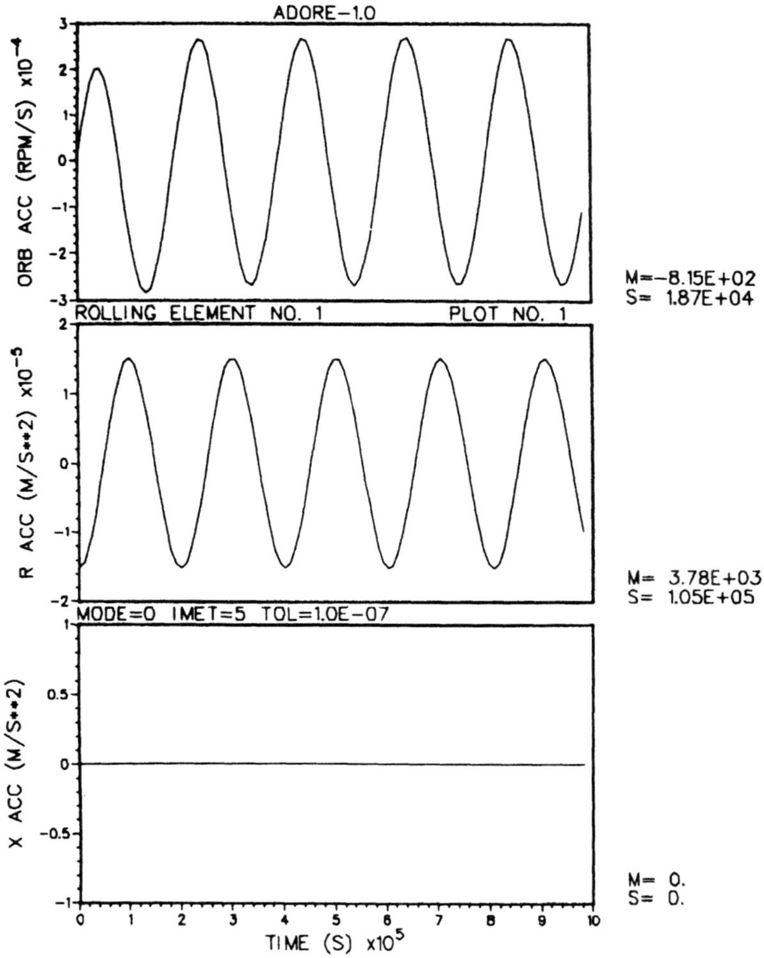

Figure 8-6 The general nature of roller acceleration and the high-frequency vibration at the roller/race interface.

$$Q = \Upsilon \delta^n \tag{8.1}$$

where Υ and n are constants.

The contact stiffness is obtained by differentiating equation (8.1):

$$\frac{\partial Q}{\partial \delta} = n\Upsilon\delta^{n-1} = n\frac{Q}{\delta} \tag{8.2}$$

and if m is the mass of the roller, the characteristic vibration frequency, ω, can be written as

$$\omega = \frac{1}{2\pi}\sqrt{\frac{1}{m}\frac{\partial Q}{\partial \delta}} = \frac{1}{2\pi}\sqrt{\frac{nQ}{\delta m}} \tag{8.3}$$

When the rolling element contacts both the inner and outer races, the two contacts are combined to given an elastic contact frequency of the rolling element ω_e as

$$\omega_e = \frac{1}{2\pi}\sqrt{\frac{n}{m}\left(\frac{Q_1}{\delta_1} + \frac{Q_2}{\delta_2}\right)} \tag{8.4}$$

where the subscripts 1 and 2 correspond to the outer and inner races, respectively.

The vibration frequency of approximately 50 kHz, shown in Figure 8-6, correlates with equation (8.4) for the particular bearing used to generate this example. It must be noted that depending on the value of n, the contact stiffness may be nonlinear, and some change in the elastic contact frequency as a function of contact load may be expected. For the line-contact configuration in roller bearings, the value of n is close to unity when a Palmgren-type of load–deflection relation [equation (3.41)] is used. Thus, the stiffnesses may be almost linear and no significant change in the vibration frequency as a function of load may be noticed.

The amplitude of vibration will depend on the external forcing functions applied on the bearing. Under an ideal case, if the two races are held fixed relative to each other, the roller vibration will be excited as the roller enters the load zone. The vibration amplitude may be somewhat reduced as it travels through the load zone due to the frictional dissipation at the roller/race contacts. When the roller exits the load zone, the vibration frequency will change, because it now contacts the outer race only under the influence of the centrifugal force. The vibrational amplitude will again reduce as a function of time, due to the frictional dissipation. If a cage is present in the bearing, the frequency roller/cage collisions may continually provide an excitation for this mode of vibration and the vibratory motion may be seen in the steady state.

8.2.2 Ball Bearings

The vibratory motion of a ball in an angular-contact ball bearing is somewhat more complicated than that discussed above for roller bearings. In addition to the elastic contact frequency, Gupta et al. [38] discovered a kinematic vibration frequency which is dependent on the bearing geometry (in particular, the contact angle and the race curvatures) and which varies as the square root of the ball/race contact load. This motion therefore constitutes a pendulum-like vibration where the ball rolls across the race groove. Since both races are curved, the exact vibration path has not yet been physically understood; however, the radius of the path has been found to be very sensitive to the race curvatures.

In order to illustrate the vibratory motion in ball bearings, consider the ball bearing described in Table 8-1. The bearing is assumed to operate at 3600 rpm (inner race rotating). Using ADORE, the vibratory motion of the

Table 8-1 Data for Ball Bearing #1

Geometry	
No. of balls	8
Ball diameter	26.9875 mm
Pitch diameter	117.5 mm
Contact angle	8.31°
Outer race curvature factor	0.52
Inner race curvature factor	0.51
Materials (both ball and race)	
Elastic Modulus	2.0×10^{11} N/m^2
Density	7.75×10^3 kg/m^3
Poisson's Ratio	0.25
Hardness	7.845×10^9 N/m^2
Wear coefficient	5.00×10^{-6}
Lubrication (relation of Figure 4-2)	
κ_u	0.
κ_m	0.010
κ_∞	0.0075
u_m	1.0 m/s

ball is investigated as a function of the applied thrust load. The size of the time step is held constant, and all frictional dissipations are excluded to highlight both the elastic contact and kinematic mode of ball vibration.

Figure 8-7 shows typical acceleration of the ball mass center. Although there are no friction forces, an orbital component of ball mass center acceleration is seen in Figure 8-7. This primarily results from the radial vibrations of the ball mass center [see equation of motion (2.18)]. The axial and radial components represent the two modes of vibration. It is seen that there are two distinct vibration frequencies. The axial and radial components of the high-frequency motion are in phase and they indicate the motion of the ball normal to the race. Thus, this high frequency corresponds to the ball/race elastic contact frequency, which is correlated to an expression similar to equation (8.4). The axial and radial components of the low-frequency motion are 180° out of phase. This clearly indicates the motion of the ball center along a circular arc (similar to the race groove) of a certain radius, and it therefore has a definite kinematic significance. It has been shown that this kinematic vibration frequency, ω_k, can be correlated with an expression of the form

$$\omega_k = \frac{1}{2\pi}\sqrt{\frac{Q}{m\ell}} \tag{8.5}$$

where m is the ball mass, Q is the ball/race contact load, and ℓ is the radius of the arc along which the ball center traverses.

The load dependence of the kinematic frequency is clear from equation

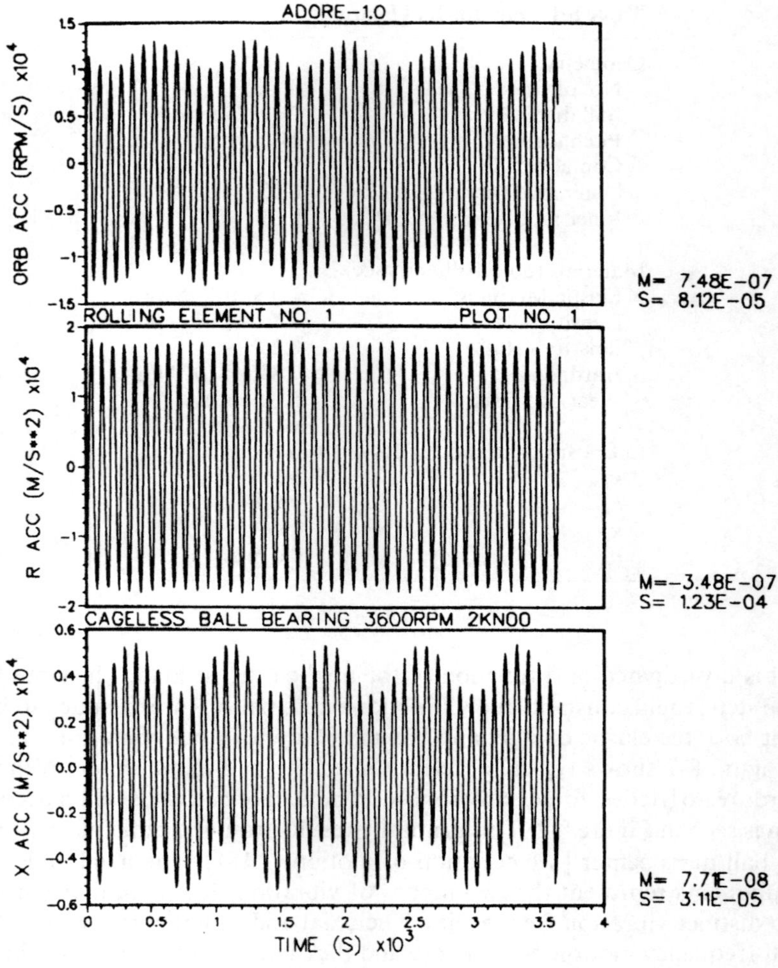

Figure 8-7 Characteristic vibrations at the ball/race contacts in angular contact ball bearings.

(8.5), while equation (8.4) establishes the load dependence of the elastic contact frequency. Since for a classical Hertzian point contact in ball bearings, the value of n is $\frac{3}{2}$, the contact stiffness is nonlinear and by combining equations (8.1) and (8.3) it can be shown that the elastic contact frequency will vary as $\frac{1}{6}$ power of the load. Such load variations were shown by Gupta et al. [38] and they are shown in Figure 8-8 for the particular bearing used in the present example. The data points used to draw the figure represent actual cases generated from ADORE simulations.

The magnitude of both the elastic contact and kinematic frequencies will clearly depend on the bearing geometry and the operating conditions. For most low-speed applications where the centrifugal force is relatively small compared to the applied load, the two frequencies will be greatly different.

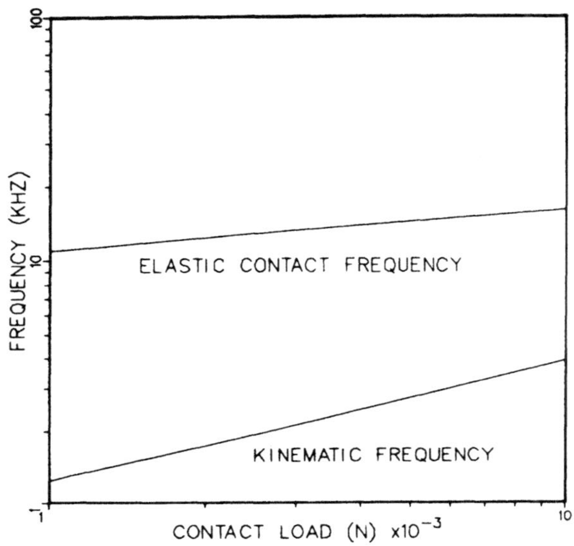

Figure 8-8 The variation of elastic contact and kinematic vibration frequencies in ball bearings as a function of the ball/race contact load.

At high speeds when the centrifugal force on the ball becomes quite large, however, the two characteristic frequencies may be close together; under such conditions, a clear distinction in the two vibration modes as seen in Figure 8.8 may not be possible.

Existence of the above vibratory motion in ball bearings has been experimentally validated by Gupta et al. [38]. It was shown that the vibrations picked up by an accelerometer mounted on the stationary outer race of ball bearings of different geometric descriptions indicate the presence of both of the characteristic frequencies discussed above.

A clear understanding of the various modes of vibration has substantial practical significance in understanding the noise generated by the bearing. For example, if the races of a ball bearing have geometric imperfections, a variation in both the elastic contact and kinematic frequencies will be observed. In fact, if the computed vibrations are transformed to a spectral density function, noticeable bands of "noise" may be observed. This establishes the strength of computer programs such as ADORE for realistic modeling of bearing noise.

8.3 General Ball Motion and Skid

Consider the ball bearing described in Table 8-1 with the parameters of the ball/race traction curve as stated. These parameters indicate a fairly well-lubricated bearing. The ball motion simulations obtained by ADORE for such conditions with pure thrust load provide significant insight into the

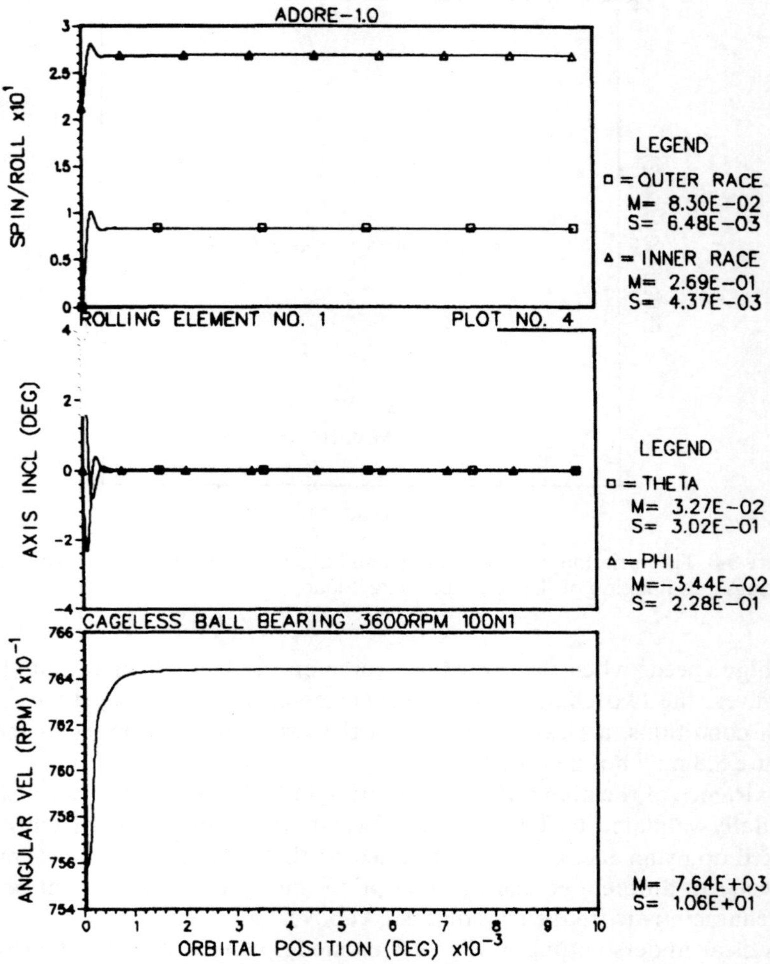

Figure 8-9 General motion of a ball in an angular-contact ball bearing with an applied thrust load of 100 N and in the absence of any churning effects.

general motion of the ball. Figures 8-9 and 8-10 show the variation in ball angular velocity for a thrust load of 100 and 1000 N, respectively. The solutions are plotted as a function of ball orbital position; it can therefore be easily seen that these solutions contain time integration of the equations of motion over a large number of shaft revolutions. The initial conditions for the dynamic solutions are estimated from the conventional quasi-static equilibrium solutions; the axial and radial positions of the ball center are constrained by equilibrium equations to eliminate the high-frequency vibratory motion. It is seen that both the magnitude and orientation of the ball angular velocity vector are significantly different from the quasi-static solution. Also, when the angular velocities are transformed into spin-to-roll ratios, it is seen

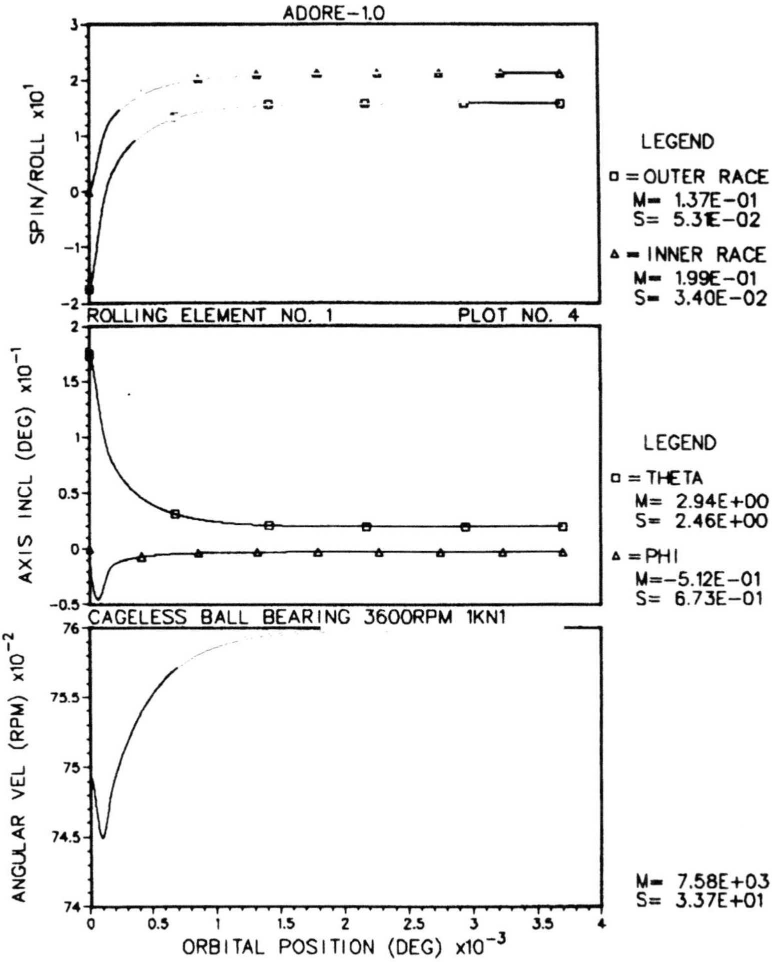

Figure 8-10 Ball motion in angular-contact ball bearing with an applied thrust load of 1000 N and in the absence of any lubricant churning effects.

in Figures 8-9 and 8-10 that the race control hypothesis (which constrains relative ball spin to one of the races) does not hold and that the steady-state solutions indicate relative ball spin on both races. Similarly, the steady-state solutions for ball orbital velocity and ball/race slip along the direction of rolling indicate noticeable deviations from the conventional quasi-static solutions. Figure 8-11 shows the ball slip and heat generation at the ball/race interface, while Figure 8-12 shows the ball orbital velocity for the 1000 N thrust condition. These findings indicate that the quasi-static solutions may not be realistic in estimating ball spin and slip under well-lubricated conditions.

The steady-state spin and slip at the ball/race interface can be used to

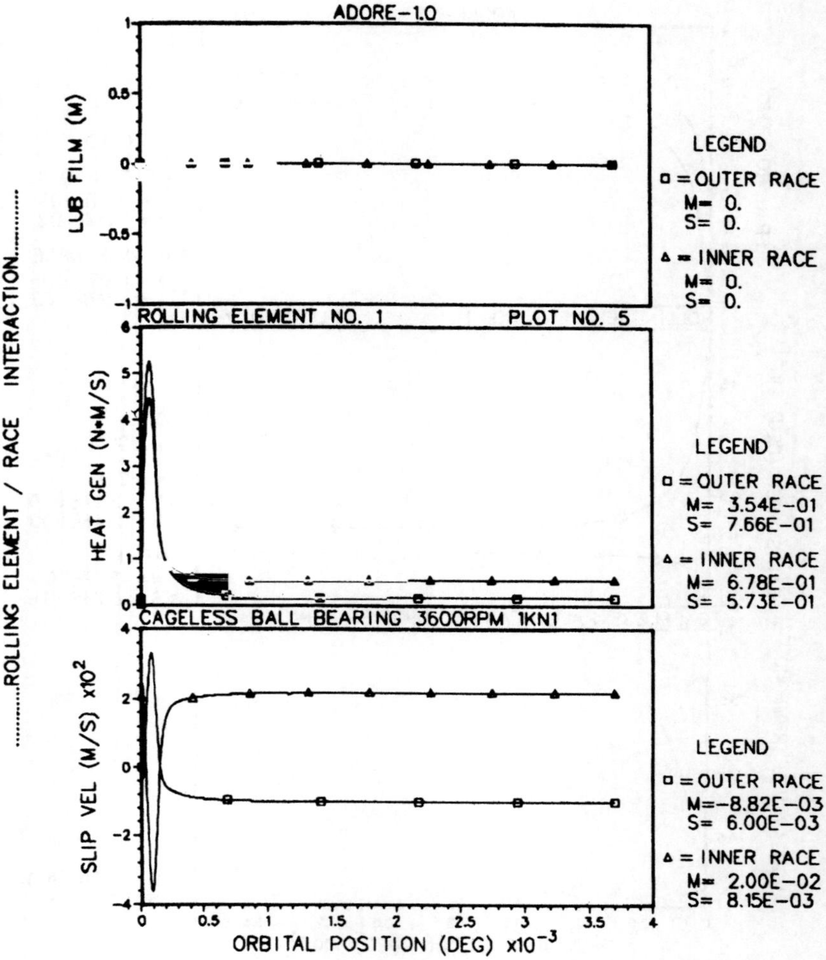

Figure 8-11 Ball/race slip and the resulting heat generation in an angular contact ball bearing with an applied thrust load of 1000 N.

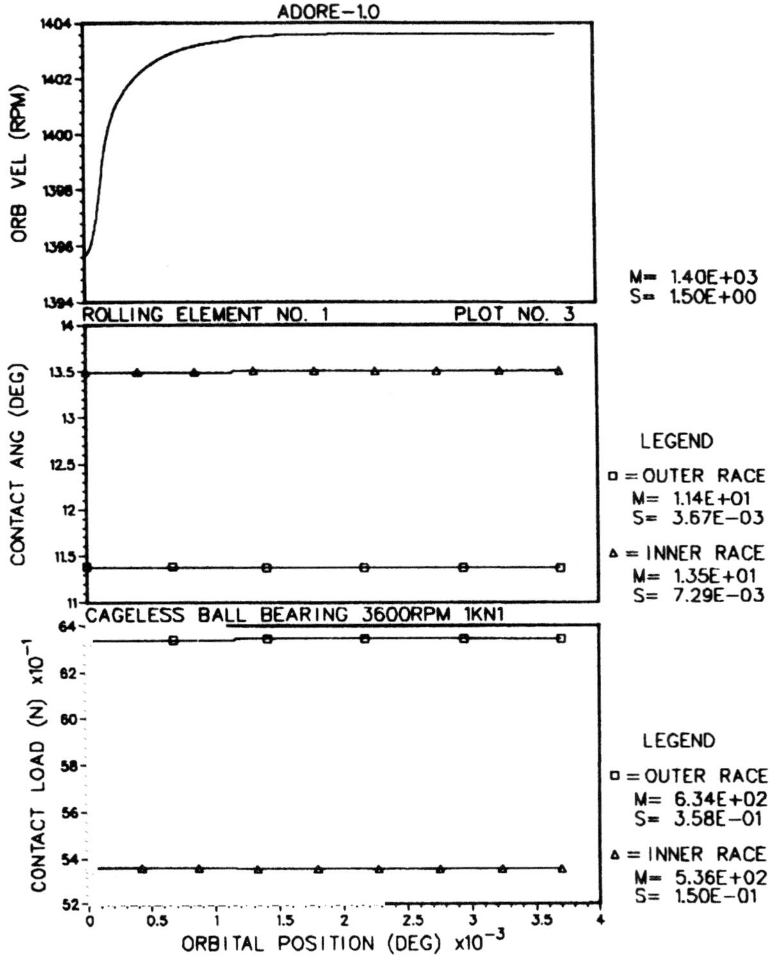

Figure 8-12 Ball/race contact loads and the orbital velocity of the ball in an angular-contact ball bearing under a thrust load of 1000 N.

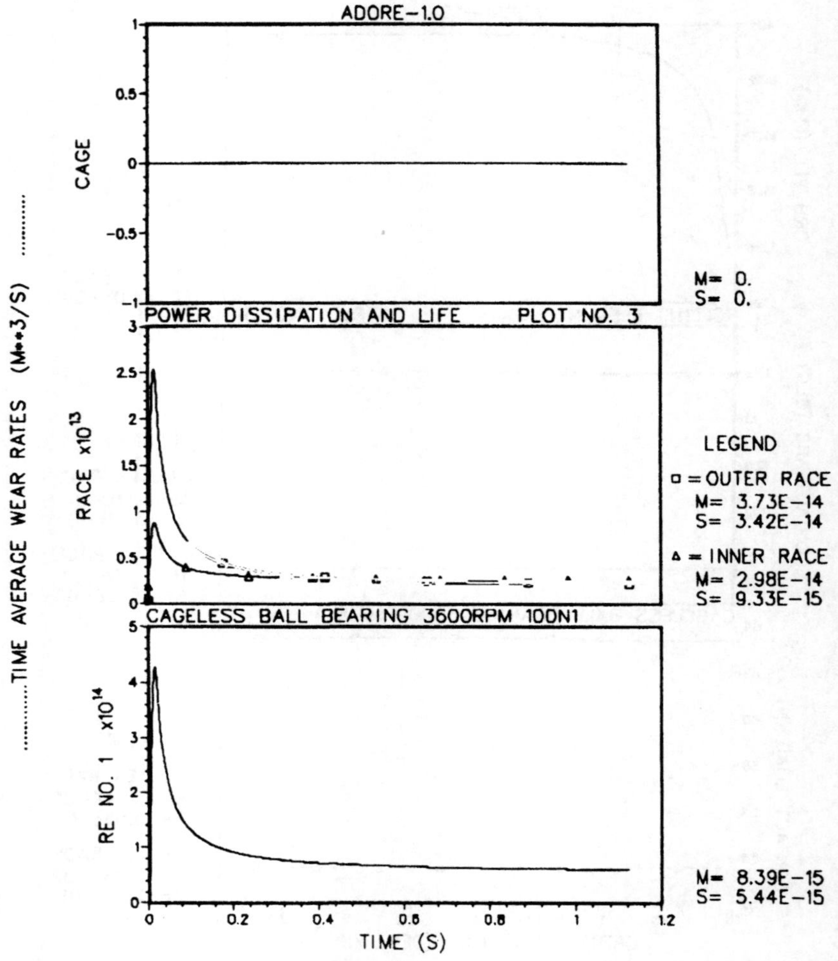

Figure 8-13 Simulated time-average ball and race wear rates under a thrust load of 100 N.

estimate wear rates of the ball and races. Corresponding to the solutions of Figures 8-9 and 8-10, the time-average wear rates are plotted in Figures 8-13 and 8-14, respectively. The conventional Archard's wear equation with a wear coefficient of 10^{-6} is used to generate these solutions. An estimate of the average wear rate is useful in estimating the expected wear life of the bearing.

Ball skid, and the determination of a realistic thrust load, normally referred to as preload, to prevent skidding are other practical problems where the dynamic solutions provide a clear understanding. Under ideal conditions, where the ball/race traction is the only means of restraining the ball in the orbital direction, the ball will reach a steady-state orbital and angular velocity

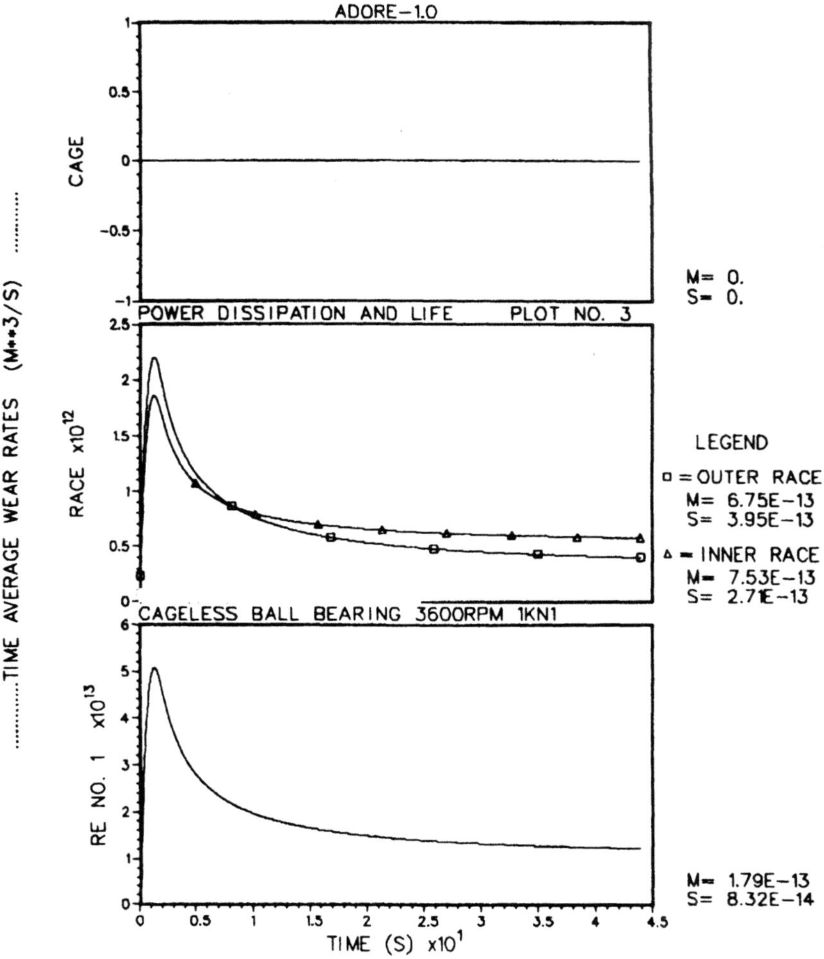

Figure 8-14 Simulated ball and race wear rates under an applied thrust load of 1000 N.

corresponding to the minimum energy dissipation. Here, a stable solution can always be obtained, provided there are no instabilities introduced by the ball/race traction curve. In more practical cases, the bearing is partially filled with oil and there are certain churning and drag forces applied to the ball, or there is a cage in the bearing and the balls are subjected to ball/cage collision forces. In order to prevent ball skid under such conditions, the ball/race traction forces must be greater than or equal to these applied forces. To illustrate this further, consider the 3600 rpm ball bearing described in the preceding section to be partially filled with oil so that the balls are subjected to a drag force and churning moment. Let the viscosity of the lubricant be 6 cp and the magnitude of the force and moment be varied as a function of the effective density of the air–oil mixture in the bearing. Figure 8-15 shows the change in ball orbital velocity for the 1000 N applied thrust load condition and the effective lubricant density of 100 kg/m^3. When this is compared to Figure 8-12, it is seen that the drag forces do result in a reduction in ball orbital velocity; similarly, the churning moments result in a reduction in ball angular velocity. The initial dip in the ball orbital velocity in Figure 8-15 is due strictly to a transient effect produced by the applied drag force which appears as a step change, when the initial conditions are determined by a quasi-static solution which neglects all such forces.

An important part of the solution in Figure 8-15 is the fact that with a thrust load of 1000 N, the ball/race traction forces during ball skid are greater than the drag force, the ball/race traction forces are therefore adequate to produce a stable operation and the ball orbital velocity rapidly stabilizes to a certain value. Now, let the applied thrust load be considerably reduced, to 75 N for example, and the drag forces are now greater than the ball/race traction forces which can be produced by this applied load and the lubrication parameters listed in Table 8-1. The ball orbital velocity now continues to decrease with time, as shown in Figure 8-16. This clearly shows a skidding instability where the ball/race traction is inadequate to balance the drag forces. The ball and race wear rates will also show an instability due to increasing sliding at the ball/race interface. This is shown in Figure 8-17.

With the effective lubricant density (or the drag force) as a parameter, the ball-orbital-to-shaft angular velocity ratio can be plotted as a function of the applied thrust load, as shown in Figure 8-18. This provides significant guidance in determining a realistic thrust load to prevent skidding. Again, it should be emphasized that Figure 8-18 represents only the bearing considered here as an example and that it is unique to the bearing geometry and the operating conditions stated above. For any practical application, it will be necessary to generate Figure 8-18 for the prescribed bearing geometry and the expected operating conditions. The example presented herein thus emphasizes the design capabilities of the dynamic analysis and the computer program ADORE.

8.4 Cage Stability

Cage stability is one of the most well-known problems that sometimes leads to rather catastrophic failure in high-speed rolling bearings. Very often the rotating shaft has a certain imbalance, which results in a rotating radial load on the support bearings. At very high speeds, this could result in a serious problem as far as the performance and life of the bearing are concerned.

As a typical illustration of the response of a rolling bearing to a rotating load, consider an angular-contact ball bearing described in Table 8-2, with the cage guided on the outer race and relatively high traction at the various interactions. Let the bearing operate at 30,000 rpm (inner race rotating) with

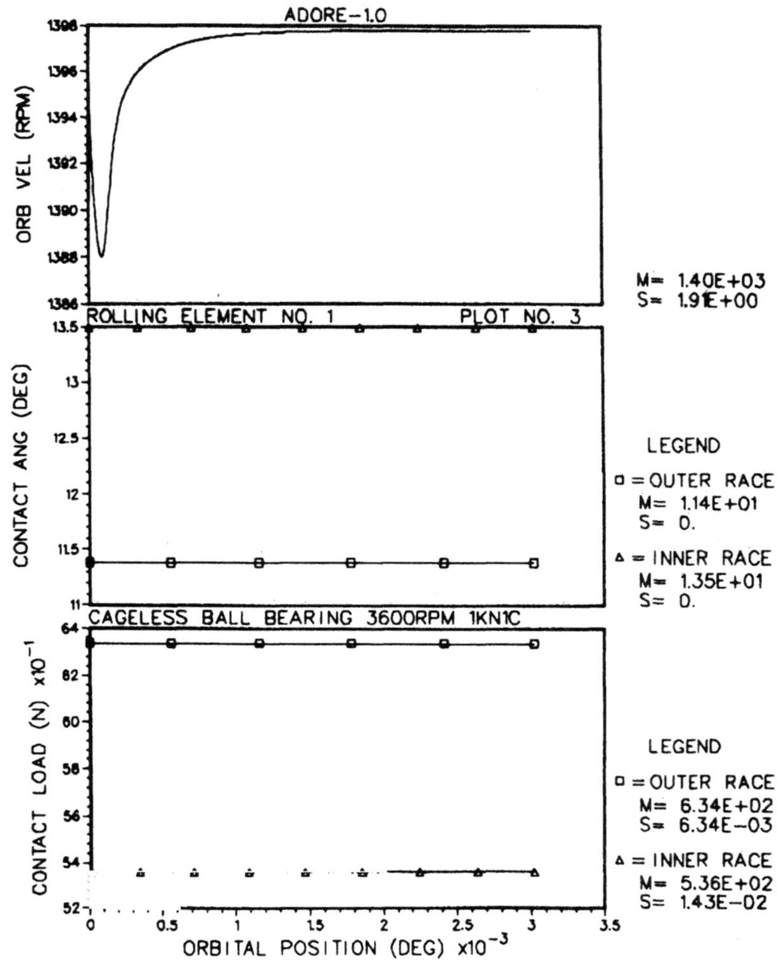

Figure 8-15 The influence of churning (with an effective density of 100 kg/m^3) on the ball orbital velocity in an angular contact ball bearing with a thrust load of 1000 N.

a thrust load of 5000 N and a rotating radial load of 4000 N. Under such conditions, the center of the inner race will just whirl in a circular orbit, the radius of which will be determined by the relative radial displacement of the races due to the combined radial and axial loads. The contact loads and angles at the ball/race interface will have a certain cyclic pattern, but it will be somewhat complicated compared to the case where the radial load is fixed. The results obtained by ADORE are shown in Figure 8-19 for ball 1, which is the most heavily loaded rolling element at the start of the simulation. As the ball travels in the orbit, the load goes through a cyclic variation corresponding to the orbital velocity of the ball relative to the rotating radial load. The variation in the contact loads results in corresponding variations in the

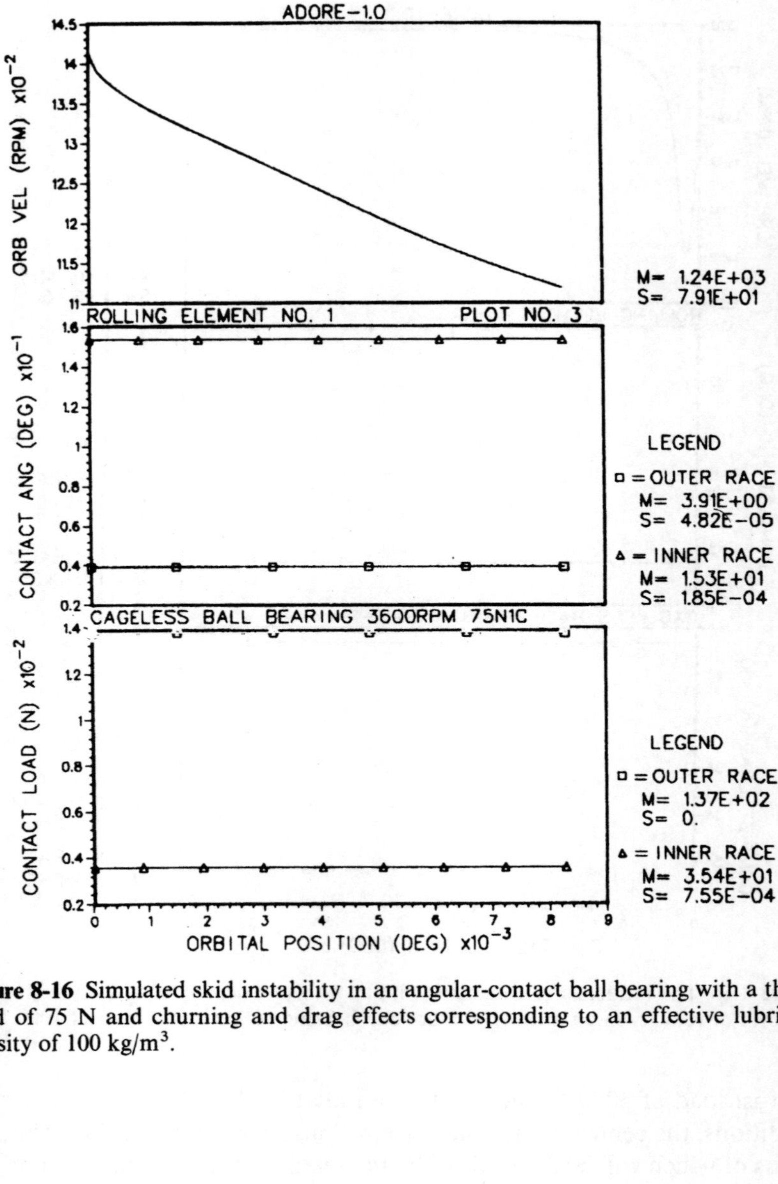

Figure 8-16 Simulated skid instability in an angular-contact ball bearing with a thrust load of 75 N and churning and drag effects corresponding to an effective lubricant density of 100 kg/m^3.

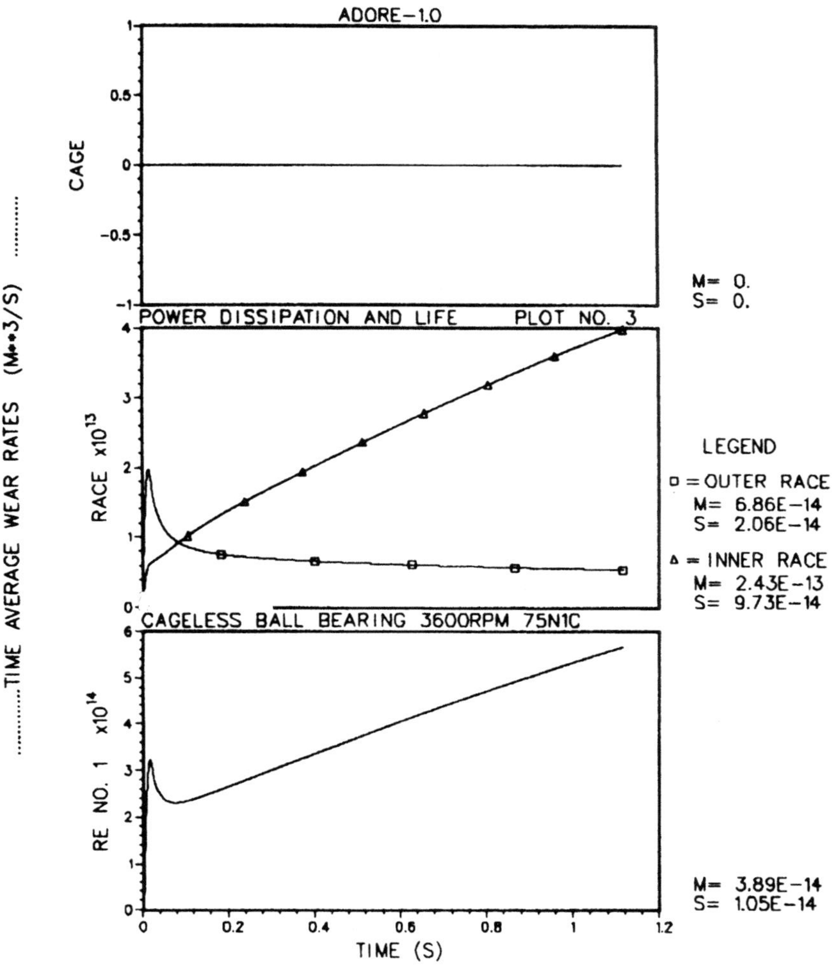

Figure 8-17 Ball/race wear resulting from the skid instability shown in Figure 8-16.

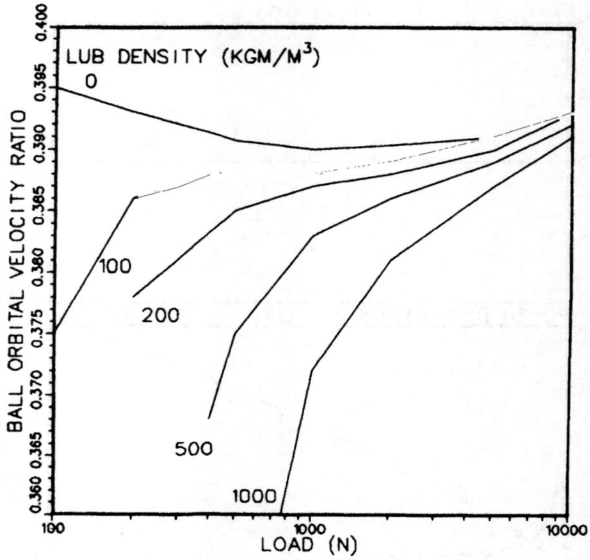

Figure 8-18 The variation of ball orbital velocity and skid in ball bearings as a function of the applied load and the level of churning and drag losses.

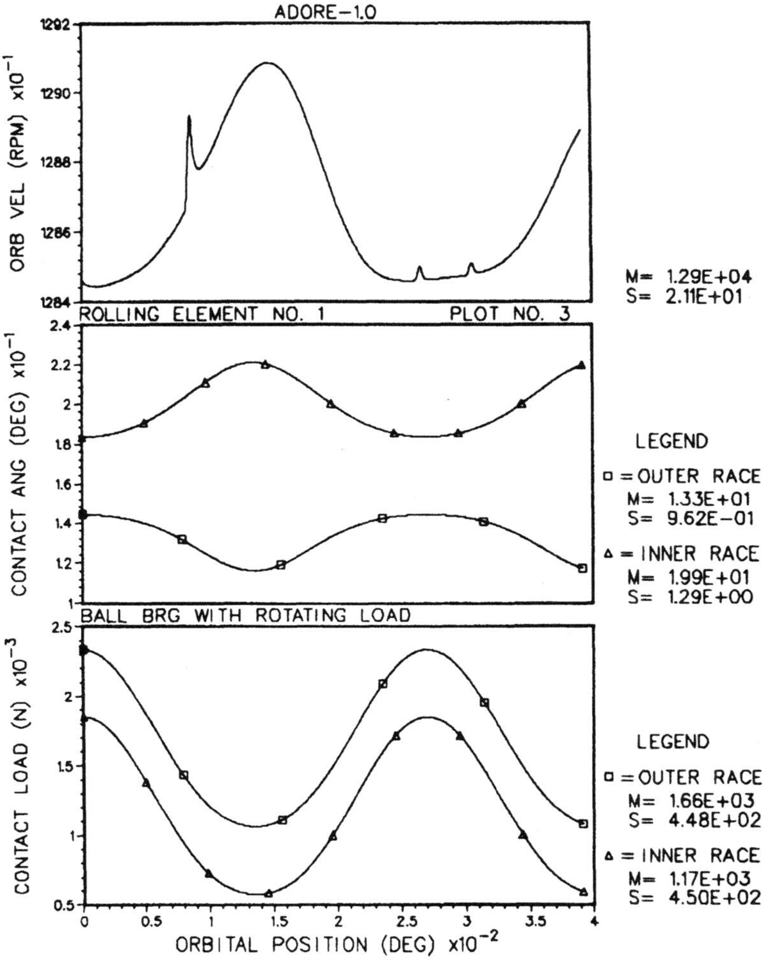

Figure 8-19 The variation in ball/race contact loads and angles in a ball bearing with rotating radial load.

Table 8-2 Data for Ball Bearing #2

Geometry	
No. of balls	13
Ball diameter	12 mm
Pitch diameter	80 mm
Contact angle	16°
Outer race curvature factor	0.53
Inner race curvature factor	0.53
Outer race fit	−0.15 mm (loose)
Inner race fit	0.03 mm
Shaft outside diameter	57 mm
Shaft inside diameter	25 mm
Cage outer diameter	87.5 mm
Cage inner diameter	76 mm
Cage/race clearance (outer)	0.25 mm (outer race guidance)
Cage/race clearance (inner)	2.00 mm
Width of guiding land	5.00 mm
Cage width	30 mm
Pocket clearances	0.25 mm

Materials

	Races and Balls	Cage	Shaft
Elastic modulus (N/m^2)	2.0×10^{11}	1.1×10^{10}	2.0×10^{11}
Density (kg/m^3)	7.75×10^{3}	2.03×10^{2}	7.75×10^{11}
Poisson's ratio	0.25	0.45	0.25
Hardness (N/m^2)	7.845×10^{9}	7.00×10^{7}	
Wear coefficient	5.00×10^{-6}	1.00×10^{-4}	

Lubrication
Ball/race (relation of Figure 4-2)
$\kappa_0 = 0, \quad \kappa_m = 0.10, \quad \kappa_\infty = 0.08, \quad u_m = 1.0$ m/s
Ball/cage and cage/race
Constant friction coefficient = 0.10

contact angles and the ball angular velocity which, in turn, lead to a variation in ball/race spin-to-roll ratio as shown in Figure 8-20. Note the slight coning motion or precession of the ball angular velocity vector. This is more pronounced for a well-lubricated bearing where the ball/race traction coefficients are much lower than those listed in Table 8-2.

The motion of the cage is a function of the interactions at the ball/cage and cage/race interfaces. Typical ball/cage interactions, as simulated by ADORE, are shown in Figure 8-21 for the first three balls located in pockets 1–3. It is noted that although the magnitude of the collision forces is relatively moderate, the time of contact may sometimes become appreciable, particularly when the collisions are close to each other, as seen for balls 2 and 3 in Figure 8-21. Typical interaction at the cage/race interface is shown in Figure 8-22. It is seen that after the initial collisions, the cage tends to establish a contact with the guiding race with a relatively moderate force but over extended periods of time.

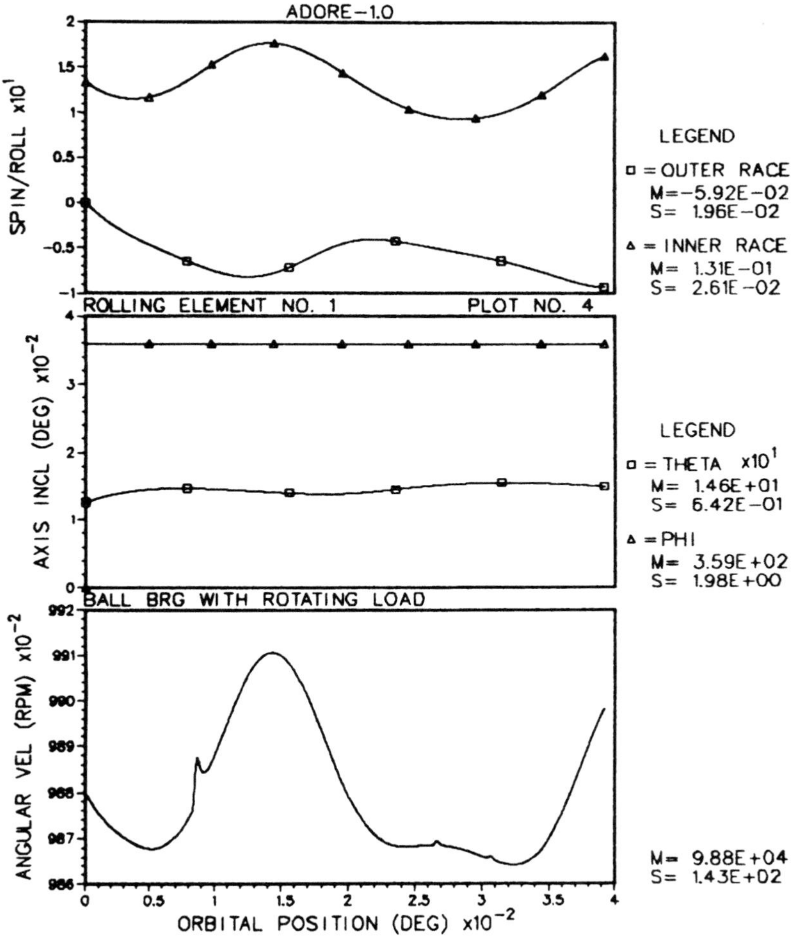

Figure 8-20 Variations in the ball angular velocity under a rotating radial load.

Figures 8-23 and 8-24 present the general motion of the cage resulting from the above ball/cage and race/cage interactions. Both the angular and the mass center whirl velocities tend to stabilize to a certain steady-state value, as seen in Figure 8-23. Also, as seen by the variation in the orientation of the angular velocity vector (angles θ and ϕ in Figure 8-23), after the initial transients the coning motion of the cage is appreciably reduced. The whirl orbit shown in Figure 8-24 reaches an almost circular shape after the effect of the initial transients has diminished. Such a cage motion may be classified as relatively stable. If either the whirl or the angular velocity continues to change as a function of time and fails to reach any steady-state value or pattern, potential cage instability is indicated.

For most practical applications, the evidence of cage instability can easily be detected by examining the variation in time-average wear rates of the cage.

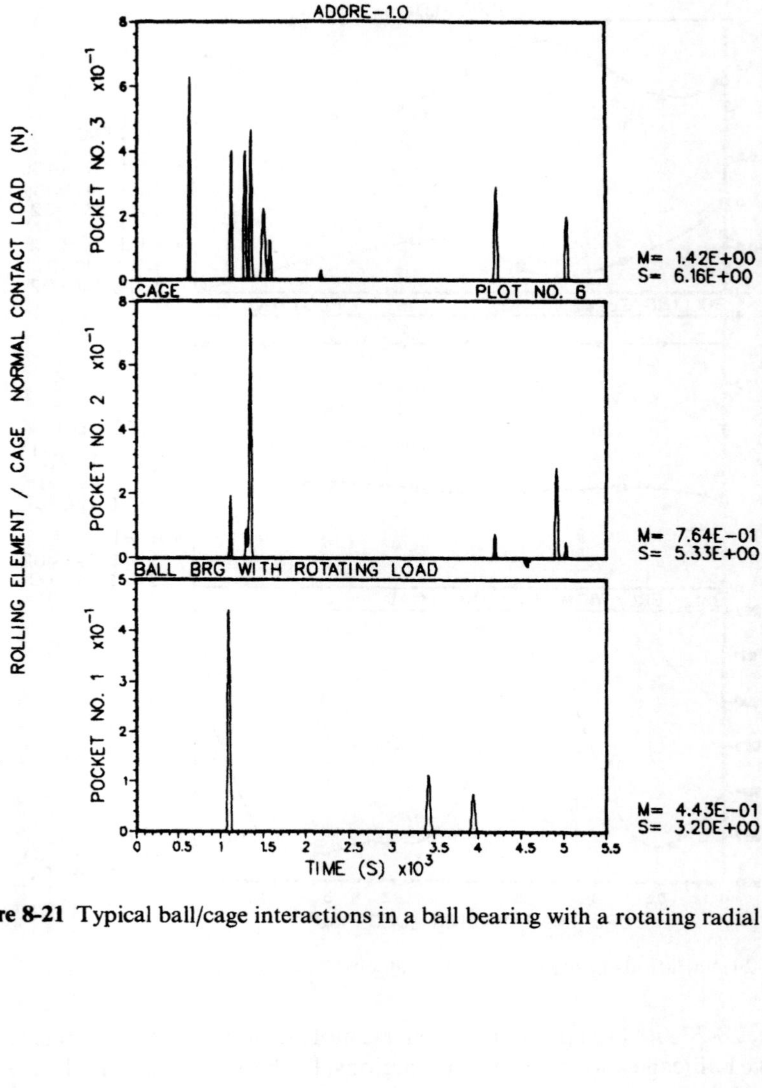

Figure 8-21 Typical ball/cage interactions in a ball bearing with a rotating radial load.

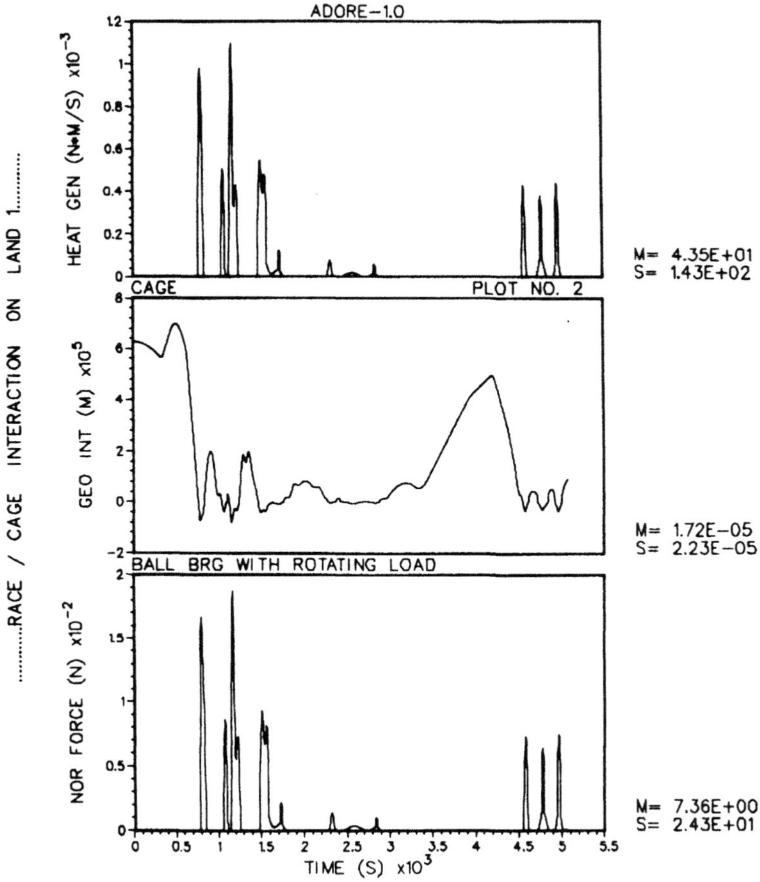

Figure 8-22 Typical interaction at the cage/race interface when the cage is guided on the stationary center race in an angular-contact ball bearing with a rotating radial load.

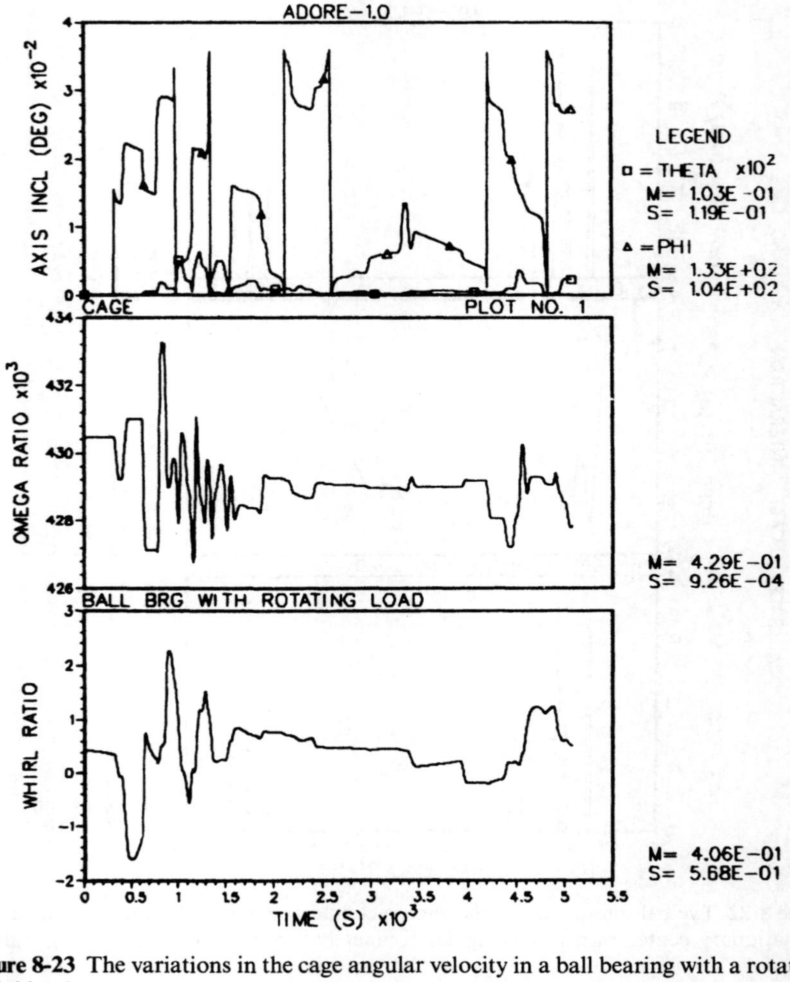

Figure 8-23 The variations in the cage angular velocity in a ball bearing with a rotating radial load.

Since the wear rate depends on both the magnitude and the time of interaction at the ball/cage and cage/race interfaces, any excessive cage interactions will result in increasing wear rates. Also, when the instability in cage motion results in increasing interactions and sliding speeds, the cage wear rates will continually increase in time. Such a condition may often lead to a catastrophic failure.

The simulated time-average wear rates, with the wear coefficients listed in Table 8-2, for the various bearing elements are shown in Figure 8-25. As expected, the cyclic variation in the ball/race contact load results in a corresponding variation in the wear rate of the ball. Wear on the races is a sum of wear at all ball/race interactions; hence, the steady-state solution approaches a certain constant value, as seen in Figure 8-25. The wear rate of the cage first increases due to the initial interactions, but after these transients

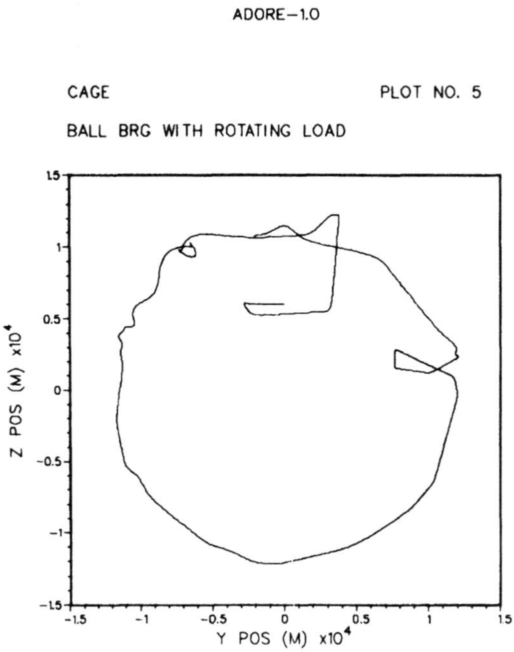

Figure 8-24 Typical orbit of the cage mass center in an angular-contact ball bearing with a rotating radial load.

have dissipated, it tends to approach a steady value. This would indicate a relatively stable motion of all of the bearing elements, the steady-state wear rates can be used as reasonable indicators of the overall expected wear life of the bearing.

It should be emphasized that the above discussion is unique to the particular example and it does not lead to any general conclusions. In fact, the nature of the traction–slip behavior at the various interactions and the ball/cage and race/cage clearances will often demonstrate a strong coupling with the operating loads and speeds in determining the overall life of the bearing. For example, Dill and Gupta [22] have shown that for well-lubricated bearings, a rotating radial load to applied thrust load ratio near unity can lead to serious instability of the cage when the rotating inner race guides the cage. It is therefore necessary to analyze each case individually. Perhaps when a large amount of data becomes available, certain trends may be established and general design guidelines can be prepared.

8.5 Roller and Cage Motion in Cylindrical Roller Bearings

Cylindrical roller bearings are generally designed to support large radial loads only and the system design is such that no axial or thrust loads are applied on the roller bearings. In such an environment the races are displaced

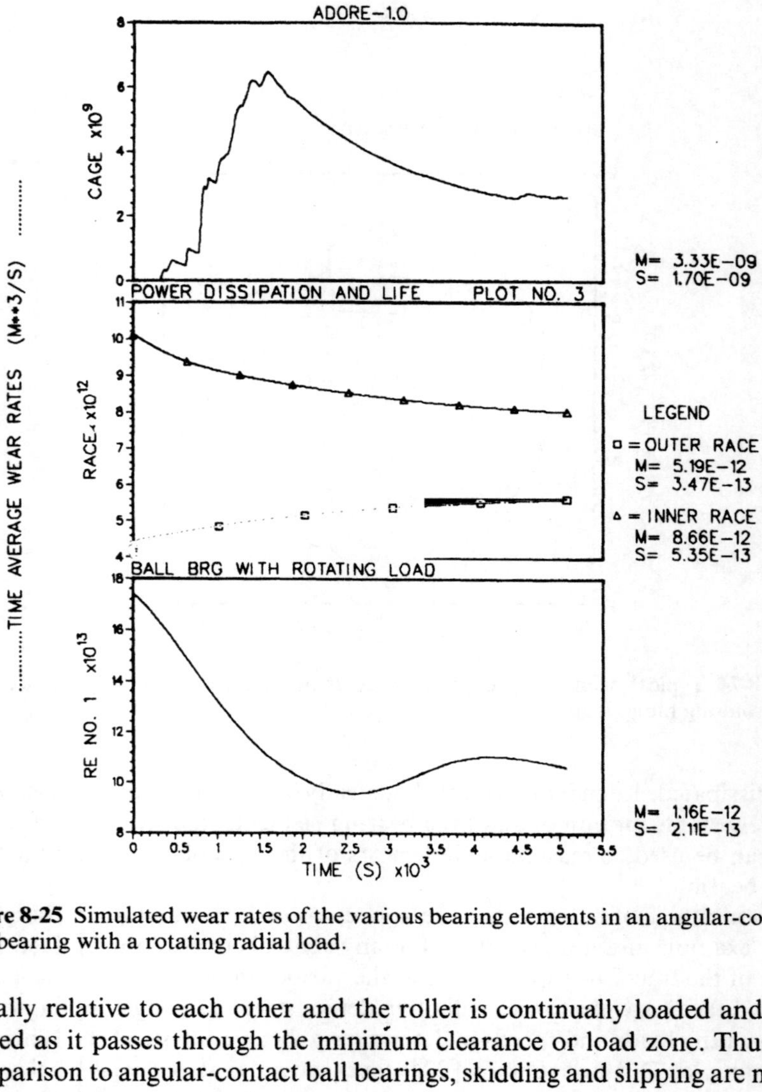

Figure 8-25 Simulated wear rates of the various bearing elements in an angular-contact ball bearing with a rotating radial load.

radially relative to each other and the roller is continually loaded and unloaded as it passes through the minimum clearance or load zone. Thus, in comparison to angular-contact ball bearings, skidding and slipping are more pronounced for such bearings.

Consider the cylindrical roller bearing described in Table 8-3 operating at 15,000 rpm with a pure radial load of 6500 N and lubricated with a MIL-L-7808 type lubricant. The variations in roller/race load as it travels around the orbit are shown in Figure 8-26; the roller/race contact angles are, of course, zero. Initially, the roller is in the most heavily loaded position and it contacts both the inner and outer races. The difference in the loads at the two races is equal to the centrifugal force exerted on the roller. As the roller travels in the orbit, the geometric interaction at the roller/race interface reduces, due to the increasing clearance, which results in a reduction in contact load. As the roller exits the load zone, the contact with the inner race is

Table 8-3 Cylindrical Roller Bearing #1

Geometry	
No. of rollers	32
Roller diameter	16 mm
Roller length	16 mm
Central land length	5 mm
Crown radius	432 mm
Corner radius	1.3 mm
Pitch diameter	196.85 mm
Diametral play	0.14 mm
Race flange clearance	0.04 mm (inner race only)
Race flange height	5 mm
Layback angle	2°
Outer race fit	0.05 mm
Inner race fit	0.10 mm
Shaft outer diameter	165 mm
Shaft inside diameter	50 mm
Bearing outside diameter	225 mm
Housing outside diameter	400 mm
Cage outer diameter	203 mm
Cage inner diameter	187 mm
Cage width	26.3 mm
Cage/race clearance (outer)	1 mm
Cage/race clearance (inner)	0.35 mm (inner race guidance)
Width of guiding land	5 mm
Pocket clearance	0.35 mm
Materials (Rollers, Races, Cage, Shaft and Housing)	
Elastic modulus	2.0×10^{11} N/m^2
Density	7.75×10^3 kg/m^3
Poisson's ratio	0.25
Hardness (rollers and races)	7.845×10^9 N/m^2
Hardness (cage)	7.845×10^8 N/m^2
Wear coefficient (rollers and races)	$5.0 \times 10^{\ 6}$
Wear coefficient (cage)	5.0×10^{-4}
Lubrication	
Roller/Race: MIL-L-7808 Traction Model	
Cage/Race and Roller/Cage: Constant friction coefficient of 0.025	

terminated and the load at the outer race becomes equal to the centrifugal force. Depending on the traction characteristics at the roller/race interface and the magnitude of roller/cage interaction, the roller may assume certain angular and orbital accelerations which will contribute to roller slipping or skidding. In fact, there is a strong coupling between the roller/race traction and the roller/cage pocket clearance. In the event of a severe collision between the roller and cage, large slip is imposed at the roller/race interface and the roller angular and orbital velocities are suddenly altered. The discontinuities in orbital velocity seen in Figure 8-26 are due to such effects. The nature of roller/race slip resulting from the collisions at the roller/cage interface is

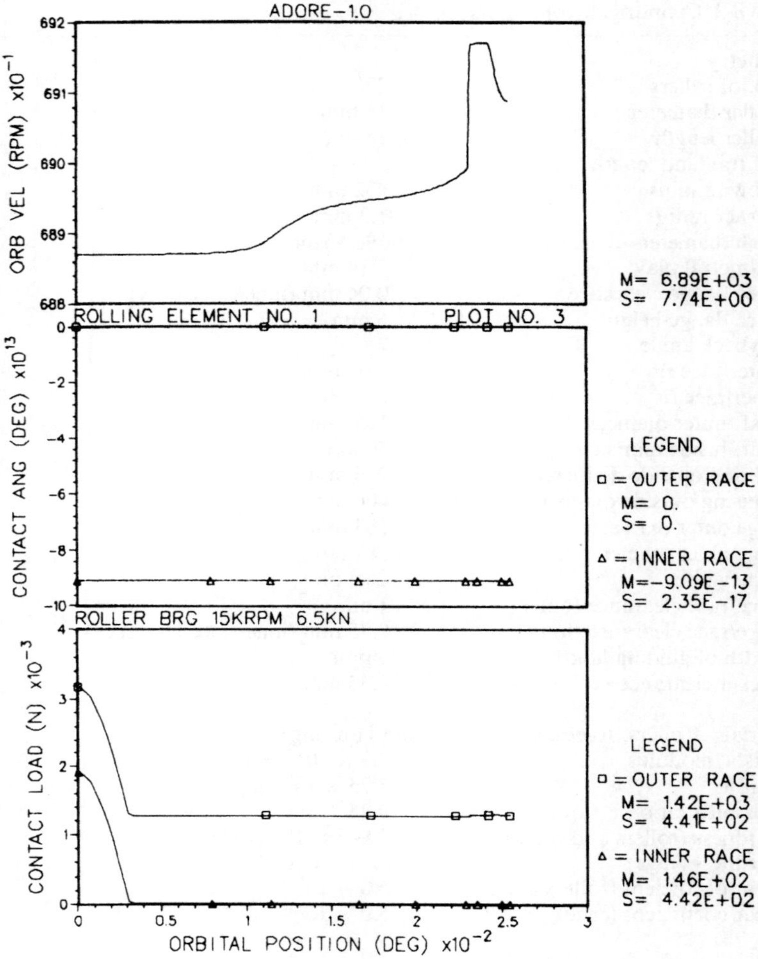

Figure 8-26 Roller/race contact load variation in a cylindrical roller bearing.

shown in Figure 8-27. The increase in roller/race slip results in increased viscous heating in the elastohydrodynamic lubricant film at the roller/race contact, which results in a somewhat reduced lubricant film and increased heat generation, as seen in Figure 8-27. Note that in this example, the general range of film thickness (0.2–0.5 μm) may not be very much larger than the roughness of the roller and race surfaces, which could result in direct metal contact and substantial damage of the bearing. The sudden drop in film thickness at the inner race contact in Figure 8-27 should be ignored, because ADORE arbitrarily sets the film thickness equal to zero when the load at the roller/race interface is zero, and any hydrodynamic effects at the unloaded interface are ignored.

The nature of roller/cage interaction is demonstrated in Figures 8-28 and

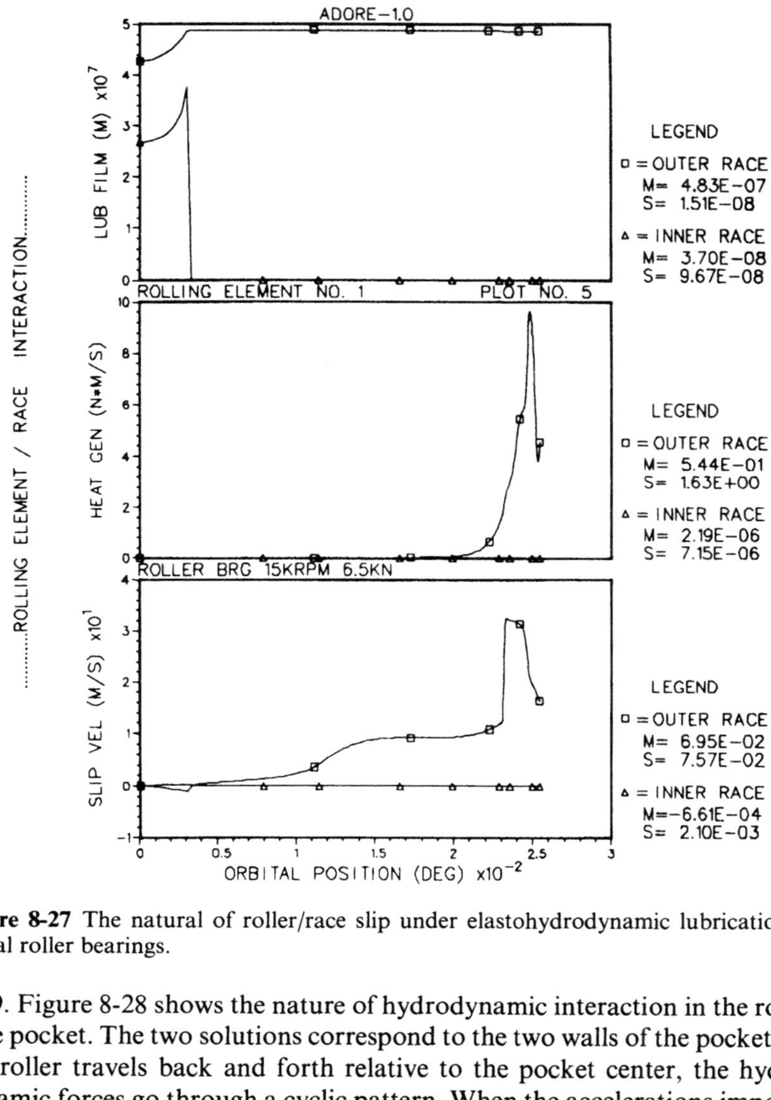

Figure 8-27 The natural of roller/race slip under elastohydrodynamic lubrication in radial roller bearings.

8-29. Figure 8-28 shows the nature of hydrodynamic interaction in the roller cage pocket. The two solutions correspond to the two walls of the pocket. As the roller travels back and forth relative to the pocket center, the hydrodynamic forces go through a cyclic pattern. When the accelerations imposed either on the roller or the cage are large, the hydrodynamic film is squeezed out and a collision between the roller and cage pocket wall takes place; this is seen in Figure 8-29. Similar to the earlier discussion on ball bearings, the frequency of these collisions will determine any damage in the cage pocket and the overall instabilities in the motion of the cage.

Typical interaction at the cage/race interface is shown in Figure 8-30. As the cage moves toward the guiding race, the geometric interaction decreases, resulting in an increased hydrodynamic force. Depending on the interaction in the cage pockets, this force may go through a cyclic pattern. If the pocket forces increase, the cage and land forces will also increase to keep the

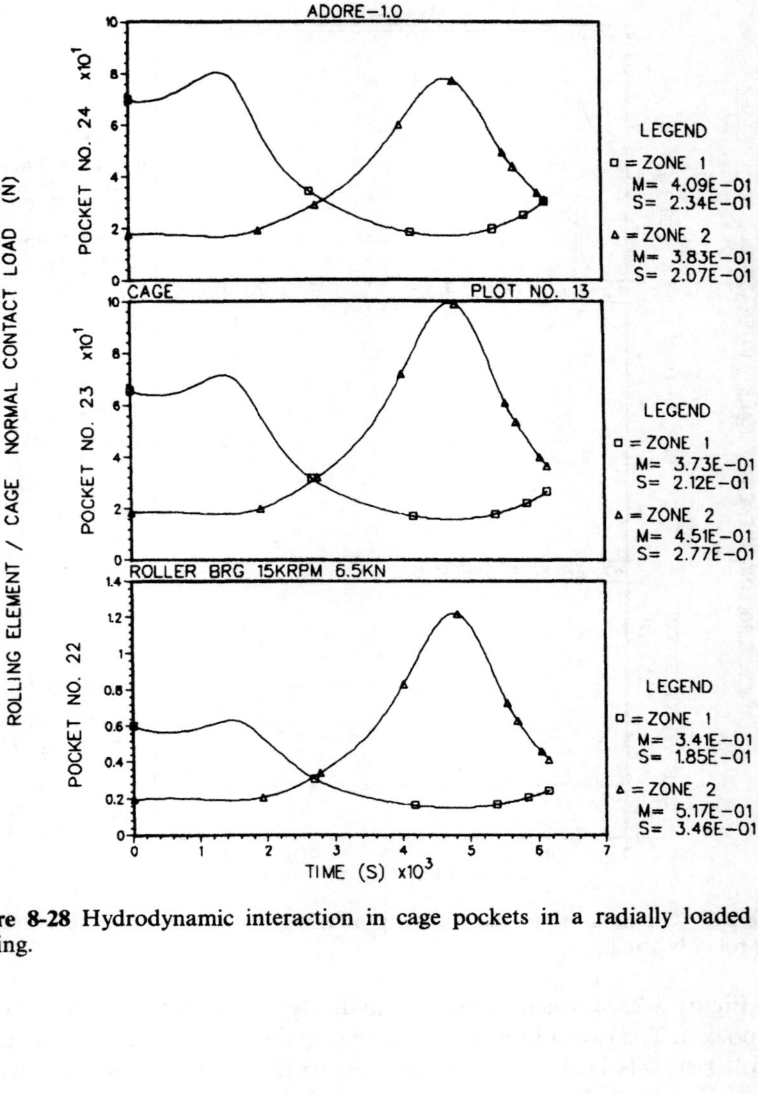

Figure 8-28 Hydrodynamic interaction in cage pockets in a radially loaded roller bearing.

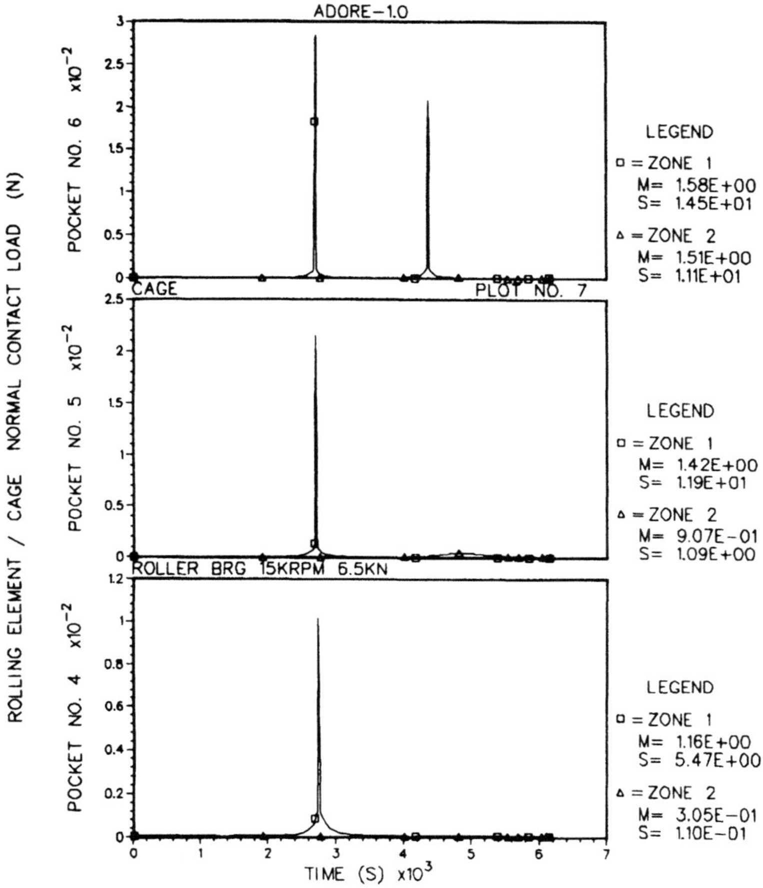

Figure 8-29 Typical roller/cage collisions in a cylindrical roller bearing.

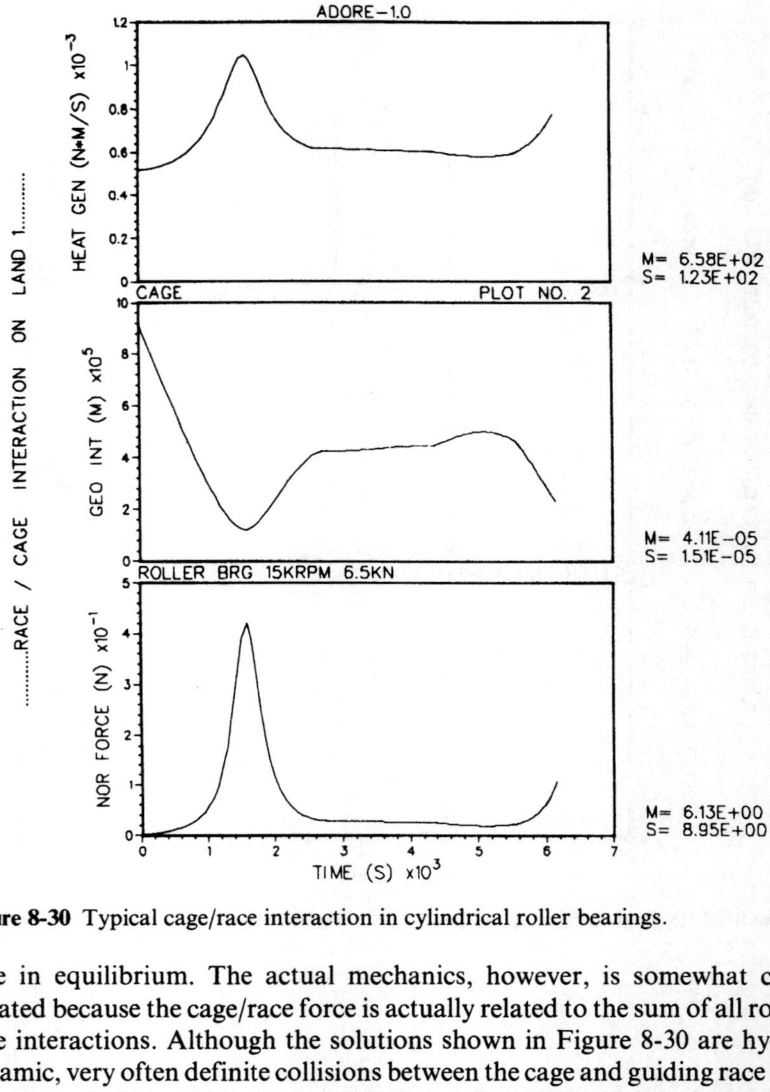

Figure 8-30 Typical cage/race interaction in cylindrical roller bearings.

cage in equilibrium. The actual mechanics, however, is somewhat complicated because the cage/race force is actually related to the sum of all roller/cage interactions. Although the solutions shown in Figure 8-30 are hydrodynamic, very often definite collisions between the cage and guiding race may be observed. This will happen particularly when the collisions in the cage pockets are severe and the hydrodynamic interaction at the cage/race interface cannot support all of the applied loads on the cage.

Severe interactions with the cage will result in substantial cage whirl and instabilities in cage motion. For the present example, however, no significant whirl is observed. The overall performance of the bearing can once again be summarized in terms of the wear rates of the various bearing elements, as shown in Figure 8-31. It is seen that the skidding of the roller on the outer race results in an increased wear rate of both the race and the roller. As the roller goes around in its orbit and the roller/race slip pattern changes, the actual wear rate will go through a cyclic pattern, and the time-average wear rate will

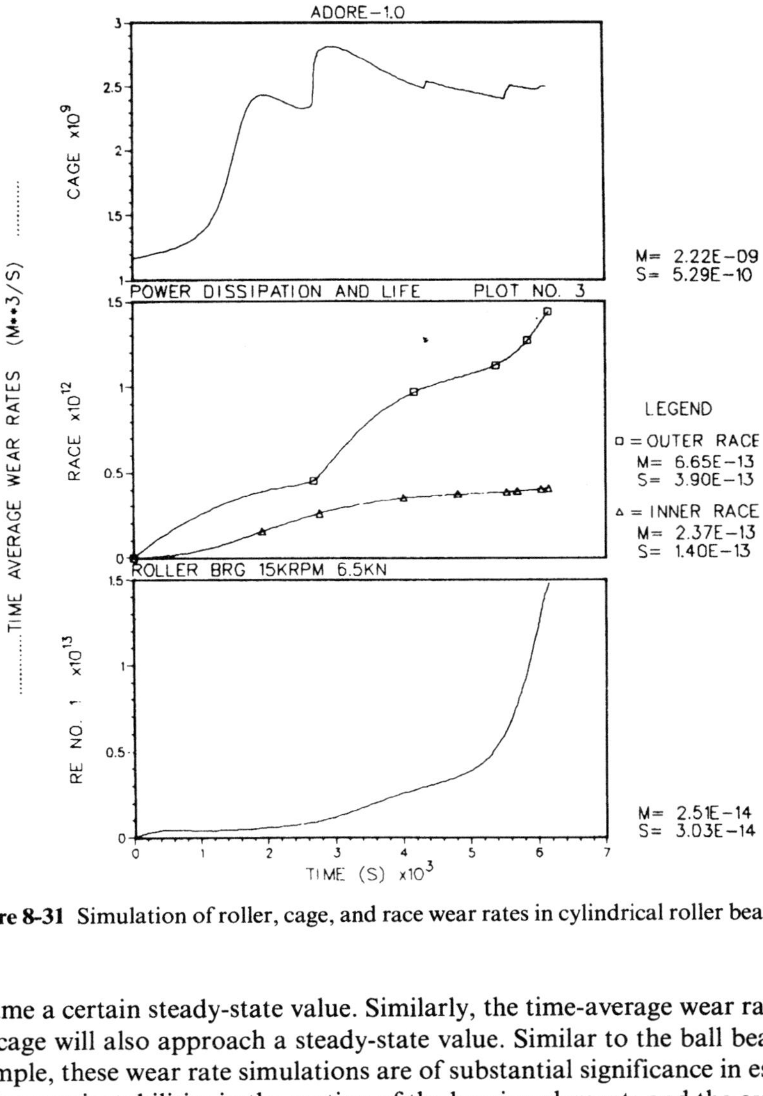

Figure 8-31 Simulation of roller, cage, and race wear rates in cylindrical roller bearings.

assume a certain steady-state value. Similarly, the time-average wear rate of the cage will also approach a steady-state value. Similar to the ball bearing example, these wear rate simulations are of substantial significance in establishing any instabilities in the motion of the bearing elements and the overall wear life of the bearing.

8.5.1 Roller Misalignment and Skew

Misalignment in cylindrical roller bearings is one of the most important parameters which influence the life and performance of the bearing. In a perfectly aligned bearing the roller simply rotates about its polar axis ($\hat{x}$), and the contact with the two races is fairly uniform and symmetric about the roller center. When the races are misaligned relative to each other, however, the

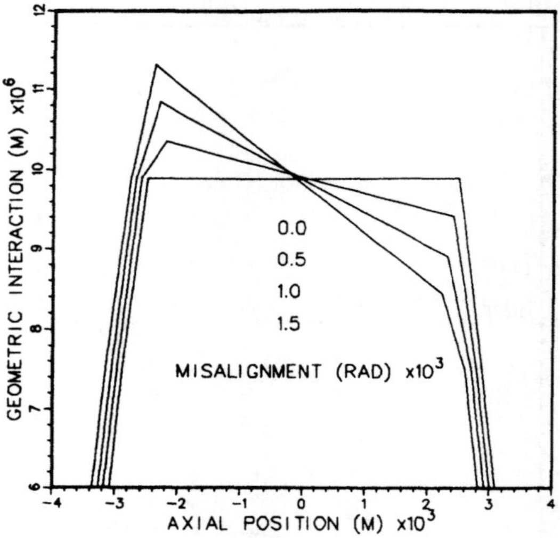

Figure 8-32 The influence of roller misalignment on the roller/race contact in cylindrical roller bearings.

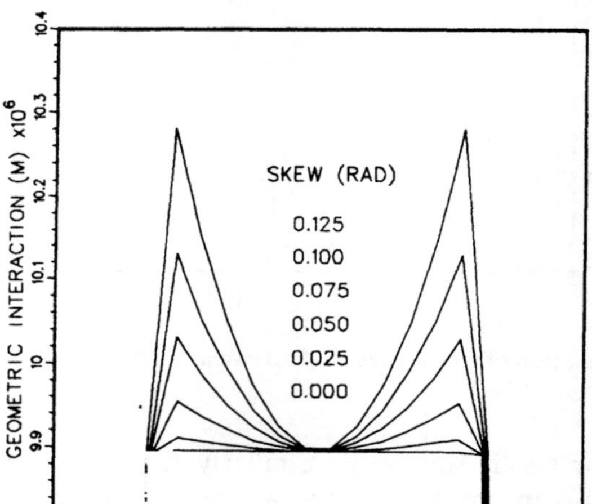

Figure 8-33 Roller skew and the resulting influence on the roller/race interaction at the outer race contact in cylindrical roller bearings.

polar axis of the roller also misaligns relative to the bearing axis. This causes certain rotation of the roller about its transverse axis. The rotation about an axis normal to the plane containing the bearing axis and the radial vector locating the roller center relative to the bearing center is referred to as misalignment, while the rotation about the transverse axis along the radial vector

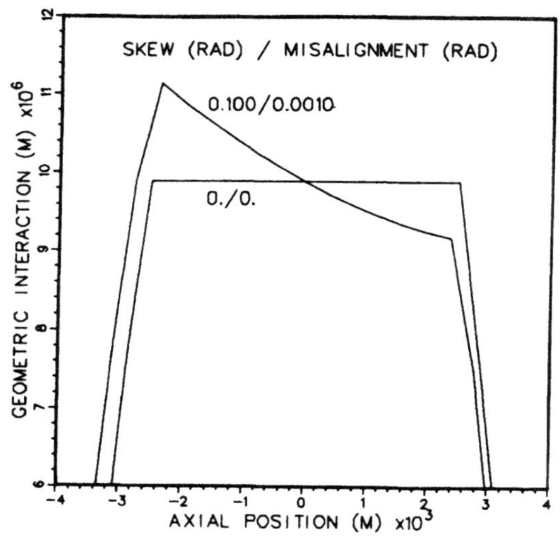

Figure 8-34 Roller/race contact (outer race) with a combined skew and misalignment.

is known as roller skew. As indicated by the general equations of motion described in Chapter 2, both misalignment and skew appear simultaneously. The roller/race contact gets somewhat complicated and if the skew angles are large, roller/flange interactions become appreciable.

To illustrate the salient features in the performance of a misaligned roller bearing, consider again the roller bearing, operating at a speed of 15,000 rpm and the applied thrust load of 6500 N, described in Table 8-3. At the outer race contact of the most heavily loaded roller, the variation in geometric interaction (which represents the contact load) along the length of the roller is shown in Figure 8-32 as a function of roller misalignment. As might be expected, the increasing misalignment results in a heavier load on one side of the roller. The influence of roller skew is somewhat more interesting, particularly at the outer race contact. As shown in Figure 8-33, an increasing angle of skew results in heavier loads at the ends of the contact zone, while the load in the center of contact remains unchanged. This can be simply explained by the geometry of interaction between the outer race and the roller, which is turned about an axis normal to the contact plane. In a stable situation this increased loading on the ends of the contact zone will result in a tractive moment that helps align the roller. However, if the direction of roller/race slip velocities is such that the net tractive moment is in the direction of skew, a potential skew instability of the roller will be indicated. When both the skew and misalignment are applied simultaneously, the contact solution is even more complicated, as shown in Figure 8-34.

The general motion of the roller in a misaligned bearing can be best understood by examining the variation in roller angular velocity as the roller travels between the misaligned races. Figure 8-35 shows the roller angular velocity

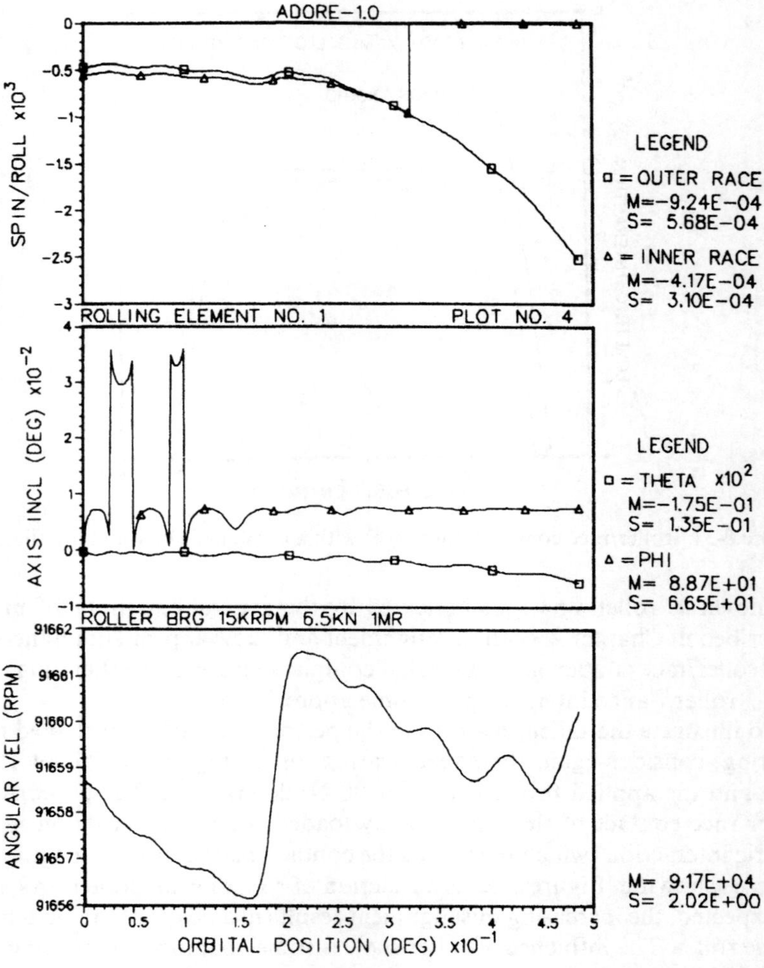

Figure 8-35 The general motion of a roller in a misaligned cylindrical roller bearing.

and the spin or roller skew component when the rotating inner race is misaligned relative to the stationary outer race by a small angle of one milliradian. As the roller travels in its orbit, the angular velocity goes through a cyclic variation superimposed on the normal decrease due to unloading of the roller. The small precession of the roller axis is indicated by the two angles which define the inclination of the roller angular velocity vector. The termination of the roller contact with the inner race is indicated by the point where the inner race spin-to-roll ratio is set to zero. The sudden increase in the roller angular velocity is due to roller interaction with the guide flange, which results from the oscillatory motion of the roller. Figure 8-36 shows the general nature of the roller/flange interaction in this example.

Table 8-4 Data for Roller Bearing #2

Geometry	
No. of rollers	16
Roller diameter	12 mm
Roller length	20 mm
Central land length	12 mm
Crown radius	1000 mm
Corner radius	1 mm
Pitch diameter	85 mm
Diametral play	0.04 mm
Outer race fit	0.02 mm
Inner race fit	0.02 mm
Shaft outer diameter	70 mm
Shaft inner diameter	20 mm
Bearing outer diameter	100 mm
Housing outer diameter	150 mm
Cage outer diameter	90 mm
Cage inner diameter	80 mm
Cage width	40 mm
Cage/race clearance (outer)	0.50 mm (outer race guidance)
Cage/race clearance (inner)	1 mm
Width of guiding land	5 mm
Pocket clearance	0.25 mm
Segmented cage, no. of segments	2
Width of segment cut	2°
Materials (Rollers, Races, Cage, Shaft and Housing)	
Elastic modulus	2.0×10^{11} N/m^2
Density	7.75×10^3 kg/m^3
Poisson's ratio	0.25
Hardness (rollers and races)	7.845×10^9 N/m^2
Hardness (cage)	7.845×10^8 N/m^2
Wear coefficient (rollers and races)	5.0×10^{-6}
Wear coefficient (cage)	5.0×10^{-4}
Lubrication	
Roller/Race: $\kappa_0 = 0$, $\kappa_m = 0.05$, $\kappa_\infty = 0.04$, $u_m = 2.0$ m/s	
Cage/Race and Roller Cage: Constant friction coefficient = 0.05	

8.5.2 Time-Varying Loads and Speeds

Very often rolling bearings are subjected to a variation in both the applied load and angular velocity of the rotating raceway. Consider the roller bearing with a segmented cage described in Table 8-4, in a connecting rod application of a reciprocating engine. As the crank turns, the bearing center travels in a circular orbit, as shown in Figure 8-37. Let the crank velocity or the orbital velocity of the bearing be 2000 rpm, and let a constant load of 5000 N be applied along a vector passing through the orbit and bearing centers and pointing radially outward; this is somewhat of a simplification of the actual application. Let the orbit radius be 0.25 m and the angular velocity of the

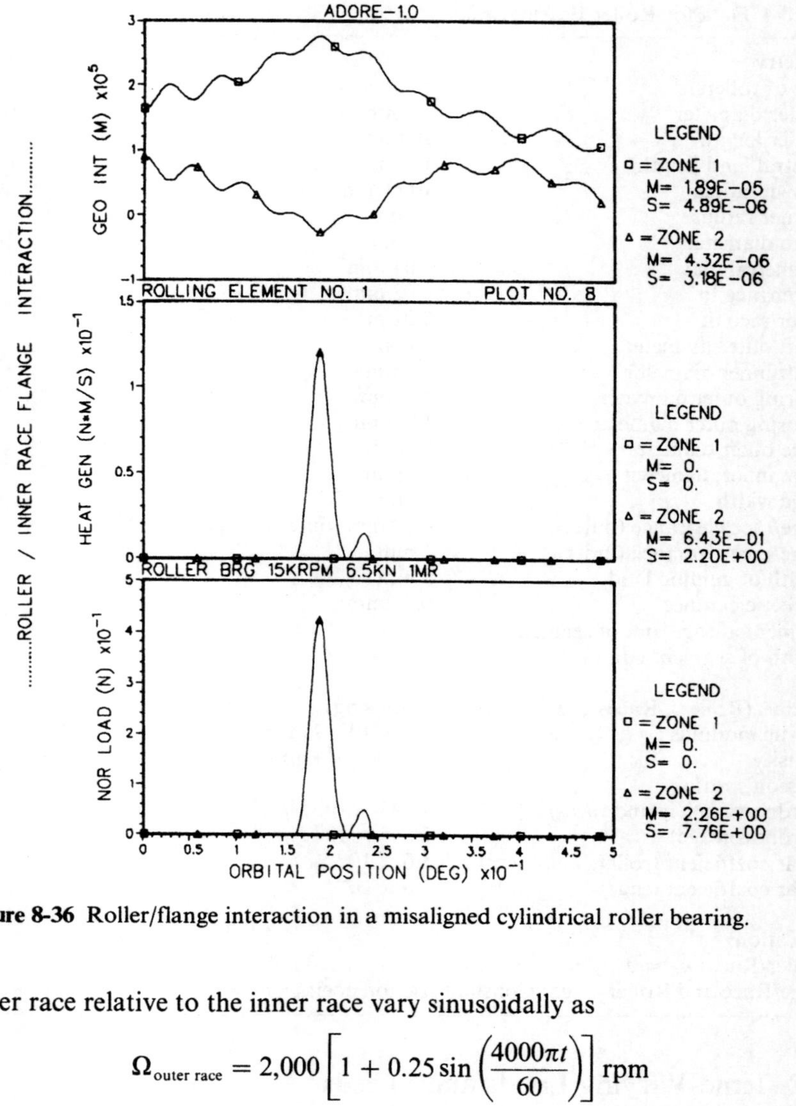

Figure 8-36 Roller/flange interaction in a misaligned cylindrical roller bearing.

outer race relative to the inner race vary sinusoidally as

$$\Omega_{\text{outer race}} = 2{,}000\left[1 + 0.25\sin\left(\frac{4000\pi t}{60}\right)\right] \text{ rpm}$$

Although the applied load has been assumed to be constant, the Coriolis forces exerted on the rollers and cage segments will contribute to a dynamic variation in the roller/race and cage/race interactions. The cyclic variation in the race angular velocity will influence the roller/race slip behavior and may subsequently effect the roller/cage interaction. The modeling of such sophisticated operating environments is accomplished in ADORE by providing appropriating coding in the variable load subroutine ADRX1.

Figures 8-38 and 8-39 show ADORE simulations of the roller/race contact load and slip variation. The high-frequency vibration is due to the roller/race elastic contact spring and this vibration is excited by the time-varying Coriolis

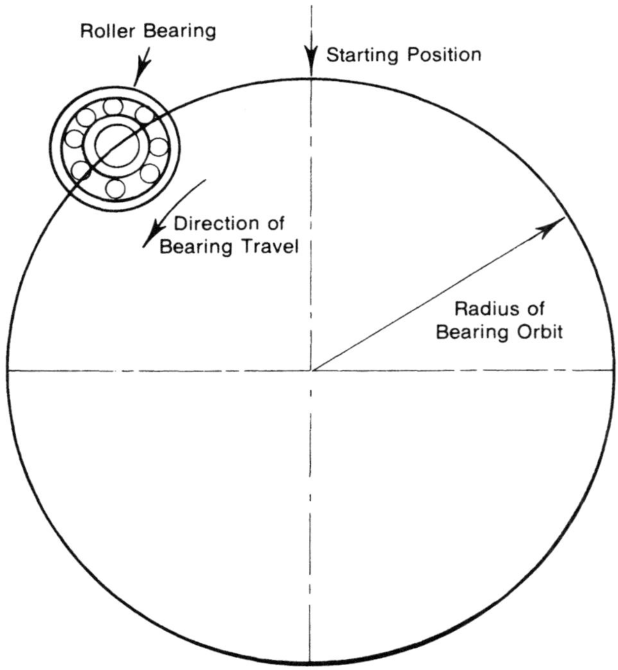

Figure 8-37 Prescribed orbit of travel for a variable-load cylindrical roller bearing.

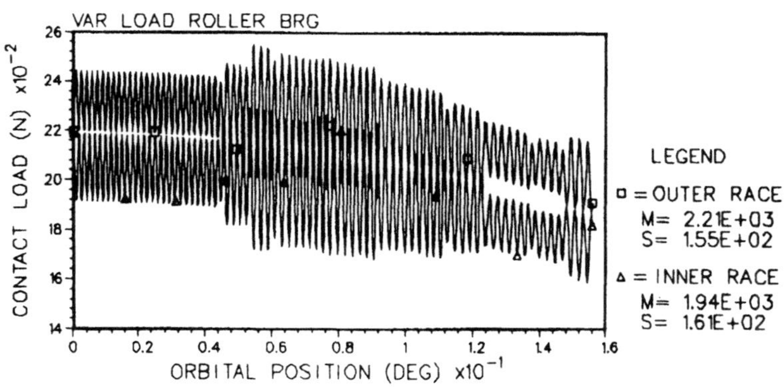

Figure 8-38 Roller/race contact load variation in the variable-load roller bearing.

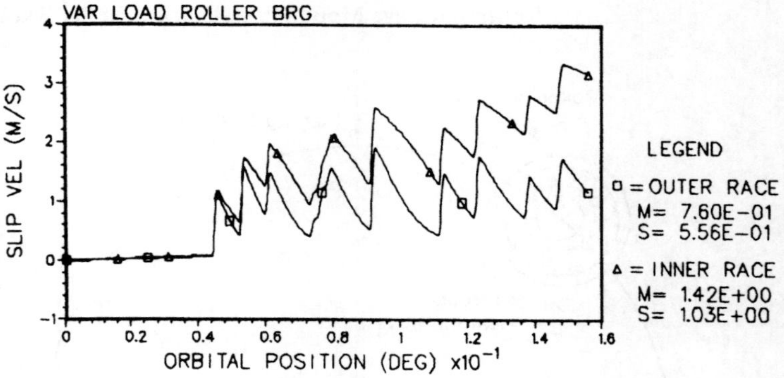

Figure 8-39 Roller/race slip variation in the variable-load roller bearing.

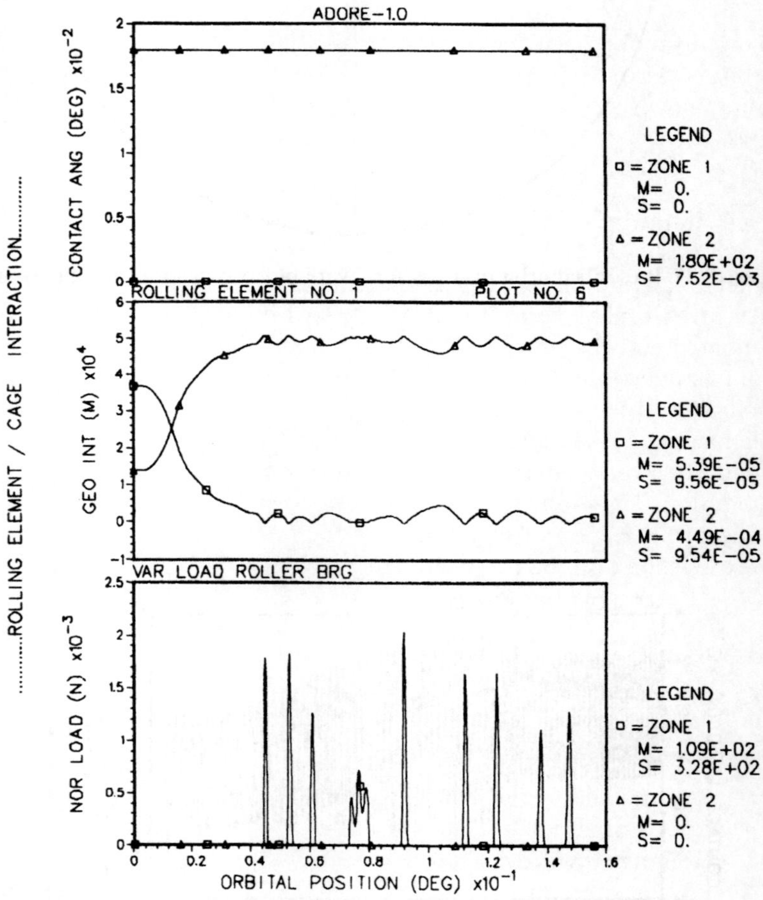

Figure 8-40 Roller/cage interaction in the variable-load roller bearing.

force exerted on the roller and the small variation in the centrifugal force resulting from the variation in the roller orbital velocity. The nominal decrease in the roller/race load is due to the unloading of the roller as it travels around the bearing. A combination of the roller unloading and the changing angular velocity of the outer race results in the nominal variation in the roller/race slip shown in Figure 8-39. The superposed saw-tooth pattern in the slip is a result of the roller/cage collisions shown in Figure 8-40. It is seen that all roller/cage interactions in Figure 8-40 correspond to "Zone 1," which represents the trailing wall of the cage pocket. Thus the cage segment is driving the roller during the time of simulation considered in this example. Note the large magnitude of roller/cage forces which is a result of roller skidding or slipping and the rather high coefficients of traction (see Table 8-4) at the roller/race interface. These large pocket forces contribute to similar forces as the cage/race interface, as shown in the cage/race contact simulation in Figure 8-41. Normally, when the pocket forces are small, the average cage/race contact force will be just equal to the centrifugal and Coriolis force exerted on the cage segment. In the present example, however, the pocket forces are very large compared to both the centrifugal and Coriolis components; thus the large forces at the cage/race interface. These large forces result in excessive cage whirl and wear, as shown in Figures 8-42 and 8-43 respectively. The base for computing the velocity ratios in Figure 8-42 is the angular velocity of the inner race relative to the outer race. In the present example, with stationary inner race and rotating outer race, the negative sign associated with the cage angular and whirl velocity ratios can be ignored. The increasing whirl and angular velocity of the cage segment, shown in Figure 8-42, is an indication of some instability in the motion of the cage segment, which may eventually result in a bearing failure. As a result of the large magnitude and high frequency of cage collisions, it is interesting to note in Figure 8-43 that the time-average cage wear rate rapidly reaches a rather high and steady value. The wear rates of the roller and raceways have clearly not stabilized during the short time of simulation considered in this example.

Aside from extending the performance simulations farther out in time, complex applications, such as the one discussed above, require an in-depth study of a number of other factors. For example, details of roller motion as the bearing travels along the orbit have to be investigated as a function of the roller/race traction behavior. This will provide significant insight into wear and skidding control and lubrication requirements. Similarly, the influence of bearing misalignment and roller skew on the already complicated conditions will be an interesting investigation, and it will have a substantial design significance. Once the various interaction models are well understood, most of such investigating can be very effectively carried out by systematic parametric studies using computer codes such as ADORE.

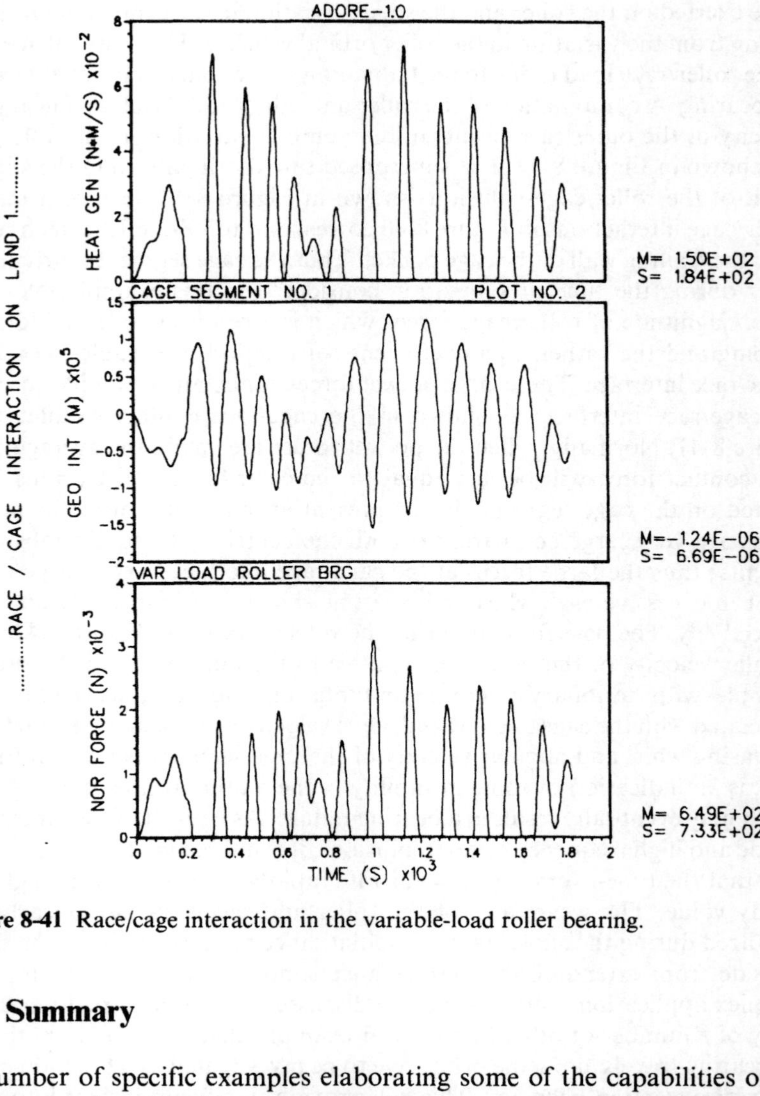

Figure 8-41 Race/cage interaction in the variable-load roller bearing.

8.6 Summary

A number of specific examples elaborating some of the capabilities of the models developed in this text and the computer program ADORE are presented. The specific areas of practical significance discussed include vibrational characteristics of rolling bearings, ball skid in an angular-contact ball bearing, cage instability in rolling bearings, roller motion, misalignment, and roller skew in cylindrical roller bearings. A roller bearing example with time-varying loads and race speeds demonstates the capabilities of ADORE to simulate rolling bearing performance under the most sophisticated operating environment.

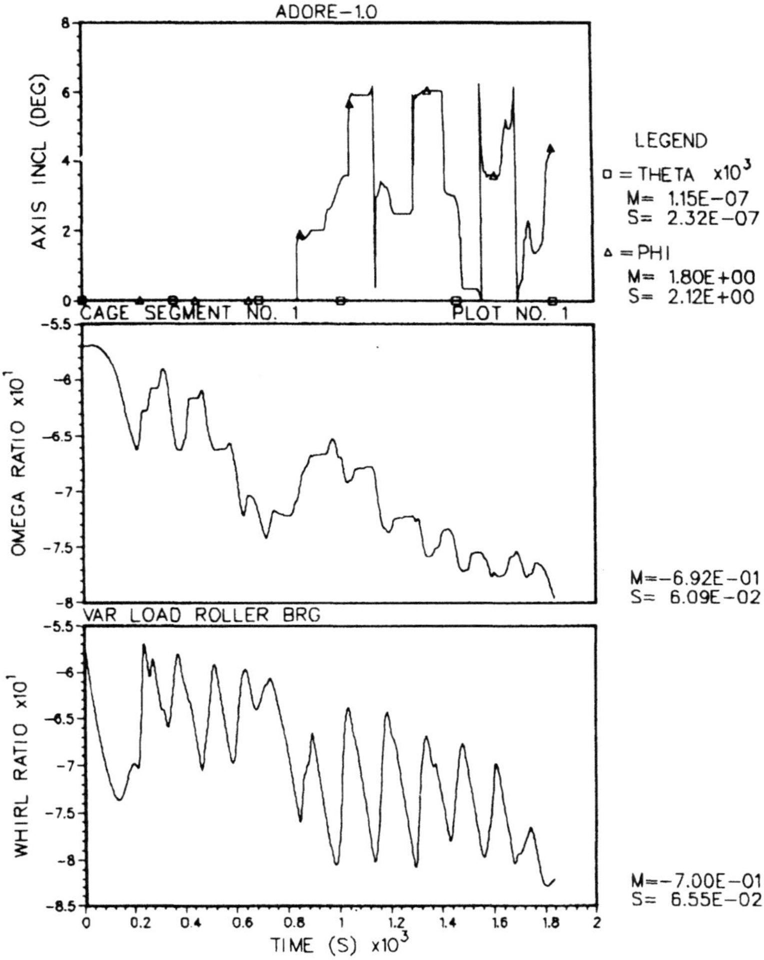

Figure 8-42 Cage segment whirl and angular velocities in the variable-load roller bearing.

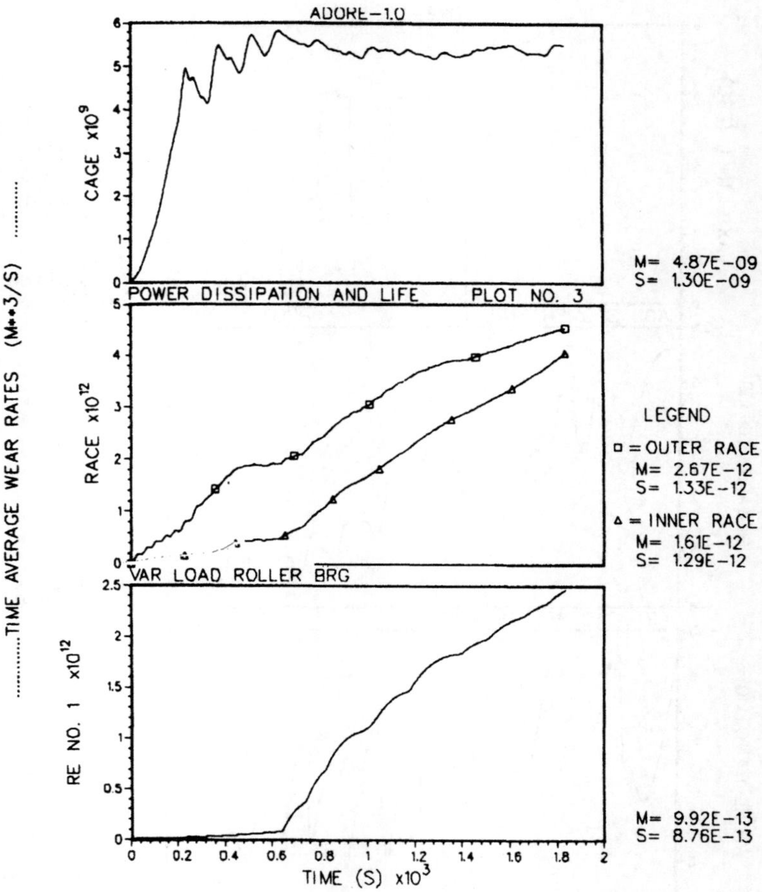

Figure 8-43 Time-average wear rates of the bearing elements in the variable-load roller bearing.

CHAPTER 9

Experimental Validation of ADORE

Bearing performance simulations obtained by ADORE are validated in the light of available experimental data on ball skid and the general motion of the cage in an angular-contact ball bearing. The reasonably good agreement between ADORE predictions and the experimental observations provide significant strength to the design capabilities of this computer program.

Experimental validation of an analytical model provides significant confidence in the predictions and design capabilities of the model. As the level of sophistication increases and the models become more complex, the need for experimental validation becomes even more important; however, the experiments required to produce the necessary data become extremely difficult and the limitations of the measurement and instrumentation systems present a real problem. For example, any precise measurement of the motion of various elements in a rolling bearing, which is necessary in order to truly validate ADORE, is very difficult, if not impossible. Furthermore, when the raw experimental data are subjected to analysis in order to derive the required parameters, experimental errors can be greatly amplified and the validation of a model with such experimental data becomes highly questionable. For instance, if a displacement is actually measured and then differentiated analytically to arrive at "measured" velocities or accelerations, a small experimental error in the displacement measurement, when combined with the noise generated by the differentiating process, can produce a very large tolerance band on the derived velocities or accelerations. Thus, considerable care is necessary in both actual measurement and analysis of the raw data. It is essential that the uncertainty introduced by the data analysis process be significantly less than the experimental error in actual measurement.

Similar to the difficulties in carrying out the necessary experiments, a realistic determination of the various input parameters to the analytical model is often a problem. For example, defining the tractive characteristic of the lubricant and realistically estimating the various drag coefficients are difficult tasks. Such input parameters can be easily varied to produce a reasonable fit to the experimental data, but the validation will hardly strengthen the model, although it may be useful in establishing certain trends. An alternate approach is to generate the analytical results over a range of input parameters and produce a "performance map." The experimental data can then be superimposed on this map and some qualitative judgement about the predictive capabilities of the model can be made. A more rigorous validation can then be deferred until both realistic inputs to the model become available and the uncertainty in the experimental data is reduced.

With particular emphasis on the analysis presented in this text and the computer program ADORE, it is essential that experimental investigations be performed on two levels. First, it is necessary to carry out simple experiments to determine realistic inputs to ADORE. This will include lubricant traction experiments and friction tests simulating the cage/race and rolling element/cage interactions. Once such data have been generated, bearing experiments will be necessary to measure the various components of motion that are predicted by ADORE. Validation of overall performance, such as for bearing failure under certain operating conditions, may be comforting, but is really inadequate to truly validate the basic models in ADORE.

In view of the above difficulties and the limited experimental data, the objective of this chapter is to present a "first step" towards a greatly needed experimental validation of advanced analytical tools such as ADORE. The subjects of ball skid and the general motion of the cage in ball bearings are considered as a starting point. It is expected that a more rigorous validation of ADORE will become possible as more experimental data become available.

9.1 Ball Motion and Skid

Ball skid in lightly loaded, angular-contact ball bearings is a problem of considerable practical significance. A number of experimental investigations aimed at the measurement of skid have been reported; however, in spite of this large amount of data, a true validation of ADORE is difficult, because neither the characteristics of the lubricant used nor adequate geometric details of the test bearing are available. Therefore, only a qualitative assessment of ADORE predictions is possible at this point.

Philips [78] measured the ball orbital velocity as a function of the applied load and the quantity of grease packed in a ball bearing having the configuration described earlier in Table 8-1. The measurement system consisted of a photonic sensor mounted on the stationary outer race to measure the deflection of the outer race as the ball traveled on the raceway. Every time the ball passed under the photonic sensor, a distinct peak in the race deflection profile was observed. Thus, the frequency of these peaks indicated the

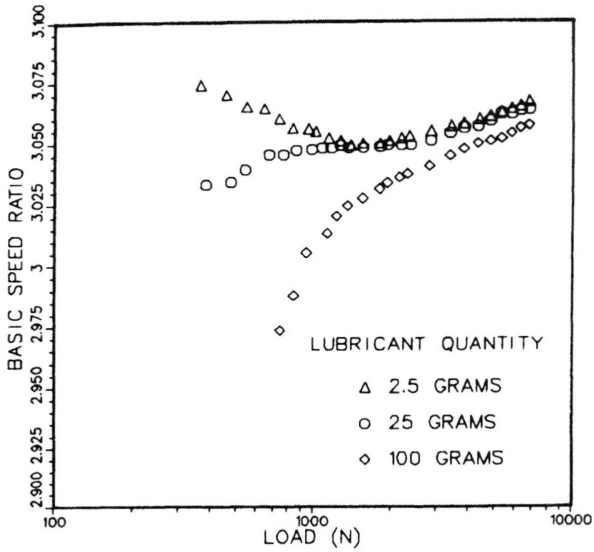

Figure 9-1 Experimental ball skid data, after Philips [78].

pass frequency. The experiment was run at a constant inner race speed of 3600 rpm and the results were presented in terms of a "basic speed ratio" which is essentially defined as the ratio of the ball pass frequency to the angular velocity of the inner race. Typical results are reproduced here in Figure 9-1.

ADORE predictions of ball orbital velocity for a varying amount of lubricant drag were presented for this bearing in Figure 8-19. Figure 9-2 shows the experimental data of Figure 9-1 reduced to ball orbital velocity ratio and compares these ratios with the ADORE predictions. It is seen that the general trends of the data are in agreement. When comparing the actual magnitude, it must be remembered that the analytical results were generated by using the simple traction curve of Figure 4-2, since there are no traction characteristics available for the grease used in the experimental investigation. Also, the test bearing did have a cage, the geometric details of which are not available. The presence of the cage will definitely increase churning and drag losses, and the drag on the cage may contribute to some reduction in the ball orbital velocity. This could, to some extent, explain the differences between the analytical and experimental results in Figure 9-2.

9.2 Cage Motion

Actual measurement of cage motion and the subsequent determination of the various components predicted by ADORE are extremely difficult tasks. Although a number of investigations have been devoted to the measurement

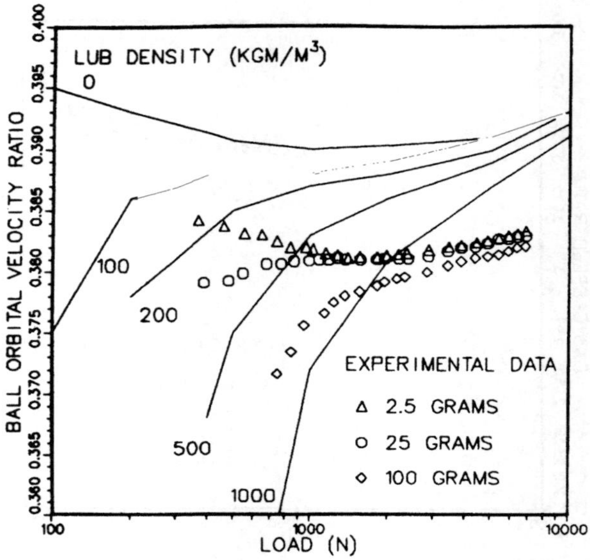

Figure 9-2 ADORE predictions versus the experimental ball skid data.

of cage slip and the average cage angular velocity, data on cage whirl and out-of-plane coning motion of the cage are extremely limited. Perhaps the very recent investigation by Gupta et al. [44] is a first attempt to the measurement of such components of cage motion. This investigation considered a 100 angular-contact ball bearing, the geometric details of which are summarized in Table 9-1.

A MIL-L-7808-type lubricant was used in the test bearing; the available traction data on such lubricants was used to determine the traction–slip behavior at the ball/race contacts. At the cage/race and ball/cage interactions, a constant coefficient of friction was assumed. The quantity of lubricant in the bearing was kept to a minimum to eliminate large churning and drag losses.

The cage width was slightly extended out of the bearing races so that its radial displacement could be measured by probes attached to the stationary outer race and directed towards the extended rim of the cage on either side of the rolling elements. As shown schematically in Figure 9-3, two probes were mounted at right angles to each other on each side of the cage. Thus, the radial motion of the cage was measured in two planes normal to the bearing axis by the two sets of radial probes labelled as (R_{11}, R_{21}) and (R_{12}, R_{22}) in Figure 9-3. The signals from the two probes in each plane were first averaged and differentiated to determine the cage mass center velocity in the two orthogonal directions. These velocities were then transformed to determine the orbital whirl component as predicted by ADORE. The measured radial displacements in the two planes can also be subtracted and divided by the distance between the two measurement planes to determine

Table 9-1 Data for Ball Bearing #3

Geometry	
No. of balls	18
Ball diameter	19.05 mm
Pitch diameter	140 mm
Contact angle	25°
Outer race curvature factor	0.52
Inner race curvature factor	0.54
Outer race fit	0.010 mm
Inner race fit	0.050 mm
Shaft outer diameter	100 mm
Shaft inner diameter	20 mm
Bearing outer diameter	180 mm
Housing outer diameter	205 mm
Cage outer diameter	146.292 mm
Cage inner diameter	130.073 mm
Cage width	36.24 mm
Cage/race clearance (outer)	4.7358 mm
Cage/race clearance (inner)	1.4605 mm (inner race guidance)
Width of guiding land	8 mm
Pocket clearance	0.8636 mm
Materials (Balls, Races, Cage, Shaft, and Housing)	
Elastic modulus	2.0×10^{11} N/m^2
Density	7.75×10^3 kg/m^3
Poisson's Ratio	0.25
Hardness (balls and races)	7.845×10^9 N/m^2
Hardness (cage)	7.845×10^8 N/m^2
Wear coefficient (balls and races)	5.0×10^{-6}
Wear coefficient (cage)	5.0×10^{-4}
Lubrication	
Ball/Race: MIL-L-7808 Traction Model	
Ball/Cage and Cage/Race: Constant friction coefficient = 0.005	

any coning motion or rotation of the cage about the transverse y and z axes. A reliable measurement of the cage angular velocity about the bearing axis and its variation in time was not possible in this investigation due to experimental limitations. An attempt was made to create a black-and-white pattern on the end face of the cage and to use two photonic sensors to measure the cage angular velocity. Unfortunately, due to the presence of lubricant in the bearing, the contrast in the black-and-white pattern was poor and the signals from the photonic sensors were rather weak, particularly at higher shaft speeds.

Before presenting a comparison between the experimental data and ADORE predictions, it is necessary to understand the expected uncertainty in the experimental data. For example, there is a definite uncertainty in the radial probe data due to the calibration and thermal compensation of the

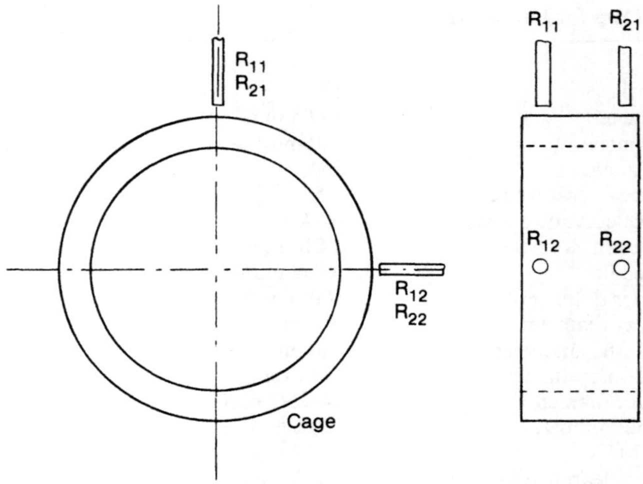

Figure 9-3 Location of the radial probes used to measure the cage motion.

probes. In addition, the rigidity of the structure on which the probes were mounted has yet to be resolved, particularly at high shaft speeds.

Although the geometry of the bearing is well established, ADORE simulations are subjected to a number of uncertainties in the lubrication data, such as the traction model for the MIL-L-7808 lubricant and the value of the friction coefficient at the ball/cage and cage/race contacts. Based on typical traction coefficients for very lightly loaded contacts, a value of 0.005 was assumed for all ball/cage and cage/race interactions. A few cases were run with a friction coefficient of 0.05 to determine the sensitivity of cage whirl to the friction coefficient.

In view of the above difficulties in both the experiments and realistic inputs to the analysis, it is fair to compare no more than general trends in the following data. A more precise validation has to be deferred until further improvements in the experimental techniques are made and the uncertainties in the input data to ADORE are reduced.

9.2.1 Cage Mass Center Orbit

Both the experimental data and ADORE predictions indicate that there is a certain critical shaft angular velocity, in the vicinity of 5000 rpm for the bearing under consideration, above which the cage mass center whirls in an almost circular orbit and below which there is no significant whirl. ADORE simulations indicate that the actual speed at which the cage starts orbiting is dependent on the friction behavior at the ball/cage and cage/race interfaces. Hence a realistic value of the friction coefficient is essential in determining the critical speed at which the cage goes into orbit. Once the cage starts orbiting, however, the shape of the orbit is insensitive to the friction coeffi-

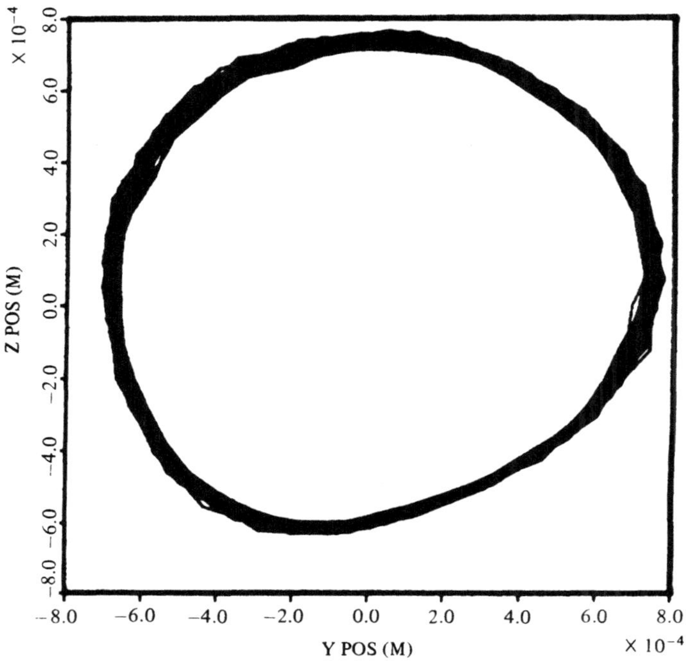

Figure 9-4 Cage mass center orbit at 10,000 rpm as derived from the radial probe data.

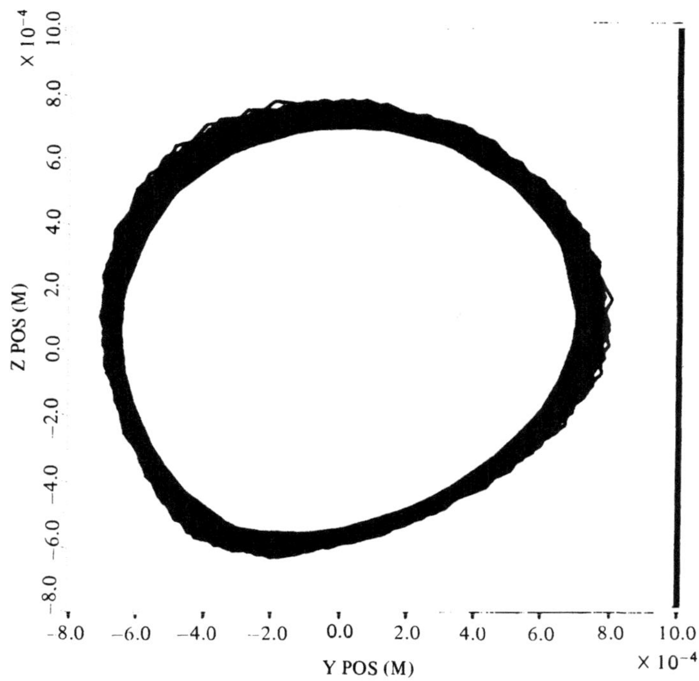

Figure 9-5 Cage mass center orbit at 20,000 rpm as determined from the experimental data.

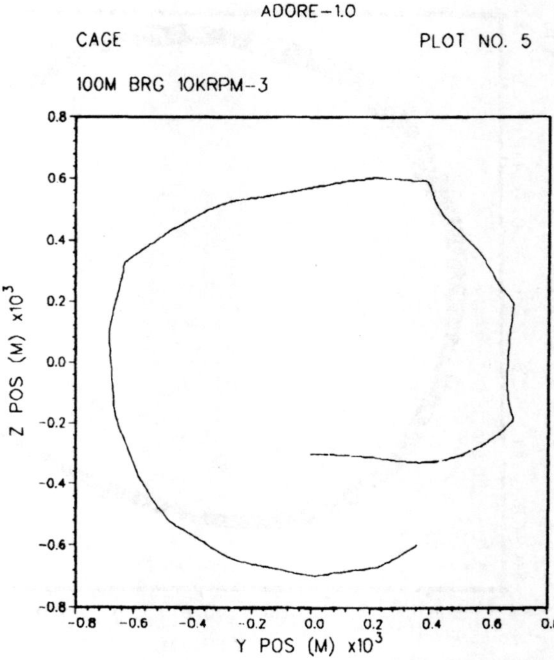

Figure 9-6 Cage mass center orbit as predicted by ADORE at 10,000 rpm.

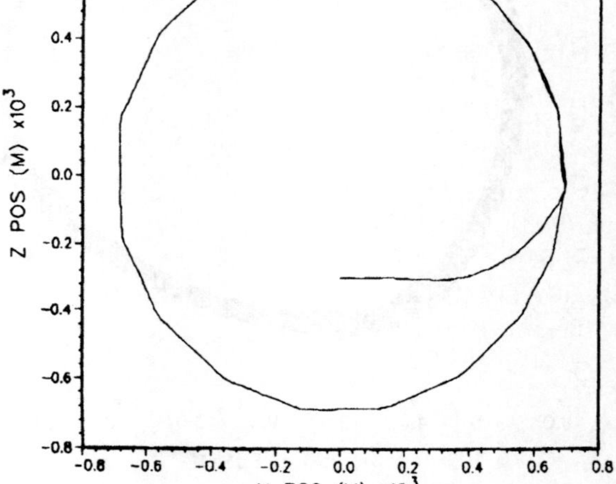

Figure 9-7 Cage mass center orbit as predicted by ADORE at 20,000 rpm.

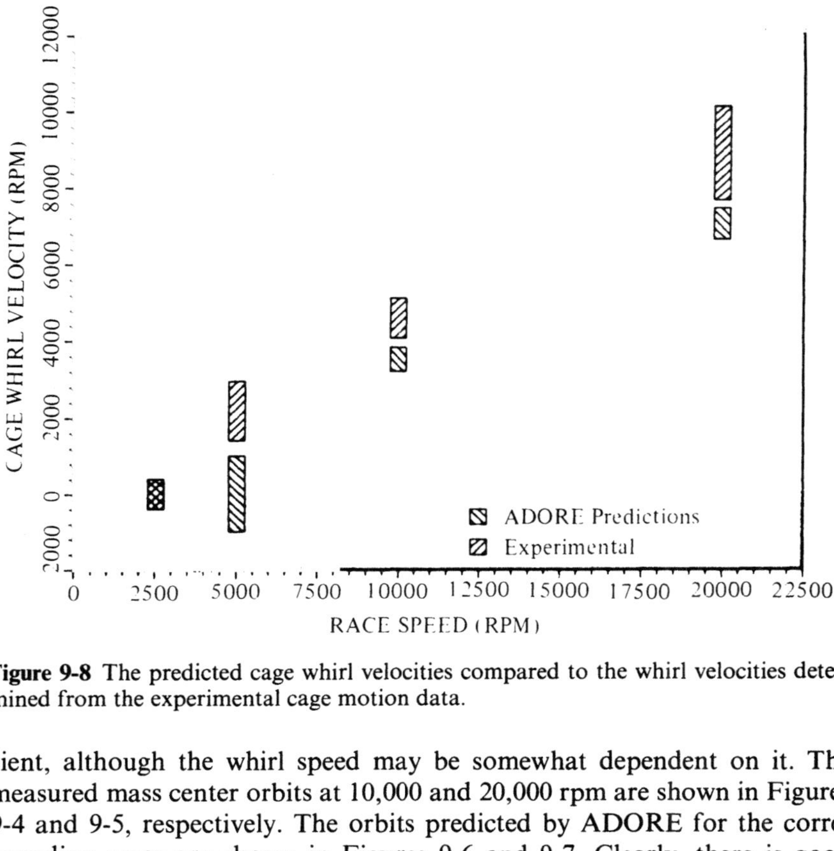

Figure 9-8 The predicted cage whirl velocities compared to the whirl velocities determined from the experimental cage motion data.

cient, although the whirl speed may be somewhat dependent on it. The measured mass center orbits at 10,000 and 20,000 rpm are shown in Figures 9-4 and 9-5, respectively. The orbits predicted by ADORE for the corresponding cases are shown in Figures 9-6 and 9-7. Clearly, there is good agreement between the shape of the experimental orbits and ADORE predictions. Once the cage goes into orbit, ADORE predictions indicate that the radius of the orbit is essentially equal to the cage/race radial clearance. Experimental observations confirm this prediction. The small differences in the orbit radii in Figures 9-5 and 9-7 are probably due to the uncertainty in the cage/race clearance.

The general nature of the whirl orbits is found to be insensitive to the applied load on the bearing, at least for the particular test bearing under consideration. Also, the critical speed at which the cage goes into orbit is independent of the applied load on the bearing. These observations are in very good agreement with ADORE predictions.

9.2.2 Whirl Velocities

The whirl angular velocities of the cage mass center are plotted as a function of shaft speed with a constant thrust load on the bearing in Figure 9-8. It is seen that the trends in both the analytical predictions and experimental

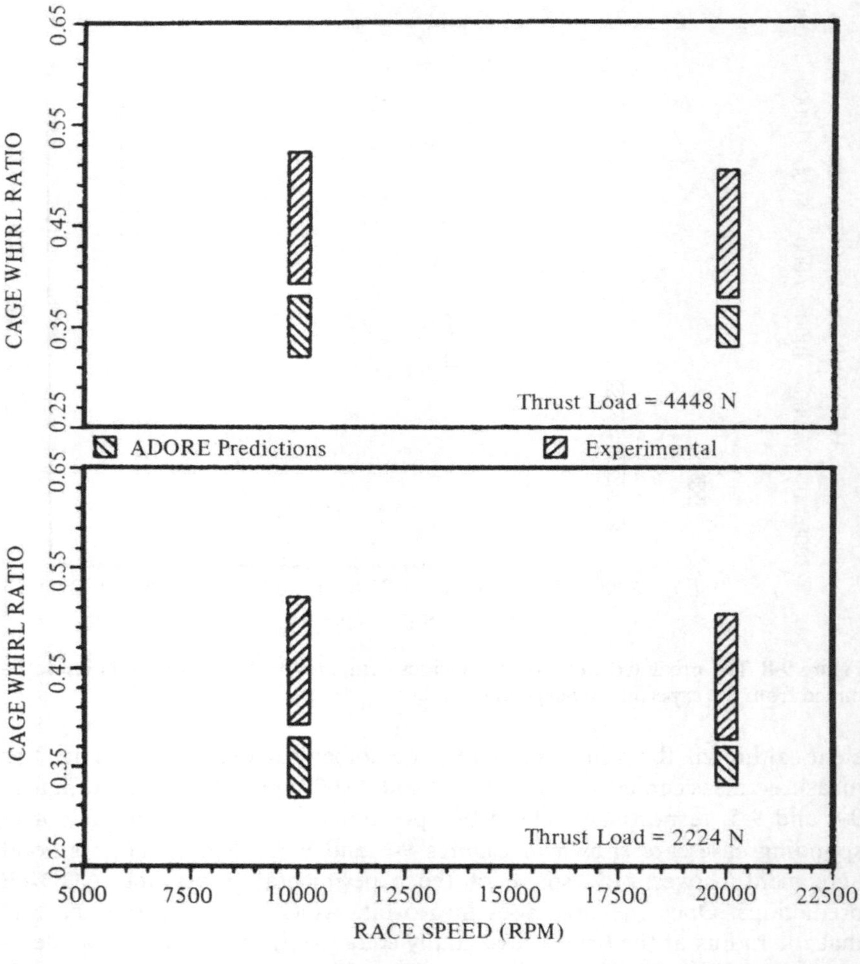

Figure 9-9 Predicted and measured cage whirl velocity ratios at two operating loads and speeds.

observations are in reasonable agreement. The differences in the critical speed at which the cage starts orbiting are probably due to the uncertainty in the ball/cage and cage/race friction coefficients. Also, once the cage goes into orbit, the deviations between the magnitudes of the predicted whirl velocities and those determined from the experimental data are perhaps a result of the uncertainties discussed above in both the experimental measurements and the input data for ADORE; in particular, the friction behavior the various interactions.

Over the range of shaft speeds (10,000–20,000 rpm) and applied thrust loads (1,000–15,000 N) the analytical results show that the ratio of the cage whirl to race angular velocity remains constant. Figure 9-9 shows that such predictions are fairly close to the experimental observations, at least in terms

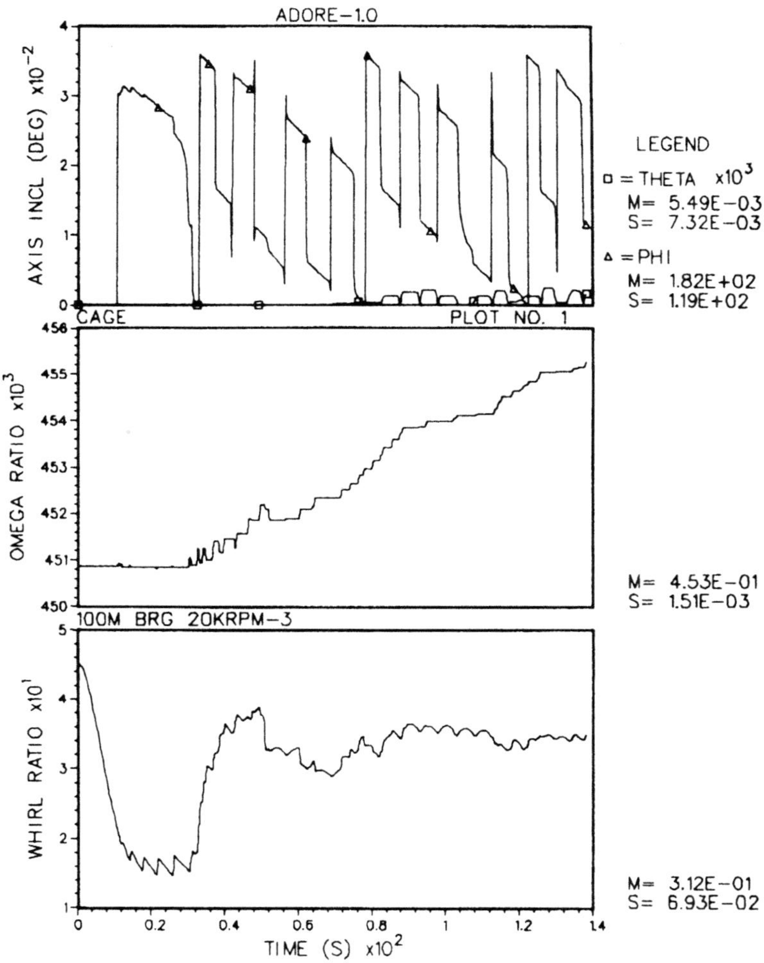

Figure 9-10 The general motion of the cage as simulated by ADORE at 20,000 rpm.

of the general trends. The whirl ratios indicate that the whirl velocity is almost equal to the angular velocity of the cage. However, the cage angular velocities could not be compared due to experimental difficulties in the measurement.

9.2.3 Coning Motion of Cage

Figure 9-10 shows a typical cage motion plot, as obtained from ADORE, which shows the cage mass center whirl velocity, and the angular velocity and the orientation of the angular velocity vector (see Figure 7-7 for definition of angles θ and ϕ) as a function of time. It is seen that even with a pure thrust load there is some coning motion indicated by the variations in angles θ and ϕ. An exact comparison of this motion with the experimental data is difficult due to the uncertainties in the measurement and subsequent analysis of the experimental data. However, the experimental data do indicate noticeable coning motion and there appears to be a qualitative agreement between the experimental observations and ADORE predictions.

9.3 Summary

ADORE predictions for the dynamic performance of angular-contact ball bearings are evaluated against experimental data on ball skid and the general motion of the cage. The capabilities to model ball skid provide considerable design strength to ADORE in determining the required preload to prevent skidding in angular-contact ball bearings. The agreement between the analytical predictions and the experimental data further strengthens such design capabilities.

The measured cage mass center whirl agrees well with ADORE predictions over the range of operating speeds and loads. It is shown that the cage mass center goes into orbit at a certain operating shaft speed and it stays in orbit at all speeds above this critical speed. Also, for the particular bearing investigated, the whirl orbit radius is essentially equal to the cage/race radial clearance and the whirl velocity is almost equal to the cage angular velocity over the entire range of loads and speeds. There is good agreement between these observations and the results predicted by ADORE. Both the experimental data and the analytical results indicate noticeable coning motion of the cage even under pure thrust conditions.

CHAPTER 10

Guidelines for Rolling Bearing Design

A general procedure for the design of rolling bearings is presented and the significance of the computer code ADORE in the design, performance diagnosis, and failure analysis of rotor–bearing systems is discussed.

A practical rotating machinery system is comprised of a shaft, a housing, and a number of bearings which support the shaft relative to the housing and allow for the relative rotation between the shaft and housing. A design procedure for such a rotor–bearing system often consists of several fundamental steps. First, a systems analysis is carried out to determine the nominal operating conditions for each bearing in the system. Once the operating environment in which the bearing must perform is known, the selection of bearing type is made and a preliminary design is produced. A rotor–bearing dynamic analysis is then carried out to determine the dynamic behavior of the rotor and the time-varying loads imposed on the bearing. Finally, a simulation of the dynamic performance of each bearing in the system is carried out to ensure acceptable system operation.

In the most general case, since both the shaft and bearings are an integral part of the system, a simultaneous solution of the governing equations for the shaft and bearings is required in order to simulate system performance. This requires that the analyses for bearing dynamics and shaft dynamics be coupled, so that the true operating conditions on the bearing may be determined. Indeed, such coupling can be very cumbersome and the resulting computer program may require prohibitively large amounts of computer time. A somewhat more acceptable approach is to analyze the performance of each bearing independently through a parametric evaluation. In essence, a "transfer function" (force–deflection response and/or stiffness–speed–load response) of the bearing is determined which can be used in the rotor dynamics analysis to simulate the system performance and response.

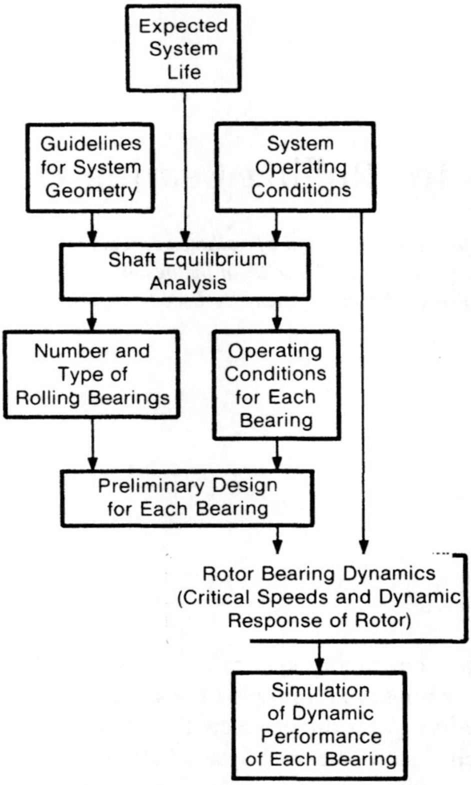

Figure 10-1 Overview of the procedure for rotor-bearing system design.

In the past, computer programs for a simultaneous analysis of a multiple bearing and shaft system have been restricted to static equilibrium analyses [17, 60]. These analyses and corresponding computer codes perform a static equilibrium or conventional bending-type analysis for the shaft, which defines the load and misalignment on each bearing. With these calculated loads and prescribed speeds, a quasi-static equilibrium analysis is carried out for each bearing to determine the overall system performance and life. The analysis, although lengthy, is quite straightforward, but the shaft and bearing problems are not really coupled. Also, since only a static equilibrium problem is considered, such an analysis is adequate only for ascertaining system fatigue life, and it does not provide any insight into the dynamic instabilities associated with either the shaft or the bearing.

Most of the shaft or rotor dynamics models, such as those credited to Lund [69, 70], Zorzi and Nelson [100], and Pan et al. [77] assume a constant stiffness for the support bearing. Although this may be reasonable in many cases, the transfer functions computed from the quasi-static portion of ADORE can provide a better representation of the rotor support bearing. Also, the improved definition of the rotor support bearing produces a more realistic estimate of the dynamic response of the rotor and the resulting time-varying loads on the bearing. In any event, the dynamic loads obtained from the rotor dynamic analyses, with either a constant bearing stiffness or with a predetermined variation of stiffness with loads and speeds, are significant inputs to ADORE for investigating any associated instabilities of the bearing elements.

10.1 System Overview

Since bearings represent integral parts of the rotor system, the interaction between the bearings and the supported rotor play a dominant role in both the design of the rotor and of the bearings. A systems overview is, therefore, a first step in the design process. The primary purpose of such an overview is to determine the types of bearings required and the operating conditions under which the bearings will operate. Figure 10-1 presents a schematic overview of the rotor–bearing system design procedure. From guidelines given for required system life, operating environment, and system size, a static equilibrium analysis for the rotor may be undertaken to determine the number and type of bearings required and the operating loads and speeds imposed on each bearing. Acceptable bearing design is then dependent on the results of a more detailed rotor–bearing interaction analysis that will ascertain the required life and a dynamically stable operation in the operating range. Assuming for the present that only rolling bearings are under consideration, details of such an interaction analysis are discussed below.

10.2 Rotor–Bearing System Interaction

The general design procedure for and system interaction of rotor–bearing systems is presented schematically in Figure 10-2. The nominal design specifications and the system performance requirements, such as nominal speeds and loads, expected bearing life, and geometric constraints, are first used to obtain a preliminary design based on recommendations by bearing manufacturers and the available design information in manufacturers' catalogs. Also, for the selected type of bearing and the available lubricant, a lubricant system has to be designed to ensure adequate lubrication at each interaction within the bearing. For the preselected design, the computer code ADORE may now be run in the quasi-static mode to determine the basic fatigue life and the radial stiffness characteristics or the transfer function of the bearing. The available life-modifying factors [7, 8, 89, 99] can be used to estimate the net expected life. If the estimated life either meets or exceeds the design specification, the computed bearing stiffness characteristics are input to a rotor dynamics simulation program, such as those described in references [69, 70, 77, 100], and a critical speed map of the rotor are generated. Failure to meet life requirements necessitates modification to the bearing design and recomputation of the life and stiffness data.

The critical speed maps are interpreted in the light of the prescribed

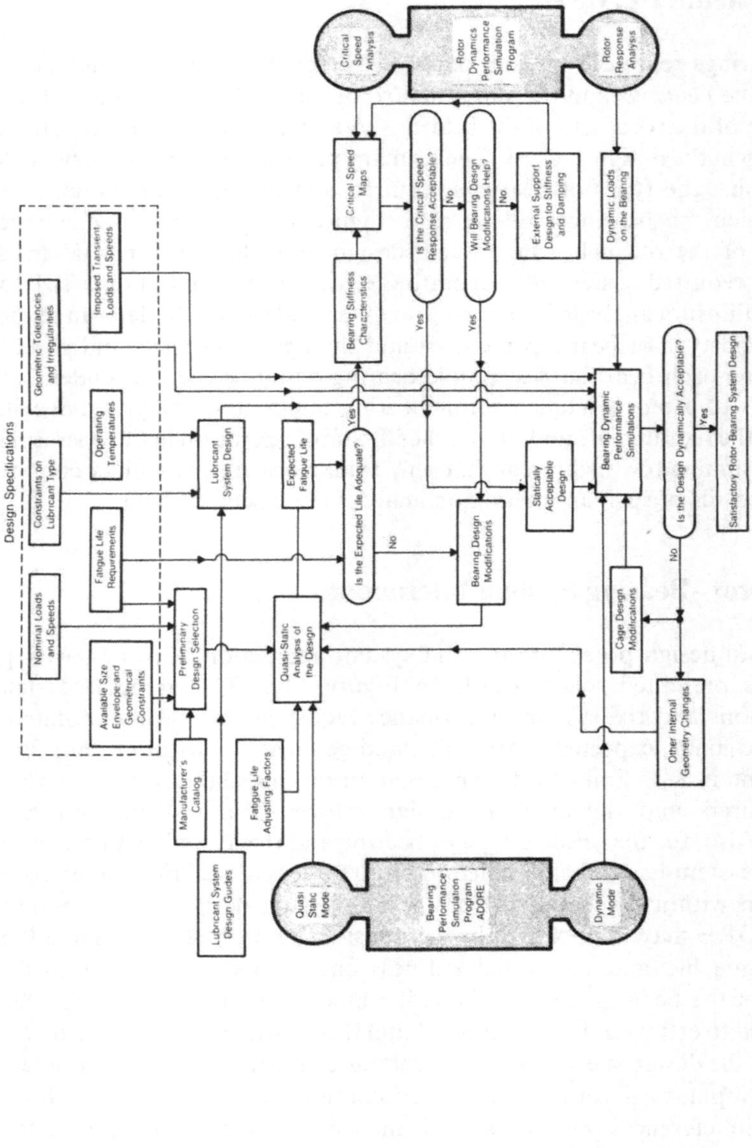

Figure 10-2 Schematic representation of the dynamic coupling in a rotor-bearing system.

operating speed range and with the objective of keeping all critical speeds out of the operating range. If this criterion is met, then the design is said to be statically acceptable; if not, the required change in stiffness that will meet the critical speed criterion must be determined. If the change required is small and no external damping is needed, then minor modifications to the bearing design may be attempted, and the quasi-static mode of ADORE can be exercised again to determine the improved stiffness and recheck the fatigue life.

When the stable rotor operation demands a greatly different stiffness and external damping seems desirable, it is necessary to design an external support for the bearing, which when combined with the bearing will meet the required stiffness and damping behavior and will ensure that no critical speed is encountered in the operating range. For example, since rolling bearings generally provide fairly high stiffness, external supports are used to obtain a reduction in the effective stiffness and to provide some damping because the rolling bearing offers negligible internal damping. In a number of applications elastomeric mounts are designed to meet these requirements.

Once a statically acceptable bearing design is obtained, the rotor dynamics code is run in the dynamic response mode to obtain the rotor response and the dynamic loads imposed on the bearing. These loads are combined with other dynamic specifications, the lubricant behavior and any geometric irregularities or manufacturer's design tolerances for the class of bearings used, to obtain the input data for the bearing dynamic performance simulation. With such data and the statically acceptable bearing design, ADORE can now be executed in the dynamic mode to ensure stable motion of each bearing element and obtain a dynamically acceptable design.

In cases where cage instabilities are observed, it may only be necessary to modify the cage geometry and repeat the dynamic performance simulation. In some cases, however, a strong coupling between lubricant behavior and cage dynamics may be observed. In such cases, it may be necessary to either choose a different lubricant or ensure a certain lubricant temperature (the iteration on lubricant behavior is not shown in Figure 10-2) at which the lubricant will provide the required tractive behavior for stable operation. In the event that the simulated dynamics of the rolling elements is undesirable, changes in the bearing internal geometry may be necessary and essentially the entire design procedure, including the quasi-static analysis, will have to be repeated. Some lubricant considerations may also be important if skidding is a problem.

10.3 ADORE: A Design and Performance Diagnosis Tool

The preceding discussion on rotor–bearing system design procedures, although abbreviated, clearly demonstrates the enhancement offered by the ADORE to system design. For most rotor–bearing system designs, ADORE will serve as a close companion to a rotor dynamics code, and the two codes together will form the analytical foundation for the rotor–bearing system design. Aside from its significance to design, ADORE is a powerful tool for bearing performance diagnosis for prescribed designs under adverse conditions. For example, lubricant failure may be simulated by changing the traction–slip characteristics and the resulting dynamic behavior can be very easily investigated. Likewise, the influence of any change in the operating parameters may be investigated. Additionally, ADORE may be extremely useful in failure analysis.

Once the computer program predictions have been experimentally validated, the general capability of simulating bearing performance as a function of a large number of operational and geometric parameters often allows the computer code to serve as a replacement for rather expensive and time-consuming experimental investigations. Also, the computer simulations provide valuable guidance for experimental investigations aimed at the development of new materials and advanced design concepts. It is expected that a general-purpose computer program such as ADORE will fulfill such practical requirements and that it will help advanced rolling bearing technology.

In the general area of computerized performance simulations for the design and diagnosis of rotor–bearing systems, it must be remembered that this text has been devoted to rolling bearings only and there has been no discussion of the dynamics of the rotor. It is well known that the rotor plays a dominant role in determining the overall dynamic performance of the rotating machinery system under question and the significance of rotor dynamics analysis and computer codes must be clearly recognized. In fact, the ultimate design tool will be a computer code which combines both the bearing and rotor dynamic analyses and simulates the performance of all system components simultaneously. Such a computer code will replace the above piecemeal-type design procedure and it will significantly advance rotor–bearing design technology. Due to computational speed limitations of present computing systems, however, such an advanced computer program would require prohibitively large amounts of computational effort. Thus, such a development must await further advancement of computer technology.

10.4 Summary

A schematic layout of the procedure used for rolling bearing design in applied rotor–bearing systems is presented. The quasi-static and dynamic modes of the computer program ADORE are explained, and the close interactions of these modes with a rotor dynamics performance program is demonstrated. It is shown that the bearing stiffness characteristics determined by the quasi-static equilibrium analysis in ADORE provide the required input to the rotor dynamics simulation program for the computation of rotor critical speed. The final rotor response simulated by the rotor dynamics codes then provides the dynamic load input to the dynamic analysis in ADORE in order to arrive at a bearing design which is free of all dynamic instabilities. It is emphasized that such a semicoupled procedure is a first step to an as-yet-unavailable generalized coupling of the rotor and bearing dynamics where all solutions and performance simulations may be obtained simultaneously.

APPENDIX I

Hertz Point-Contact Solution

The Hertzian point-contact solution is well-known and it is well documented in a number of textbooks, such as the work of Timoshenko and Goodier [91]. With its particular emphasis on rolling bearings, however, Harris [47] is an excellent reference. The following summary of the various expressions for contact half-widths and elastic deflection as a function of the applied loads and ball bearing geometry is after Harris [47].
Semimajor axis of the contact ellipse,

$$a = a^* \left(\frac{3Q}{2E' \sum \bar{\rho}} \right)^{1/3}$$

Semiminor axis of the contact ellipse,

$$b = b^* \left(\frac{3Q}{2E' \sum \bar{\rho}} \right)^{1/3}$$

Elastic deflection or geometric interaction,

$$\delta = \delta^* \left(\frac{3Q}{2E' \sum \bar{\rho}} \right)^{2/3} \frac{\sum \bar{\rho}}{2}$$

where

$$a^* = \left(\frac{2k^2 \mathscr{E}_2}{\pi} \right)^{1/3}$$

$$b^* = \left(\frac{2\mathscr{E}_2}{2k} \right)^{1/3}$$

$$\delta^* = \frac{2\mathscr{E}_1}{\pi} \left(\frac{\pi}{2k^2 \mathscr{E}_2} \right)^{1/3}$$

$$\mathscr{E}_1 = \int_0^{\pi/2} [1 - (1 - 1/k^2) \sin^2 \phi]^{-1/2} d\phi$$

$$\mathscr{E}_2 = \int_0^{\pi/2} [1 - (1 - 1/k^2) \sin^2 \phi]^{1/2} d\phi$$

$$\sum \bar{\rho} = \bar{\rho}_{I1} + \bar{\rho}_{I2} + \bar{\rho}_{II1} + \bar{\rho}_{II2}$$

$$\frac{1}{E} = \frac{1 - \sigma_1^2}{E_1} + \frac{1 - \sigma_2^2}{E_2}$$

and the ratio, $k = a/b$, is determined by solving the equation

$$\frac{(k^2 + 1)\mathscr{E}_2 - 2\mathscr{E}_1}{(k^2 - 1)\mathscr{E}_2} = \frac{(\bar{\rho}_{I1} - \bar{\rho}_{I2}) + (\bar{\rho}_{II1} - \bar{\rho}_{II2})}{\sum \bar{\rho}}$$

The subscripts I and II denote the race and ball, while 1 and 2 denote the direction of rolling and the direction normal to rolling respectively. E is the elastic modulus, σ is the Poisson ratio, Q is the contact load and $\bar{\rho}$ is the curvature of the interacting bearing element.

APPENDIX II

Shrink Fit and Thermal Expansion of Races

The influence of shrink fits, thermal gradients, and centrifugal expansion on the geometry of the races and cage can be easily determined from the expressions available for a cylindrical element [20]. The geometric changes in the rolling element will generally be very small compared to those in the races and therefore they may be neglected.

For a cylindrical element with the outer and inner radii as r_0 and r_i respectively, the elastic expansions δ_i, δ_0, and δ_c at radius γ due to an internal pressure p_i, external pressure p_0, and rotation with an angular velocity ω are written as [20]

$$\delta_i = p_i \frac{r}{E} \frac{r_i^2}{r_0^2 - r_i^2} \left[(1 - \sigma) + (1 + \sigma) \frac{r_0^2}{r^2} \right]$$

$$\delta_0 = -p_0 \frac{r}{E} \frac{r_0^2}{r_0^2 - r_i^2} \left[(1 - \sigma) + (1 + \sigma) \frac{r_i^2}{r^2} \right]$$

$$\delta_c = \rho \omega^2 \frac{r}{E} \frac{(3 + \sigma)(1 - \sigma)}{8} \left[r_0^2 + r_i^2 + \frac{1 + \sigma}{1 - \sigma} \frac{r_0^2 r_i^2}{r^2} - \frac{1 + \sigma}{3 + \sigma} r^2 \right]$$

where E is the elastic modulus, σ is the Poisson's ratio, and ρ is the density of the material.

The thermal growth δ_t can be simply expressed in terms of a thermal strain

$$\frac{\delta_t}{r} = \mathrm{w} \Delta T$$

where a is the thermal coefficient of expansion and ΔT is the temperature rise.

The above simple expressions may be linearly superimpose to determine the combined effects on the internal geometry of the bearing.

APPENDIX III

Fatigue Life Computation

Once the normal contact load and the size of the contact zone are known at the various rolling element/race interactions, the expected fatigue life of the bearing can be easily computed by using the available semi-empirical formulas. The following is a summary of such formulas after Jones [61] Harris [47], and Anderson [7].

Fatigue Life of Ball Bearings

For a point contact subjected to a load of Q, the fatigue life of the contact is defined as

$$L = \left(\frac{Q_c}{Q}\right)^3 \text{ million revolutions} \tag{III.1}$$

where Q is the basic dynamic load capacity of the contact defined by the following relation for a ball bearing:

$$Q_c = A_p \lambda_p \left(\frac{2f}{2f-1}\right)^{0.41} \frac{(1 \mp \gamma)^{1.39}}{(1 \pm \gamma)^{1/3}} \left(\frac{\gamma}{\cos\bar{\alpha}}\right)^{0.3} D^{1.8} N^{-1/3} \tag{III.2}$$

$$\gamma = \frac{d\cos\bar{\alpha}}{d_m} \tag{III.3}$$

where the upper sign refers to the inner race while the lower sign denotes outer race contact.

The variables, d, d_m, $\bar{\alpha}$, f, and N are respectively the ball diameter, pitch diameter, contact angle, race curvature factor, and the number of balls in the bearing. A_p is an empirical constant which depends on the material of the balls and races. For conventional 52100 steel the suggested value [47] of A_p is 2.3615×10^7 N/m$^{1.8}$ (7450 lb$_f$/in$^{1.8}$). The constant λ_p is a factor intended to account for a reduction in fatigue life due to ball/race spin, i.e., relative angular velocity normal to the plane of contact. Using a race control hypothesis, Jones [61] has suggested the following relation for the ball/race contact with nonzero spin:

$$\lambda_p = 1 - \tfrac{1}{3}\sin\bar{\alpha} \tag{III.4}$$

Until a more rigous evaluation of ball/race spin effect on fatigue life becomes available the above expression can be used to compute λ_p. In the event spin exists on both races and the race control hypothesis does not apply, equation (III.4) may be applied to the contact having a higher spin velocity.

In order to compute the life of the raceway, an equivalent load is shown to be

$$Q_{e\mu} = \left(\frac{1}{Z}\sum_{j=1}^{Z} Q_j^3\right)^{1/3} \tag{III.5}$$

for a rotating raceway and

$$Q_{ev} = \left(\frac{1}{Z}\sum_{j=1}^{Z} Q_j^{10/3}\right)^{0.3} \tag{III.6}$$

for a stationary raceway.

The lives of the rotating and stationary races are thus computed as

$$L_\mu = \left(\frac{Q_{c\mu}}{Q_{e\mu}}\right)^3 \text{ million revolutions} \tag{III.7}$$

and

$$L_v = \left(\frac{Q_{cv}}{Q_{ev}}\right)^3 \text{ million revolutions} \tag{III.8}$$

and the net life of the bearing is expressed as

$$L = (L_\mu^{-1.11} + L_v^{-1.11})^{-0.90} \text{ million revolutions} \tag{III.9}$$

Fatigue Life for Roller Bearings

Similar to the above expressions for ball bearings, the life of a line contact in a roller bearing is defined as

$$L = \left(\frac{Q_c}{Q}\right)^4 \text{ million revolutions} \tag{III.10}$$

where the basic dynamic capacity for the line contact is written as

$$Q_c = A_r \lambda_r \frac{(1 \mp d/d_m)^{29/27}}{(1 \pm d/d_m)^{1/4}} \left(\frac{d}{d_m}\right)^{2/9} d^{29/27} \ell^{7/9} N^{-1/4} \tag{III.11}$$

where the additional parameter ℓ is the roller race contact length and the empirical material constant A_r has a value of 1.9806 $\text{N/m}^{50/27}$ (49,500 $\text{lb}_\text{f}/\text{in.}^{50/27}$) for conventional 52100 steel [47].

The constant λ_r is suggested by Jones [61] to allow for a reduction in fatigue life due to stress concentration in roller bearings. A value of 0.45 is suggested for uncrowned rollers while a value of 0.61 is proposed for the crowned rollers.

Similar to the case of ball bearings, if the dynamic capacity Q_c is a constant and the load along the roller length is fairly uniform, the effective raceway load can be computed from the individual loads at the various roller/race interactions:

$$Q_{e\mu} = \left(\frac{1}{N}\sum_{j=1}^{N} Q_j^4\right)^{1/4} \tag{III.12}$$

for a rotating raceway, and

$$Q_{ev} = \left(\frac{1}{N}\sum_{j=1}^{N} Q_j^{4.5}\right)^{2/9} \tag{III.13}$$

for a stationary raceway.

The expressions for the raceway and bearing lives are also very similar:

$$L_\mu = \left(\frac{Q_{c\mu}}{Q_{e\mu}}\right)^4 \text{ million revolutions} \tag{III.14}$$

for the rotating raceway,

$$L_v = \left(\frac{Q_{cv}}{Q_{ev}}\right)^4 \text{ million revolutions} \tag{III.15}$$

for the stationary raceway, and

$$L = (L_\mu^{-9/8} + L_v^{-9/8})^{-8/9} \text{ million revolutions} \tag{III.16}$$

for the total bearing.

When a misalignment exists in a roller bearing, the load along the roller length is not uniform, and the contact length may also vary from roller to roller. This requires that the dynamic capacity be determined for each roller/race contact and effective values of Q_j for equations (III.12) and (III.13) are determined from the local variation in load along the roller length by an averaging procedure of the form

$$Q_{ej} = \left[\frac{1}{\ell}\int_0^\ell q^n dx\right]^{1/n} \tag{III.17}$$

where q is the load intensity along the contact length and n will assume the values of 4.0 and 4.5 for the rotating and stationary raceways.

The effect of varying contact length from roller to roller can be easily accounted for by computing the load capacity Q_c (equation III.11) for each contact separately and using the relevant value of the contact length ℓ. The summations of equation (III.12) and (III.13) can now be replaced by an equation for the form

$$\left(\frac{Q_c}{Q}\right) = \left\{\frac{1}{N}\sum_{j=1}^{N}\left(\frac{Q_{cj}}{Q_{ej}}\right)^n\right\}^{1/n} \tag{III.18}$$

where the values of n will again by 4.0 and 4.5 for the rotating and stationary raceways.

These equivalent load ratios can now be substituted in equations (III.14) to (III.16) to determine the fatigue life of the bearing.

It should be noted that the above expressions, for both ball and roller bearings, determine a "basic" life. Often it will be necessary to modify this life by certain multiplying factors to account for practical considerations such as lubricant film thickness, surface roughness, slip velocities, material hardness, manufacturing processes, etc. Based on experimental data, estimates for some of these factors are available in the literature [7, 8, 89, 99].

APPENDIX IV

Source Listing of ADORE

The following is the FORTRAN listing of the main program ADORE and the user-programmable subroutines ADRX1, ADRX2, ADRX3, ADRX4, ADRX5, and ADRX6:

```
      PROGRAM ADORE(INPUT,OUTPUT,MASTER,FINAL,SOL1,SOL2,                ADORE
     1SOL3,SOL4,SOL5,SOL6,SOL7,TAPE5=INPUT,TAPE6=OUTPUT,                ADORE
     2TAPE7=MASTER,TAPE8=FINAL,TAPE11=SOL1,TAPE12=SOL2,                 ADORE
     3TAPE13=SOL3,TAPE14=SOL4,TAPE15=SOL5,TAPE16=SOL6,TAPE17=SOL7)      ADORE
C                                                                       ADORE
C---------------------------------------------------------------------- ADORE
C                                                                       ADORE
C                                ADORE                                  ADORE
C                                =====                                  ADORE
C                                                                       ADORE
C                 ADVANCED DYNAMICS OF ROLLING ELEMENTS                 ADORE
C                 =====================================                 ADORE
C                         (VERSION ADORE-1.0)                           ADORE
C                                                                       ADORE
C                                 BY                                    ADORE
C                          PRADEEP K. GUPTA                             ADORE
C                                                                       ADORE
C                                                                       ADORE
C                                                                       ADORE
C              *********************************************            ADORE
C              *                                           *            ADORE
C              *    ADORE IS A PROPRIETARY SOFTWARE OF     *            ADORE
C              *          PRADEEP K. GUPTA, INC.           *            ADORE
C              *     REPRODUCTION IN WHOLE OR IN PART      *            ADORE
C              *          IS STRICTLY PROHIBITED.          *            ADORE
C              *                                           *            ADORE
C              *  THIS LISTING IS PROVIDED FOR TECHNICAL   *            ADORE
C              *  EVALUATION AND REFERENCE PURPOSES ONLY.  *            ADORE
C              *                                           *            ADORE
C              *********************************************            ADORE
C                                                                       ADORE
C                                                                       ADORE
C  THIS COMPUTER PROGRAM, ADORE (ADVANCED DYNAMICS OF ROLLING           ADORE
C  ELEMENTS), PROVIDES A REAL TIME SIMULATION OF THE DYNAMIC PER-       ADORE
C  FORMANCE OF A ROLLING BEARING BY INTEGRATING THE CLASSICAL DIFFER-   ADORE
C  ENTIAL EQUATIONS OF MOTION OF EACH ELEMENT (ROLLING ELEMENT, CAGE    ADORE
C  AND THE RACES) OF THE BEARING.  THE PROGRAM PROVIDES CONSIDERABLE    ADORE
C  FLEXIBILITY IN TERMS OF BOTH THE GEOMETRY OF THE BEARING AND THE     ADORE
C  IMPOSED CONDITIONS OF OPERATION.  THE TYPES OF ROLLING BEARING CON-  ADORE
C  SIDERED ARE ANGULAR CONTACT OR RADIAL BALL BEARINGS, CYLINDRICAL     ADORE
C  ROLLER BEARINGS, SPHERICAL ROLLER BEARINGS, TAPERED ROLLER BEARINGS  ADORE
C  AND SPHERICAL TAPERED ROLLER BEARINGS.  THE GEOMETRY OF EACH BEARING ADORE
C  ELEMENT CAN BE INDPENDENTLY VARIED AND THE APPLIED LOADS AND SPEEDS  ADORE
C  MAY ARBITRARILY VARY AS A FUNCTION OF TIME.  ALSO, THE BEARING AS A  ADORE
C  WHOLE MAY ASSUME ANY PRESCRIBED MOTION WHICH MAY GENERATE SUBSTAN-   ADORE
C  TIAL INERTIAL FORCES CONSISTING OF BOTH CENTRIFUGAL AND CORIOLIS     ADORE
C  COMPONENTS.   IT IS ANTICIPATED THAT WITH SUCH A FLEXIBILITY, ADORE  ADORE
C  WILL PROVIDE A REALISTIC MODEL FOR THE MOST SOPHISTICATED OPERATING  ADORE
C  ENVIRONMENT AND IT WILL PROVIDE SIGNIFICANT INSIGHT INTO THE         ADORE
C  INFLUENCE OF GEOMETRICAL VARIATIONS (OR MANUFACTURING TOLERANCES)    ADORE
C  ON THE OVERALL DYNAMIC PERFORMANCE OF THE BEARING.                   ADORE
C                                                                       ADORE
C  ASIDE FROM TREATING THE SOPHISTICATED CASES, THE INPUT/OUTPUT OF     ADORE
C  ADORE IS DESIGNED SUCH THAT THE SOLUTION TO THE CONVENTIONAL QUASI-  ADORE
C  STATIC PROBLEMS CAN ALSO BE EASILY OBTAINED.   THUS THE COMPUTATION  ADORE
C  OF FATIGUE LIFE AND BEARING STIFFNESS AS A FUNCTION OF THE OPERATING ADORE
C  CONDITIONS IS QUITE STRAIGHT FORWARD.                                ADORE
```

```
C
C  ALTHOUGH ADORE HAS BEEN DEVELOPED ON A CDC CYBER SYSTEM, IT EMPLOYS
C  THE WELL ESTABLISHED FORTRAN 77 LANGUAGE AND THE CODE CLOSELY
C  CONFORMS TO THE ANSI STANDARDS.  THUS IMPLEMENTATION OF ADORE ON
C  OTHER COMPUTER SYSTEMS CAN BE EASILY ACCOMPLISHED.  ASIDE FROM THE
C  GRAPHIC SOFTWARE, ADORE IS SELF CONTAINED IN TERMS OF ALL THE
C  REQUIRED MATHEMATICAL SUBROUTINES.  THE GRAPHIC SOFTWARE USED IS
C  THE WELL KNOWN PACKAGE CALLED DISSPLA WHICH IS DEVELOPED AND
C  SUPPORTED BY ISSCO.
C
C
C  COORDINATE FRAMES
C  =================
C
C  ADORE CONSIDERS THE GENERAL MOTION OF EACH BEARING ELEMENT IN A
C  SIX DEGREES OF FREEDOM SYSTEM.  THE FIRST THREE COORDINATES
C  LOCATE THE MASS CENTER OF A BEARING ELEMENT IN AN INERTIAL (SPACE
C  FIXED) COORDINATE FRAME AND THE LATTER THREE COORDINATES DEFINE
C  THE ANGLES WHICH LOCATE THE BODY FIXED COORDINATE FRAME RELATIVE
C  TO THE INERTIAL FRAME AND THEREBY DEFINE THE ANGULAR ORIENTATION OF
C  THE BEARING ELEMENT.  MOTION OF THE MASS CENTER IS NORMALLY CONSI-
C  DERED IN THE INERTIAL FRAME WHILE THE ROTATIONAL MOTION IS CONSI-
C  DERED IN A BODY FIXED COORDINATE FRAME.  HOWEVER, FOR SPECIALIZED
C  APPLICATIONS WHERE THE WHOLE BEARING MOVES IN SPACE, THE MASS
C  CENTER MOTION MAY BE CONSIDERED IN A MOVING COORDINATE FRAME.
C
C  THE INERTIAL REFERENCE FRAME IS DEFINED WITH THE ORIGIN AT THE
C  BEARING CENTER, X-AXIS ALONG THE BEARING AXIS AND Y AND Z AXES
C  NORMAL TO THE BEARING AXIS AND AS DETERMINED BY THE RIGHT HAND
C  SCREW RULE.
C
C  NORMALLY THE INITIAL (TIME=0) THRUST LOAD ACTS ALONG THE X-AXIS,
C  THE RADIAL LOAD ACTS ALONG THE Z-AXIS AND THE MISALIGNMENT IS
C  PRESCRIBED BY A ROTATION ABOUT THE Y-AXIS.
C
C
C  SYSTEM OF UNITS
C  ===============
C
C  BY APPROPRIATE SELECTION OF AN INPUT PARAMETER, EITHER THE ENGLISH
C  OR THE SI SYSTEM OF UNITS MAY BE SELECTED.  IT IS NECESSARY THAT
C  ALL THE INPUTS BE PRESCRIBED IN THE APPROPRIATE UNITS USING THE
C  FOLLOWING FUNDAMENTAL QUANTITIES --
C
C    QUANTITY            ENGLISH SYSTEM      SI SYSTEM
C    --------            --------------      ---------
C
C    MASS (MA)           POUND MASS (LBM)    KILOGRAM MASS (KGM)
C    LENGTH (L)          INCH (IN)           METER (M)
C    FORCE (F)           POUND FORCE (LBF)   NEWTON (N)
C    TIME (T)            SECOND (S)          SECOND (S)
C    TEMPERATURE (TM)    DEGREE RANKINE (R)  DEGREE KELVIN (K)
C
C  THE QUANTITIES IN PARENTHESIS INDICATE THE ABBREVIATIONS USED FOR
C  THE VARIOUS PARAMETERS.  IN THE INPUT DESCRIPTION ALL UNITS ARE
C  INDICATED BY THE FUNDAMENTAL QUANTITIES (MA, L, F, T AND TM) AND
C  APPROPRIATE UNITS MUST BE SUBSTITUTED WHILE PREPARING THE INPUT
C  DATA.  THE TWO SYSTEMS OF UNITS MUST NOT BE MIXED.
C
C
C  INPUT DATA
C  ==========
C
C  INPUT TO ADORE CONSISTS OF TWO PARTS, THE INPUT DATA AND THE
C  OPTIONAL CODE SUPPLIED IN ANY OF THE USER PROGRAMMABLE SUBROUTINES.
C  AS WILL BE DISCUSSED LATER, THESE OPTIONAL SUBROUTINES PERMIT
C  CUSTOMIZING ADORE TO CLOSELY MEET THE NEEDS OF HIGHLY SPECIALIZED
C  AND SOPHISTICATED BEARING APPLICATIONS.
C
C  ALL THE INPUT DATA IS SUPPLIED IN FREE FORMAT AND THE VARIABLES MAY
C  BE SEPARATED BY EITHER A COMMA OR A SPACE ON THE INPUT RECORD.  THE
C  FIRST VARIABLE IN EACH RECORD IS A RECORD IDENTIFIER WHICH IS ONLY
```

```
C  USED TO IDENTIFY THE RECORD TO THE USER FOR CONVENIENT EDITING OF     ADORE
C  THE DATA, WHEN ONLY A FEW VARIABLES HAVE TO BE MODIFIED IN AN EXIST-  ADORE
C  ING DATA FILE TO CREATE A FILE FOR A NEW RUN.  THE RECORD IDENTI-     ADORE
C  FIER IS A STRING OF A MAXIMUM OF 12 ALPHANUMERIC CHARACTERS AND IT    ADORE
C  MUST BE ENCLOSED IN APHOSTROPHES.                                      ADORE
C                                                                         ADORE
C  VARIABLES NAMES BEGINNING WITH I, J, K, L, M AND N ARE INTEGERS AND   ADORE
C  THE REMAINING ARE FLOATING POINT NUMBERS.                              ADORE
C                                                                         ADORE
C  FOR ANY GIVEN SET OF DATA, WHEN ADORE IS FIRST EXECUTED ALL THE       ADORE
C  INPUT DATA IS STORED IN THE DATA FILE, -MASTER- (SEE DATA FILES       ADORE
C  SECTION OF THIS DOCUMENTATION).   FOR SUBSEQUENT RUNS MOST OF THE     ADORE
C  INPUT DATA MAY, THEREFORE, BE READ FROM THE FILE -MASTER- AND         ADORE
C  FOR SUCH RUNS IT WILL ONLY BE NECESSARY TO SUPPLY RECORDS 1 AND 2     ADORE
C  IN THE INPUT, FOLLOWED BY ANY OPTIONAL INPUT REQUIRED BY ANY OF       ADORE
C  THE USER PROGRAMMABLE SUBROUTINES (RECORDS 12.K) AND ANY OPTIONAL     ADORE
C  PLOT INPUT (RECORDS 13.K) WHEN THE PLOT OUTPUT IS DESIRED.            ADORE
C                                                                         ADORE
C  RECORD 1                                                               ADORE
C  --------                                                               ADORE
C                                                                         ADORE
C  JREC    RECORD IDENTIFIER.                                             ADORE
C  MODE    PROGRAM MODE DEFINED AS FOLLOWS --                             ADORE
C          -2 = QUASI-STATIC EQUILIBRIUM SOLUTION WITH GYROSCOPIC RES-    ADORE
C               TRAINTS AS USED IN RACE CONTROL HYPOTHESIS FOR BALL       ADORE
C               BEARINGS (KBT MUST BE NEGATIVE ON RECORD 3.2).            ADORE
C          -1 = QUASI-STATIC EQUILIBRIUM SOLUTION WITHOUT GYROSCOPIC      ADORE
C               RESTRAINTS (KBT MUST BE NEGATIVE ON RECORD 3.2).          ADORE
C           0 = GENERALIZED DYNAMIC SIMULATION.                           ADORE
C           1 = DYNAMIC SIMULATION WITH EQUILIBRIUM CONSTRAINTS           ADORE
C               ON ROLLING ELEMENTS WHERE THE RADIAL AND AXIAL EQUILI-    ADORE
C               BRIUM IS PERFORMED ONLY FOR THE ROLLING ELEMENTS AND      ADORE
C               THE POSITION OF THE RACE CENTERS IS EITHER FIXED OR       ADORE
C               PRESCRIBED IN ACCORDANCE TO ANY PREDETERMINED PATH.       ADORE
C           2 = DYNAMIC SIMULATION WITH EQUILIBRIUM CONSTRAINTS           ADORE
C               ON BOTH THE ROLLING ELEMENTS AND THE RACES.  THE EQUILI-  ADORE
C               BRIUM EQUATIONS DETERMINE THE POSITION OF ALL ROLLING     ADORE
C               ELEMENTS AND ALSO THE RELATIVE POSITION OF THE TWO        ADORE
C               RACES.  GENERALLY THE OUTER RACE WILL BE HELD FIXED       ADORE
C               WHILE RELATIVE POSITION OF THE INNER RACE IS DETERMINED   ADORE
C               BY THE EQUILIBRIUM EQUATIONS.                             ADORE
C  JPT     PRINT OUTPUT CONTROL (FOR DEFINITION OF THE VARIOUS OUTPUT     ADORE
C          SECTIONS SEE DISCUSSION OF PRINT OUTPUT PRESENTED BELOW IN     ADORE
C          THIS DOCUMENTATION) --                                         ADORE
C          -2 = PRINT SECTION 4 OUTPUT ONLY.                              ADORE
C          -1 = PRINT SECTIONS 3 AND 4 ONLY.                              ADORE
C           0 = PRINT OUTPUT SECTIONS 2, 3 AND 4.                         ADORE
C           N = (N.GT.0) PRINT ALL SECTIONS BUT PRINT SOLUTIONS FOR       ADORE
C               EVERY NTH ROLLING ELEMENT.  N=1 WILL PRINT ALL ROLLING    ADORE
C               ELEMENT SOLUTIONS, N=2 WILL PRINT SOLUTIONS FOR EVERY     ADORE
C               OTHER ROLLING ELEMENT AND SO ON.                          ADORE
C  KSTEP   FREQUENCY OF TIME STEPS FOR PRINT OUTPUT.  KSTEP=1 WILL PRINT  ADORE
C          SOLUTIONS AT EVERY STEP, KSTEP=2 WILL PRINT AT EVERY OTHER     ADORE
C          STEP ETC.  NOTE THAT TIME=0 CORRESPONDS TO STEP NO. 0.         ADORE
C  LSTEP   FREQUENCY OF TIME STEPS FOR PLOT OUTPUT WHICH IS STORED IN     ADORE
C          DATA FILES.  LSTEP=1 WILL STORE ALL SOLUTIONS, LSTEP=2, WILL   ADORE
C          STORE SOLUTIONS AT EVERY OTHER STEP AND SO ON.                 ADORE
C  JPLT    PLOT CONTROL DEFINED AS FOLLOWS --                             ADORE
C           0 = NO PLOTS REQUIRED IN THIS RUN.                            ADORE
C           1 = PLOT THE DATA AT THE END OF THIS RUN.                     ADORE
C           2 = PLOTS ARE REQUIRED FOR THE PREVIOUSLY COMPUTED SOLUTIONS  ADORE
C               TIONS ONLY.  UNDER THIS OPTION ALL INPUT IN RECORDS 2 TO  ADORE
C               12 MUST BE SKIPPED.                                       ADORE
C  MAX     MAXIMUM NUMBER OF STEPS FOR THIS RUN.                          ADORE
C  IMET    INTEGRATION ALGORITHM DEFINED AS FOLLOWS --                    ADORE
C           1 TO  7 EXPLICIT RUNGA-KUTTA-FEHLBERG METHOD OF ORDER IMET.   ADORE
C          11 TO 18 PREDICTOR-CORRECTOR METHOD OR ORDER (IMET-10).  AN    ADORE
C                   EXPLICIT METHOD OF ORDER 4 IS USED TO START THE       ADORE
C                   PREDICTOR-CORRECTOR PROCESS.                          ADORE
C          AN INITIAL TRIAL VALUE OF IMET=5 IS SUGGESTED.                 ADORE
C                                                                         ADORE
C  RECORD 2   (ONLY FOR DYNAMIC SIMULATIONS, MODE.GE.0 ON RECORD 1)       ADORE
```

```
C  --------
C
C  JREC    RECORD IDENTIFIER.
C  DINT    STARTING SIZE OF DIMENSIONLESS TIME STEP (AN INITIAL TRIAL
C          VALUE OF 0.050 IS SUGGESTED).
C  DMIN    MINIMUM PERMISSIBLE SIZE OF DIMENSIONLESS TIME STEP (AN INI-
C          TIAL TRIAL VALUE OF 5.0E-04 IS SUGGESTED).
C  DMAX    MAXIMUM PERMISSIBLE SIZE OF DIMENSIONLESS TIME STEP (AN INI-
C          TRIAL VALUE OF 0.50 IS SUGGESTED).
C  TF      FINAL VALUE OF DIMENSIONLESS TIME (AN INITIAL TRIAL VALUE OF
C          100.  IS SUGGESTED, ALSO SEE NOTE 1).
C
C
C  RECORD 3.1
C  ----------
C
C  JREC    RECORD IDENTIFIER.
C  ISPEC   BEARING SPECIFICATION CODE - A STRING OF CHARACTERS (MAXIMUM
C          36 CHARACTERS) WHICH WILL BE USED TO IDENTIFY THIS RUN.  THE
C          STRING MUST BE ENCLOSED IN APHOSTROPHES.
C
C  RECORD 3.2
C  ----------
C
C  JREC    RECORD IDENTIFIER.
C  LU      CODE FOR SYSTEM OF UNITS DEFINED AS FOLLOWS --
C           1 = ENGLISH UNITS.
C           2 = SI UNITS.
C  KBT     BEARING TYPE.  A POSITIVE VALUE INDICATES A BEARING WITH
C          CAGE WHILE A NEGATIVE VALUE DENOTES A CAGELESS BEARING.  THE
C          VALID BEARING TYPES ARE --
C           1 = BALL BEARING (ANGULAR CONTACT OR RADIAL).
C           2 = CYLINDRICAL ROLLER BEARING.
C           3 = SPHERICAL ROLLER BEARING (SEE NOTE 2).
C           4 = TAPERED ROLLER BEARING.
C           5 = SPHERICAL TAPERED ROLLER BEARING.
C  KS      WHEN THE CAGE IS SEGMENTED IN SEVERAL EQUAL AND SYMETRIC
C          SEGMENTS, KS DENOTES THE NUMBER OF CAGE SEGMENTS.  FOR A
C          POSITIVE VALUE SEGMENTATION WILL TAKE PLACE THROUGH THE
C          POCKETS WHILE A NEGATIVE VALUE WILL INDICATE SEGMENTATION
C          BETWEEN THE POCKETS.  KS=0 WILL DENOTE NORMAL ONE PIECE
C          CAGE.  SINCE NO HYDRODYNAMIC INTERACTION IS PERMISSIBLE WITH
C          A SEGMENTED CAGE, THE HYDRODYNAMICS CODES, IHBC AND IHRC,
C          MUST BE SET TO ZERO ON THIS RECORD.  SEE NOTE 3 FOR FURTHER
C          DISCUSSION OF THE SEGMENTED CAGE OPTION.
C  NB      NUMBER OF ROLLING ELEMENTS.  NB.LE.40 FOR A ONE PIECE
C          CAGE OR A CAGELESS BEARING AND (NB+ABS(KS)).LE.41 FOR A
C          SEGMENTED CAGE.
C  IRG1    TYPE OF CAGE GUIDANCE ON THE NEGATIVE X-AXIS --
C           0 = NO RACE GUIDANCE.
C           1 = OUTER RACE GUIDANCE.
C           2 = INNER RACE GUIDANCE.
C  IRG2    TYPE OF CAGE GUIDANCE ON THE POSITIVE X-AXIS --
C           0 = NO RACE GUIDANCE.
C           1 = OUTER RACE GUIDANCE.
C           2 =INNER RACE GUIDANCE.
C  IHBC    HYDRODYNAMICS CODE AT ROLLING ELEMENT/CAGE INTERACTION --
C           0 = NEGLECT HYDRODYNAMICS.
C           1 = INCLUDE HYDRODYNAMICS.
C  IHRC    HYDRODYNAMICS CODE AT CAGE/RACE INTERFACE --
C           0 = NEGLECT HYDRODYNAMICS.
C           1 = INCLUDE HYDRODYNAMICS.
C  KTC     TRACTION CODE AT ROLLING ELEMENT/RACE INTERACTION --
C           0 = AN ARBITRARY TRACTION CURVE (SEE NOTES 4 AND 5).
C           1 = MINERAL OIL, SAE 30 OR MOBIL DTE.
C           2 = 5P4E POLYPHENYL ETHER.
C           3 = MIL-L-7808 TYPE OIL.
C           4 = MIL-L-23699 OR MOBIL JET II.
C           5 = MIL-L-27502 OR MCS 1780, A HIGH TEMP VERSION OF 23699.
C           6 = TRACTION FLUID SANTOTRAC 30.
C           7 = TRACTION FLUID SANTOTRAC 50.
C           8 = TRACTION MODEL WITH USER DEFINED COEFFICIENTS (NOTE 6).
```

```
C               THIS CASE IS DIFFERENT FROM THE CASE KTC=0 IN THE SENSE   ADORE
C               THAT THE TRACTION/SLIP BEHAVIOR IS COMPUTED BY THE MODEL  ADORE
C               USED UNDER KTC = 1 TO 7, HOWEVER, THE VARIOUS COEFFI-     ADORE
C               CIENTS OF THE CONSTITUTIVE EQUATION OF THE LUBRICANT ARE  ADORE
C               SUPPLIED BY THE USER ON RECORDS 10.4.K.                   ADORE
C  ICHR   CHURNING CODE --                                                ADORE
C          0 = NEGLECT CHURNING.                                          ADORE
C          1 = INCLUDE CHURNING WITH SPECIFIED LUBRICANT PROPERTIES.      ADORE
C          2 = INCLUDE CHURNING WITH LUBRICANT PROPERTIES DERIVED         ADORE
C              FROM THE LUBRICANT MODEL SPECIFIED BY PARAMETER KTC.       ADORE
C  IFL11  EXISTENCE OF ROLLER GUIDE FLANGE ON THE NEGATIVE X-AXIS OF      ADORE
C         THE OUTER RACE --                                               ADORE
C          0 = NO GUIDE FLANGE PRESENT.                                   ADORE
C          1 = GUIDE FLANGE EXISTS.                                       ADORE
C  IFL21  EXISTENCE OF ROLLER GUIDE FLANGE ON THE POSITIVE X-AXIS OF      ADORE
C         THE OUTER RACE PER DEFINITION OF IFL11.                         ADORE
C  IFL12  EXISTENCE OF ROLLER GUIDE FLANGE ON THE NEGATIVE X-AXIS OF      ADORE
C         THE INNER RACE PER DEFINITION OF IFL11.                         ADORE
C  IFL22  EXISTENCE OF ROLLER GUIDE FLANGE ON THE POSITIVE X-AXIS OF      ADORE
C         THE INNER RACE PER DEFINITION OF IFL11.                         ADORE
C  MBR1   MATERIAL CODE FOR THE ROLLING ELEMENTS --                       ADORE
C          0 = STANDARD MATERIALS (BEARING STEEL).                        ADORE
C          1 = MATERIALS WITH PROPERTIES SPECIFIED ON RECORD 8.1.         ADORE
C  MBR2   MATERIAL CODE FOR THE RACES --                                  ADORE
C          0 = STANDARD MATERIALS (BEARING STEEL).                        ADORE
C          1 = RACE MATERIAL PROPERTIES SPECIFIED ON RECORD 8.2.          ADORE
C  MHS    MATERIAL CODE FOR THE SHAFT AND HOUSING --                      ADORE
C          0 = STANDARD MATERIALS (STEEL).                                ADORE
C          1 = MATERIAL PROPERTIES SPECIFIED ON RECORDS 8.3 AND 8.4.      ADORE
C  MCR    MATERIAL CODE FOR THE CAGE --                                   ADORE
C          0 = STANDARD MATERIAL (STEEL).                                 ADORE
C          1 = CAGE MATERIAL PROPERTIES SPECIFIED ON RECORD 8.5.          ADORE
C                                                                         ADORE
C  FOR TAPERED BEARINGS THE RACE FLANGE MAY EXIST ONLY ON THE NEGATIVE    ADORE
C  X-AXIS.  THEREFORE, IFL21 AND IFL22 MUST BE ZERO.                      ADORE
C                                                                         ADORE
C  RECORD 3.3                                                             ADORE
C  ----------                                                             ADORE
C                                                                         ADORE
C  JREC   RECORD IDENTIFIER.                                              ADORE
C  JF1    AXIAL RACE CONSTRAINT FOR QUASI-STATIC SOLUTION --              ADORE
C          0 = PRESCRIBED AXIAL FORCE.                                    ADORE
C          1 = PRESCRIBED AXIAL DISPLACEMENT.                             ADORE
C  JF2    RADIAL RACE CONSTRAINT FOR QUASI-STATIC SOLUTION --             ADORE
C          0 = PRESCRIBED RADIAL FORCE.                                   ADORE
C          1 = PRESCRIBED RADIAL DISPLACEMENT.                            ADORE
C  NDQ    NUMBER OF POINTS IN THE STIFFNESS SPEED TABLE.  SPECIFY ZERO    ADORE
C         IF NO STIFFNESS COMPUTATION IS DESIRED.                         ADORE
C  ICOD   INITIAL CONDITIONS FOR DYNAMIC SOLUTIONS --                     ADORE
C          0 = EITHER COMPUTE INITIAL CONDITIONS FROM QUASI-STATIC        ADORE
C              SOLUTION (IF THIS THE FIRST RUN) OR READ INITIAL CONDI-    ADORE
C              TIONS FROM THE DATA FILE -MASTER- CREATED BY AN            ADORE
C              EARLIER RUN.                                               ADORE
C          1 = INITIAL CONDITIONS SPECIFIED IN THE DATA FILE -FINAL-      ADORE
C              (SEE DISCUSSION OF DATA FILES).                            ADORE
C  IROT   DYNAMIC SIMULATION WITH A ROTATING LOAD WITH RACE ORBITING      ADORE
C         WITH A RADIUS EQUAL TO THE STATIC RADIAL DEFLECTION.            ADORE
C          0 = NO ROTATING LOAD.                                          ADORE
C          1 = OUTER RACE ORBITING.                                       ADORE
C          2 = INNER RACE ORBITING.                                       ADORE
C  IBXM   CODE FOR INERTIAL PARAMETERS OF THE ROLLING ELEMENTS --         ADORE
C          0 = STANDARD PARAMETERS (IDEAL GEOMETRY).                      ADORE
C          M = PARAMETERS SUPPLIED FOR M ROLLING ELEMENTS (M.GT.0) ON     ADORE
C              RECORDS 6.1.                                               ADORE
C  ICXM   INERTIAL PARAMETERS FOR THE CAGE OR CAGE SEGMENTS --            ADORE
C          0 = STANDARD PARAMETERS (IDEAL GEOMETRY).                      ADORE
C          1 = INERTIAL PARAMETERS SPECIFIED ON RECORD 7.5.               ADORE
C  IRXM   INERTIAL PARAMETERS ASSOCIATED WITH ACCELERATING RACES --       ADORE
C          0 = RACES CONSTRAINED, NO INERTIAL PARAMETERS REQUIRED.        ADORE
C          1 = INERTIAL PARAMETERS SPECIFIED ON RECORDS 6.2.K.            ADORE
C  NBW    NUMBER OF ELEMENTS (MAXIMUM 6) FOR WHICH THE PLOT               ADORE
C         OUTPUT WILL BE SAVED.                                           ADORE
```

```
C   IBW     VECTOR OF LENGTH NBW CONTAINING THE INDICES OF THE NBW         ADORE
C           ELEMENTS IN INCREASING ORDER (SEE NOTE 7).                     ADORE
C                                                                          ADORE
C   RECORD 4.1                                                             ADORE
C   ----------                                                             ADORE
C                                                                          ADORE
C   JREC    RECORD IDENTIFIER.                                             ADORE
C   BOR     BEARING BORE (L).                                              ADORE
C   OD      OUTSIDE DIAMETER OF BEARING (L).                               ADORE
C   SDA     SHAFT INSIDE DIAMETER (L) FOR A HOLLOW SHAFT.                  ADORE
C   HDA     HOUSING OUTSIDE DIAMETER (L).                                  ADORE
C   SF1     SHRINK FIT ALLOWANCE ON OUTER RACE (L).                        ADORE
C   SF2     SHRINK FIT ALLOWANCE ON INNER RACE (L).                        ADORE
C                                                                          ADORE
C   FOR AN INTERFERENCE FIT THE SHRINK FIT ALLOWANCE IS POSITIVE AND A     ADORE
C   NEGATIVE VALUE INDICATES A LOOSE FIT.                                  ADORE
C                                                                          ADORE
C   RECORD 4.2                                                             ADORE
C   ----------                                                             ADORE
C                                                                          ADORE
C   ALTHOUGH ADORE DOES NOT CARRY OUT ANY THERMAL ANALYSIS, THE GEOMETRI-  ADORE
C   CAL DISTORTION OF THE VARIOUS ELEMENTS CAN BE ACCOUNTED FOR BY SPEC-   ADORE
C   IFYING THE ESTIMATED TEMPERATURES ON THIS RECORD.  NOTE THAT ALL       ADORE
C   GEOMETRICAL PARAMETERS IN THE INPUT DATA ARE ASSUMED TO BE AT          ADORE
C   ROOM TEMPERATURE.                                                      ADORE
C                                                                          ADORE
C   JREC    RECORD IDENTIFIER.                                             ADORE
C   ROT     ROOM TEMPERATURE (TM).                                         ADORE
C   HOT     HOUSING TEMPERATURE (TM).                                      ADORE
C   SHT     SHAFT TEMPERATURE (TM).                                        ADORE
C   RAT1    OUTER RACE TEMPERATURE (TM).                                   ADORE
C   RAT2    INNER RACE TEMPERATURE (TM).                                   ADORE
C   RET     ROLLING ELEMENT TEMPERATURE (TM).                              ADORE
C   CAT     CAGE TEMPERATURE (TM).                                         ADORE
C                                                                          ADORE
C   RECORD 5A   (FOR BALL BEARINGS ONLY)                                   ADORE
C   ---------                                                              ADORE
C                                                                          ADORE
C   JREC    RECORD IDENTIFIER.                                             ADORE
C   DF      BALL DIAMETER (L).  IF BALLS ARE NOT OF A UNIFORM SIZE, THEN   ADORE
C           SET DF = -D, WHERE D IS THE NOMINAL DIAMETER (L) AND SPECIFY   ADORE
C           THE VARIATION IN BALL DIAMETER ON RECORD 5G.1.                 ADORE
C   DP      PITCH DIAMETER (L).                                            ADORE
C   ALFA    FREE CONTACT ANGLE (DEG).  IF THE DIAMETRAL PLAY IS KNOWN IN   ADORE
C           LIEU OF THE CONTACT ANGLE THEN ALFA = -DPLAY, WHERE DPLAY IS   ADORE
C           THE DIAMETRAL PLAY (L).                                        ADORE
C   RCUR1   OUTER RACE CURVATURE FACTOR.                                   ADORE
C   RCUR2   INNER RACE CURVATURE FACTOR.                                   ADORE
C                                                                          ADORE
C   RACE CURVATURE FACTOR IS DEFINED AS THE RATIO OF THE RADIUS OF CURVA-  ADORE
C   TURE OF THE RACE GROOVE TO THE NOMINAL BALL DIAMETER, DF.              ADORE
C                                                                          ADORE
C   RECORD 5B   (FOR CYLINDRICAL ROLLER BEARINGS ONLY)                     ADORE
C   ---------                                                              ADORE
C                                                                          ADORE
C   JREC    RECORD IDENTIFIER.                                             ADORE
C   DF      ROLLER DIAMETER (L).  USE DEFINITION OF DF ON RECORD 5A FOR    ADORE
C           SPECIFYING A VARIATION IN ROLLER DIAMETER.                     ADORE
C   CRN     CROWN RADIUS (L).  FOR A VARIATION IN CROWN RADIUS SET CRN     ADORE
C           = -C, WHERE C IS THE NOMINAL CROWN RADIUS (L) AND SPECIFY THE  ADORE
C           VARIATION IN CROWN RADIUS ON RECORD 5G.2.                      ADORE
C           FOR INFINITE RADIUS, SPECIFY 1.0E+10 AND SET XL2=XL1 IN        ADORE
C           THIS RECORD.                                                   ADORE
C   XL1     ROLLER LENGTH (L).  FOR A VARIATION IN ROLLER LENGTH SPECIFY   ADORE
C           XL1 = -L, WHERE L IS THE NOMINAL ROLLER LENGTH (L) AND SPECIFY ADORE
C           LENGTH VARIATION ON RECORD 5G.3.                               ADORE
C   XL2     LENGTH OF CENTRAL LAND (L).  FOR EITHER A VARIATION IN THE     ADORE
C           LENGTH OF LAND OR ITS CENTRALITY SET XL2 = -XX, WHERE XX       ADORE
C           IS THE NOMINAL LENGTH OF LAND (L) AND SPECIFY THE VARIATION    ADORE
C           IN LENGTH OF LAND ON RECORD 5G.4 AND THE VARIATION IN          ADORE
C           CENTRALITY ON LAND ON RECORD 5G.5.                             ADORE
C   CRAD1   CORNER RADIUS ON THE NEGATIVE X-AXIS OF ROLLER (L).            ADORE
```

```
C  CRAD2  CORNER RADIUS ON THE POSITIVE X-AXIS OF ROLLER (L).          ADORE
C  DP     PITCH DIAMETER (L).                                          ADORE
C  DCL    DIAMETRAL CLEARANCE OR PLAY (L).                             ADORE
C                                                                      ADORE
C  RECORD 5C   (SPHERICAL ROLLER BEARINGS ONLY, SEE NOTE 2)            ADORE
C  ---------                                                           ADORE
C                                                                      ADORE
C  DF     ROLLER DIAMETER (L).  USE DEFINITION OF DF ON RECORD 5A FOR  ADORE
C         SPECIFYING A VARIATION IN ROLLER DIAMETER.                   ADORE
C  CRN    CROWN RADIUS (L).  USE DEFINITION OF CRN ON RECORD 5B FOR    ADORE
C         SPECIFYING A VARIATION IN CROWN RADIUS.                      ADORE
C  XL1    ROLLER LENGTH (L).  USE DEFINITION OF XL1 ON RECORD 5B FOR   ADORE
C         SPECIFYING A VARIATION IN ROLLER LENGTH.                     ADORE
C  CRAD1  CORNER RADIUS ON THE NEGATIVE X-AXIS ON ROLLER (L).          ADORE
C  CRAD2  CORNER RADIUS ON THE POSITIVE X-AXIS OF THE ROLLER (L).      ADORE
C  DP     PITCH DIAMETER (L).                                          ADORE
C  DCL    DIAMETRAL CLEARANCE OR PLAY (L).                             ADORE
C  RCUR1  OUTER RACE CURVATURE FACTOR.                                 ADORE
C  RCUR2  INNER RACE CURVATURE FACTOR.                                 ADORE
C                                                                      ADORE
C  RACE CURVATURE FACTOR IS DEFINED AS THE RATIO OF THE RADIUS OF CUR- ADORE
C  VATURE OF THE RACE GROOVE TO THE DIAMETER OF ROLLER CROWN, 2*CRN.   ADORE
C                                                                      ADORE
C  RECORD 5D   (TAPERED ROLLER BEARINGS ONLY)                          ADORE
C  ---------                                                           ADORE
C                                                                      ADORE
C  DF     ROLLER DIAMETER AT THE LARGE END (L).  FOR VARIATION IN      ADORE
C         ROLLER SIZE USE DEFINITION OF DF ON RECORD 5A.               ADORE
C  XL1    ROLLER LENGTH (L).  FOR A VARIATION IN ROLLER LENGTH USE     ADORE
C         DEFINITION OF XL1 ON RECORD 5B.                              ADORE
C  ERAD   END RADIUS AT THE LARGER END OF ROLLER (L).                  ADORE
C  RANG1  OUTER RACE SEMI CONE ANGLE (DEG).                            ADORE
C  RANG2  INNER RACE SEMI CONE ANGLE (DEG).                            ADORE
C                                                                      ADORE
C  RECORD 5E   (SPHERICAL TAPERED ROLLER BEARINGS ONLY)                ADORE
C  ---------                                                           ADORE
C                                                                      ADORE
C  JREC   RECORD IDENTIFIER.                                           ADORE
C  DF     ROLLER DIAMETER (L) AT THE LARGE END.  USE DEFINITION OF DF  ADORE
C         ON RECORD 5A FOR SPECIFYING A SIZE VARIATION.                ADORE
C  CRN    CROWN RADIUS (L).  FOR A VARIATION IN CROWN RADIUS USE       ADORE
C         DEFINITION OF CRN ON RECORD 5B.                              ADORE
C  XL1    ROLLER LENGTH (L).  FOR A VARIATION IN ROLLER LENGTH, USE DEF- ADORE
C         INITION OF XL1 ON RECORD 5B.                                 ADORE
C  ERAD   END RADIUS AT THE LARGE END OF THE ROLLER (L)                ADORE
C  RANG1  OUTER RACE SEMI CONE ANGLE (DEG).                            ADORE
C  RANG2  INNER RACE SEMI CONE ANGLE (DEG).                            ADORE
C  RCUR1  OUTER RACE CURVATURE FACTOR.                                 ADORE
C  RCUR2  INNER RACE CURVATURE FACTOR.                                 ADORE
C                                                                      ADORE
C  RACE CURVATURE FACTOR IS DEFINED AS THE RATIO OF THE RADIUS OF CUR- ADORE
C  VATURE OF THE RACE GROOVE TO THE DIAMETER OF ROLLER CROWN, 2*CRN.   ADORE
C                                                                      ADORE
C  RECORD 5F   (ONLY IF ANY RACE FLANGE IS SPECIFIED ON RECORD 3.2)    ADORE
C  ---------                                                           ADORE
C                                                                      ADORE
C  JREC   RECORD IDENTIFIER.                                           ADORE
C  FAN11  FLANGE LAYBACK ANGLE (DEG), OUTER RACE, NEGATIVE X-AXIS.     ADORE
C  FAN21  FLANGE LAYBACK ANGLE (DEG), OUTER RACE, POSITIVE X-AXIS.     ADORE
C  FAN12  FLANGE LAYBACK ANGLE (DEG), INNER RACE, NEGATIVE X-AXIS.     ADORE
C  FAN22  FLANGE LAYBACK ANGLE (DEG), INNER RACE, POSITIVE X-AXIS.     ADORE
C  FLH11  FLANGE HEIGHT (L), OUTER RACE, NEGATIVE X-AXIS.              ADORE
C  FLH21  FLANGE HEIGHT (L), OUTER RACE, POSITIVE X-AXIS.              ADORE
C  FLH12  FLANGE HEIGHT (L), INNER RACE, NEGATIVE X-AXIS.              ADORE
C  FLH22  FLANGE HEIGHT (L), INNER RACE, POSITIVE X-AXIS.              ADORE
C  FCL1   ROLLER/FLANGE AXIAL CLEARANCE (L), OUTER RACE.               ADORE
C  FCL2   ROLLER/FLANGE AXIAL CLEARANCE (L), INNER RACE.               ADORE
C                                                                      ADORE
C  THE FLANGE LAYBACK ANGLE IS MEASURED FROM A PLANE NORMAL TO THE     ADORE
C  BEARING AXIS.  THE FLANGE SURFACE RELATIVE TO THE CYLINDRICAL SUR-  ADORE
C  FACE OF THE RACE IS LOCATED BY AN ANGLE (90 + LAYBACK ANGLE).       ADORE
C  FOR TAPERED ROLLER BEARINGS, WHERE THE FLANGE EXISTS ONLY ON THE    ADORE
```

```
C  NEGATIVE X-AXIS, SET FCL1=FCL2=0.                                     ADORE
C                                                                        ADORE
C  RECORD 5G.1   (ONLY IF DF IS NEGATIVE ON ONE OF THE ABOVE RECORDS)    ADORE
C  -----------                                                           ADORE
C                                                                        ADORE
C  JREC    RECORD IDENTIFIER                                             ADORE
C  RR      VECTOR OF LENGTH NB, CONTAINING THE VARIATION IN DIAMETER (L) ADORE
C          OF NB ROLLING ELEMENTS.                                       ADORE
C                                                                        ADORE
C  RECORD 5G.2   (FOR ROLLER BEARINGS ONLY WHEN CRN IS NEGATIVE)         ADORE
C  -----------                                                           ADORE
C                                                                        ADORE
C  JREC    RECORD IDENTIFIER.                                            ADORE
C  BRAD    VECTOR OF LENGTH NB, CONTAINING THE CROWN RADIUS VARIATION    ADORE
C          (L) FOR THE NB ROLLERS.                                       ADORE
C                                                                        ADORE
C  RECORD 5G.3   (FOR ROLLER BEARINGS ONLY WHEN XL1 IS NEGATIVE)         ADORE
C  -----------                                                           ADORE
C                                                                        ADORE
C  JREC    RECORD IDENTIFIER.                                            ADORE
C  BLEN1   VECTOR OF LENGTH NB, CONTAINING THE LENGTH VARIATION (L) OF   ADORE
C          THE NB ROLLERS.                                               ADORE
C                                                                        ADORE
C  RECORD 5G.4   (FOR ROLLER BEARINGS ONLY WHEN XL2 IS NEGATIVE)         ADORE
C  -----------                                                           ADORE
C                                                                        ADORE
C  JREC    RECORD IDENTIFIER.                                            ADORE
C  BLEN2   VECTOR OF LENGTH NB, CONTAINING THE VARIATION IN LENGTH       ADORE
C          OF THE FLAT LAND (L) ON THE NB ROLLERS.                       ADORE
C                                                                        ADORE
C  RECORD 5G.5   (FOR ROLLER BEARINGS ONLY WHEN XL2 IS NEGATIVE)         ADORE
C  -----------                                                           ADORE
C                                                                        ADORE
C  JREC    RECORD IDENTIFIER.                                            ADORE
C  D       VECTOR OF LENGTH NB, CONTAINING THE VARIATION IN CENTRALITY   ADORE
C          OF THE FLAT LAND (L) ON THE NB ROLLERS.                       ADORE
C                                                                        ADORE
C  RECORD 6.1   (ONLY FOR ARBITRARY INERTIAL PARAMETERS FOR THE          ADORE
C  ----------    ROLLING ELEMENTS, IBXM.GT.0 ON RECORD 3.2)              ADORE
C                                                                        ADORE
C  JREC    RECORD IDENTIFIER.                                            ADORE
C  IBX     VECTOR OF LENGTH IBXM, CONTAINING THE INDICES OF THE          ADORE
C          IBXM ROLLING ELEMENTS FOR WHICH INERTIAL PARAMETERS WILL      ADORE
C          PRESCRIBED ON THE FOLLOWING IBXM RECORDS.                     ADORE
C                                                                        ADORE
C  RECORDS 6.1.1 TO 6.1.IBXM   (ONLY IF RECORD 6.1 IS PRESENT)           ADORE
C  -------------------------                                             ADORE
C                                                                        ADORE
C  JREC    RECORD IDENTIFIER                                             ADORE
C  XMB     ROLLING ELEMENT MASS (MA).                                    ADORE
C  XIX     MOMENT OF INERTIA ABOUT (POLAR) X-AXIS (MA*L**2).             ADORE
C  XIY     MOMENT OF INERTIA ABOUT Y-AXIS (MA*L**2).                     ADORE
C  XIZ     MOMENT OF INERTIA ABOUT Z-AXIS (MA*L**2).                     ADORE
C  BGC     VECTOR OF LENGTH THREE CONTAINING THE COORDINATES LOCATING    ADORE
C          THE GEOMETRIC CENTER OF THE ROLLING ELEMENT RELATIVE TO THE   ADORE
C          MASS CENTER IN THE ROLLING ELEMENT FRAME OF REFERENCE (L).    ADORE
C                                                                        ADORE
C          NOTE THAT THE ABOVE DATA WILL BE REPEATED FOR THE IBXM        ADORE
C          ROLLING ELEMENTS.                                             ADORE
C                                                                        ADORE
C  RECORD 6.2.1   (ONLY FOR THE INERTIAL PARAMETERS OF THE RACES,        ADORE
C  ------------    IROT=0 AND IRXM=1 ON RECORD 3.3)                      ADORE
C                                                                        ADORE
C  JREC    RECORD IDENTIFIER.                                            ADORE
C  XMR1    EFFECTIVE MASS (MA) OF OUTER RACE.                            ADORE
C  XJX1    VECTOR OF LENGTH THREE CONTAINING THE PRINCIPAL MOMENT OF     ADORE
C          INERTIA (MA*L**2) OF THE OUTER RACE (ABOUT THE X,Y,Z AXES)    ADORE
C  RGC1    VECTOR OF LENGTH THREE CONTAINING THE COORDINATES OF THE      ADORE
C          GEOMETRIC CENTER OF OUTER RACE RELATIVE TO THE MASS CENTER    ADORE
C          (L) IN THE RACE FRAME.                                        ADORE
C  IRF1    VECTOR OF LENGTH THREE DEFINING THE CONSTRAINTS ON THE MASS   ADORE
C          CENTER MOTION OF OUTER RACE.  THE THREE COMPONENTS CORRESPOND ADORE
```

```
C         RESPECTIVELY TO THE THREE AXES X, Y AND Z.  THE DEFINITION IS ADORE
C         AS FOLLOWS --                                                 ADORE
C           IRF1(I) = 0, SPECIFIED FORCE ALONG THE ITH AXIS.  THE X,    ADORE
C                        Y,Z AXES ARE NUMBERED AS 1,2,3 RESPECTIVELY.   ADORE
C           IRF1(I) = 1, SPECIFIED ACCELERATION ALONG THE ITH AXIS.     ADORE
C  IRM1   CONSTRAINT DEFINING THE ROTATIONAL MOTION OF THE OUTER RACE   ADORE
C         AS FOLLOWS --                                                 ADORE
C           IRM1    = 0, SPECIFIED MOMENT ALONG THE BEARING AXIS.       ADORE
C           IRM1    = 1, SPECIFIED ANGULAR ACCELERATION ALONG THE       ADORE
C                        BEARING AXIS.                                  ADORE
C                                                                       ADORE
C  RECORD 6.2.2    (ONLY FOR THE INERTIAL PARAMETERS OF THE RACES,      ADORE
C  ------------    IROT=0 AND IRXM=1 ON RECORD 3.3)                     ADORE
C                                                                       ADORE
C  JREC  RECORD IDENTIFIER.                                             ADORE
C  XMR2   EFFECTIVE MASS (MA) OF THE INNER RACE.                        ADORE
C  XJX2   VECTOR OF LENGTH THREE CONTAINING THE PRINCIPAL MOMENT OF     ADORE
C         INERTIA (MA*L**2) OF THE INNER RACE (ABOUT THE X,Y,Z AXES).   ADORE
C  RGC2   VECTOR OF LENGTH THREE CONTAINING THE COORDINATES OF THE      ADORE
C         GEOMETRIC CENTER OF THE INNER RACE RELATIVE TO THE MASS CENTER ADORE
C         (L) IN THE RACE FRAME.                                        ADORE
C  IRF2   CONSTRAINTS ON THE MASS CENTER MOTION OF THE INNER RACE PER   ADORE
C         DEFINITION OF IRF1 ON RECORD 6.2.1.                           ADORE
C  IRM2   CONSTRAINTS ON THE ANGULAR MOTION OF THE INNER RACE PER DEFI- ADORE
C         NITION OF IRM1 ON RECORD 6.2.1.                               ADORE
C                                                                       ADORE
C  RECORD 7.1   (ONLY IF A CAGE IS PRESENT, KBT.GT.0 ON RECORD 3.2)     ADORE
C  ----------                                                           ADORE
C                                                                       ADORE
C  JREC   RECORD IDENTIFIER.                                            ADORE
C  CD1    CAGE OUTER DIAMETER (L).                                      ADORE
C  CD2    CAGE INNER DIAMETER (L).                                      ADORE
C  CWC    CAGE WIDTH (L).                                               ADORE
C  CL1    CAGE/RACE OUTER DIAMETRAL CLEARANCE (L).                      ADORE
C  CL2    CAGE/RACE INNER DIAMETRAL CLEARANCE (L).                      ADORE
C  CWT    ANGULAR WIDTH OF CUT (DEG) IF CAGE IS SEGMENTED.              ADORE
C                                                                       ADORE
C  THE DIAMETRAL CLEARANCES SPECIFIED ON THIS RECORD ARE ONLY USED FOR  ADORE
C  COMPUTING THE CHURNING EFFECTS.                                      ADORE
C                                                                       ADORE
C  RECORD 7.2.1    (ONLY IF THE CAGE IS RACE GUIDED ON THE LAND ON NEG- ADORE
C  ------------    ATIVE X-AXIS, KBT.GT.0 AND IRG1.GT.0 ON RECORD 3.2)  ADORE
C                                                                       ADORE
C  JREC   RECORD IDENTIFIER.                                            ADORE
C  CRL1   CAGE GUIDE LAND DIAMETER (L) FOR LAND ON NEGATIVE X-AXIS.     ADORE
C         THIS LAND IS REFERED TO AS LAND NO. 1.                        ADORE
C  CW1    LAND WIDTH (L) FOR THE LAND ON NEGATIVE X-AXIS.               ADORE
C  CLS1   DISTANCE (L) OF OUTER EDGE OF LAND ON NEGATIVE X-AXIS FROM    ADORE
C         THE GEOMETRIC CENTER OF CAGE.                                 ADORE
C  CLC1   DIAMETRAL CLEARANCE (L) ON THE LAND ON NEGATIVE X-AIXS.       ADORE
C                                                                       ADORE
C  RECORD 7.2.2    (ONLY IF THE CAGE IS RACE GUIDED ON THE LAND ON POS- ADORE
C  ------------    ITIVE X-AXIS, KBT.GT.0 AND IRG2.GT.0 ON RECORD 3.2)  ADORE
C                                                                       ADORE
C  JREC   RECORD IDENTIFIER.                                            ADORE
C  CRL2   CAGE GUIDE LAND DIAMETER (L) FOR LAND ON POSITIVE X-AXIS.     ADORE
C         THIS LAND IS REFERED TO AS LAND NO. 2.                        ADORE
C  CW2    LAND WIDTH (L) FOR THE LAND ON POSITIVE X-AXIS.               ADORE
C  CLS2   DISTANCE OF OUTER EDGE OF LAND ON POSITIVE X-AXIS FROM        ADORE
C         THE GEOMETRIC CENTER OF CAGE.                                 ADORE
C  CCL2   DIAMETRAL CLEARANCE (L) ON THE LAND ON POSITIVE X-AXIS.       ADORE
C                                                                       ADORE
C  RECORD 7.3   (ONLY IF CAGE IS PRESENT, KBT.GT.0 ON RECORD 3.2)       ADORE
C  ----------                                                           ADORE
C                                                                       ADORE
C  JREC   RECORD IDENTIFIER.                                            ADORE
C  RCL    DIAMETRAL CLEARANCE (L) IN CAGE POCKET.  IF CLEARANCE VARIES  ADORE
C         FROM POCKET TO POCKET, SET RCL = -RC, WHERE RC IS THE NOMINAL ADORE
C         CLEARANCE (L) AND SPECIFY VARIATION ON RECORDS 7.4.           ADORE
C  RCLH   MAXIMUM LUBRICANT FILM (L) IN CAGE POCKET.                    ADORE
C  AXPS   INITIAL AXIAL POSITION OF THE CAGE (L) RELATIVE TO THE LOCUS  ADORE
C         OF THE CENTERS OF THE ROLLING ELEMENTS.                       ADORE
```

```
C  RPOS    INITIAL RADIAL POSITION OF THE CAGE (L) RELATIVE TO THE          ADORE
C          CENTER OF THE GUIDING RACE.                                      ADORE
C  APOS    INITIAL ANGULAR ROTATION OF THE CAGE ABOUT THE BEARING AXIS,     ADORE
C          SPECIFIED IN TERMS OF A DISPLACEMENT (L) OF THE POCKET CENTER    ADORE
C          RELATIVE TO THE ROLLING ELEMENT CENTER.                          ADORE
C                                                                           ADORE
C  RECORDS 7.4.1 TO 7.4.NB  (ONLY IF RCL IS NEGATIVE ON RECORD 7.3)         ADORE
C  -----------------------                                                  ADORE
C                                                                           ADORE
C  EACH OF THESE RECORDS CORRESPOND TO THE NB POCKETS IN THE CAGE AND       ADORE
C  THEY SPECIFY THE POCKET GEOMETRY IN TERMS OF THE FOLLOWING --            ADORE
C                                                                           ADORE
C  JREC    RECORD IDENTIFIER.                                               ADORE
C  RCLI    DIAMETRAL POCKET CLEARANCE (L).                                  ADORE
C  WLI     POCKET THICKNESS (L).  WLI IS NORMALLY THE CAGE RIM THICK-       ADORE
C          NESS OR THE DIFFERENCE BETWEEN THE OUTER AND INNER RADII.        ADORE
C  RPS     VECTOR OF LENGTH THREE CONTAINING THE POSITION OF ACTUAL         ADORE
C          POCKET CENTER (L) RELATIVE TO THE CORRESPONDING IDEAL            ADORE
C          POCKET CENTER IN A GEOMETRICALLY IDEAL CAGE.                     ADORE
C  TCP     VECTOR OF LENGTH TWO CONTAINING THE ANGLES (DEG) DEFINING        ADORE
C          THE ORIENTATION OF THE ACTUAL POCKET RELATIVE TO THE CORR-       ADORE
C          ESPONDING POCKET IN A GEOMETRICALLY IDEAL CAGE.  THE TWO         ADORE
C          ANGLES ARE THE TRANSFORMATION ANGLES ABOUT THE Y AND Z           ADORE
C          AXES RESPECTIVELY.                                               ADORE
C                                                                           ADORE
C  RECORD 7.5    (ONLY FOR ARBITRARY INERTIAL PARAMETERS FOR THE            ADORE
C  ----------    CAGE, ICXM=1 ON RECORD 3.3)                                ADORE
C                                                                           ADORE
C  JREC    RECORD IDENTIFIER.                                               ADORE
C  XMC     CAGE MASS (MA).                                                  ADORE
C  XKX     VECTOR OF LENGTH THREE CONTAINING THE PRINCIPAL MOMENT OF        ADORE
C          INERTIA (MA*L**2) OF THE CAGE (ABOUT THE X,Y,Z AXES).            ADORE
C  CGC     VECTOR OF LENGTH THREE CONTAINING THE X,Y,Z COORDINATES OF       ADORE
C          THE CAGE GEOMETRIC CENTER RELATIVE TO ITS MASS CENTER IN THE     ADORE
C          CAGE FIXED FRAME.                                                ADORE
C                                                                           ADORE
C  RECORD 8.1    (ONLY FOR SPECIAL ROLLING ELEMENT MATERIAL, MBR1=1 ON      ADORE
C  ----------    RECORD 3.2)                                                ADORE
C                                                                           ADORE
C  JREC    RECORD IDENTIFIER.                                               ADORE
C  DEN     ROLLING ELEMENT DENSITY (MA/L**3).                               ADORE
C  EB      ROLLING ELEMENT ELASTIC MODULUS (F/L**2).                        ADORE
C  POB     ROLLING ELEMENT POISSON-S RATIO.                                 ADORE
C  CTEB    COEFFICIENT OF THERMAL EXPANSION OF ROLLING ELEMENT (1/TM).      ADORE
C  HRB     ROLLING ELEMENT HARDNESS (F/L**2).                               ADORE
C  WB      ROLLING ELEMENT WEAR COEFFICIENT.                                ADORE
C                                                                           ADORE
C  RECORD 8.2    (ONLY FOR SPECIAL RACE MATERIAL, MBR2=1 ON RECORD 3.2)     ADORE
C  ----------                                                               ADORE
C                                                                           ADORE
C  JREC    RECORD IDENTIFIER.                                               ADORE
C  DEN     MATERIAL DENSITY (MA/L**3) FOR THE RACES.                        ADORE
C  ER      ELASTIC MODULUS (F/L**2) FOR THE RACES.                          ADORE
C  POR     POISSON-S RATIO FOR THE RACES.                                   ADORE
C  CTER    COEFFICIENT OF THERMAL EXPANSION (1/TM) FOR THE RACES.           ADORE
C  HRR     HARDNESS (F/L**2) OF THE RACES.                                  ADORE
C  WR      WEAR COEFFICIENT FOR THE RACES.                                  ADORE
C                                                                           ADORE
C  RECORD 8.3    (ONLY FOR SPECIAL SHAFT MATERIAL, MHS=1 ON RECORD 3.2)     ADORE
C  ----------                                                               ADORE
C                                                                           ADORE
C  JREC    RECORD IDENTIFIER.                                               ADORE
C  DENSF   MATERIAL DENSITY (MA/L**3) FOR THE SHAFT.                        ADORE
C  ES      ELASTIC MODULUS (F/L**2) FOR THE SHAFT.                          ADORE
C  POS     POISSON-S RATIO FOR THE SHAFT.                                   ADORE
C  CTES    COEFFICIENT OF THERMAL EXPANSION (1/TM) FOR THE SHAFT.           ADORE
C                                                                           ADORE
C  RECORD 8.4    (ONLY FOR SPECIAL HOUSING MATERIAL, MHS=1 ON RECORD        ADORE
C  ----------    3.2)                                                       ADORE
C                                                                           ADORE
C  JREC    RECORD IDENTIFIER.                                               ADORE
C  DENH    MATERIAL DENSITY (MA/L**3) FOR THE HOUSING.                      ADORE
```

```
C  EH      ELASTIC MODULUS (F/L**2) FOR THE HOUSING.
C  POH     POISSON-S RATIO FOR THE HOUSING.
C  CTEH    COEFFICIENT OF THERMAL EXPANSION (1/TM) FOR THE HOUSING.
C
C  RECORD 8.5   (ONLY FOR SPECIAL CAGE MATERIAL, MCR=1 ON RECORD 3.2)
C  ----------
C
C  JREC    RECORD IDENTIFIER.
C  DENC    MATERIAL DENSITY (MA/L**3) FOR THE CAGE.
C  EC      CAGE ELASTIC MODULUS (F/L**2).
C  POC     POISSON-S RATION OF THE CAGE MATERIAL.
C  CTEC    COEFFICIENT OF THERMAL EXPANSION (1/TM) FOR THE CAGE.
C  HRC     HARDNESS (F/L**2) OF THE CAGE MATERIAL.
C  WC      WEAR COEFFICIENT FOR THE CAGE.
C
C  RECORD 9.1
C  ----------
C
C  THE OPERATING DATA SUPPLIED ON THIS RECORD IS USED FOR COMPUTING
C  THE QUASI-STATIC SOLUTION WHICH MAY BE USED FOR COMPUTING THE
C  INITIAL CONDITIONS FOR THE DYNAMIC SOLUTIONS.  ANY TIME DEPENDENT
C  OPERATING CONDITIONS MUST BE PROGRAMMED IN THE OPTIONAL LOAD AND
C  SPEED SUBROUTINE ADRX1.
C
C  JREC    RECORD IDENTIFIER.
C  QF1     AXIAL LOAD (F), IF JF1=0 ON RECORD 3.3 OR AXIAL DISPLACEMENT
C          (L) OF THE INNER RACE RELATIVE TO THE OUTER, IF JF1=1.
C  QF2     RADIAL LOAD (F), IF JF2=0 ON RECORD 3.3, OR RADIAL  DISPLACE-
C          MENT (L) OF INNER RACE RELATIVE TO THE OUTER, IF JF2=1.
C  GAMA11 OUTER RACE MISALIGNMENT-Y, ROTATION ABOUT Y-AXIS (RAD).
C  GAMA21 OUTER RACE MISALIGNMENT-Z, ROTATION ABOUT Z-AXIS (RAD).
C  GAMA12 INNER RACE MISALIGNMENT-Y, ROTATION ABOUT Y-AXIS (RAD).
C  GAMA22 INNER RACE MISALIGNMENT-Z, ROTATION ABOUT Z-AXIS (RAD).
C  RPM1    ANGULAR VELOCITY OF OUTER RACE (RPM).
C  RPM2    ANGULAR VELOCITY OF THE INNER RACE (RPM)
C
C  RECORD 9.2   (ONLY IF NDQ.GT.0 ON RECORD 3.3)
C  ----------
C
C  JREC    RECORD IDENTIFIER.
C  DMRI    PERCENT DISPLACEMENT INCREMENT FOR STIFFNESS COMPUTATION.
C  OMG11   INITIAL OUTER RACE VELOCITY (RPM) IN STIFFNESS-SPEED TABLE.
C  OMG21   FINAL OUTER RACE VELOCITY (RPM) IN STIFFNESS-SPEED TABLE.
C  OMG12   INITIAL INNER RACE VELOCITY (RPM) IN STIFFNESS-SPEED TABLE.
C  OMG22   FINAL INNER RACE VELOCITY (RPM) IN STIFFNESS-SPEED TABLE.
C
C  RECORD 10.1
C  -----------
C
C  THIS RECORD SPECIFIES TRACTION PARAMETERS FOR ROLLING ELEMENT/RACE
C  INTERACTION FOR EITHER KTC=0 ON RECORD 3.2 OR WHEN THE LUBRICANT
C  MODEL BREAKS DOWN UNDER KTC.GT.0 (NOTES 4 AND 5).
C
C  JREC    RECORD IDENTIFIER.
C  FCFB1   TRACTION COEFFICIENT AT ZERO SLIP UNDER OPTION 1.
C          COEFFICIENT -A- UNDER OPTION 2.
C  FCFB2   MAXIMUM TRACTION COEFFICIENT UNDER OPTION 1.
C          COEFFICIENT -B- (T/L) UNDER OPTION 2.
C  FCFB3   TRACTION COEFFICIENT AT INFINITE SLIP UNDER OPTION 1.
C          COEFFICIENT -C- (T/L) UNDER OPTION 2.
C  FCFB4   SLIP (L/T) AT MAXIMUM TRACTION COEFFICIENT UNDER OPTION 1.
C          MAXIMUM PERMISSIBLE TRACTION COEFFICIENT WITH A NEGATIVE SIGN
C          (FCFB4.LT.0.) UNDER OPTION 2.
C
C  RECORD 10.2    (FOR ROLLER/FLANGE INTERACTION ONLY, ABS(KBT).GT.1
C  -----------    AND IFL11, IFL21, IFL12 OR IFL22 =1 ON RECORD 3.2)
C
C  THIS RECORD SPECIFIES TRACTION PARAMETERS FOR ROLLER/FLANGE INT-
C  ERACTION FOR ROLLER BEARINGS (NOTES 4 AND 5).
C
C  JREC    RECORD IDENTIFIER.
C  FCFF1   TRACTION COEFFICIENT AT ZERO SLIP UNDER OPTION 1.
```

```
C         COEFFICIENT -A- UNDER OPTION 2.
C   FCFF2 MAXIMUM TRACTION COEFFICIENT UNDER OPTION 1.
C         COEFFICIENT -B- (T/L) UNDER OPTION 2.
C   FCFF3 TRACTION COEFFICIENT AT INFINITE SLIP UNDER OPTION 1.
C         COEFFICIENT -C- (T/L) UNDER OPTION 2.
C   FCFF4 SLIP (L/T) AT MAXIMUM TRACTION COEFFICIENT UNDER OPTION 1.
C         MAXIMUM PERMISSIBLE TRACTION COEFFICIENT WITH A NEGATIVE
C         SIGN (FCFF4.LT.0.) UNDER OPTION 2.
C   HCRF  CRITICAL FILM (L) WHICH ESTABLISHES ROLLER/RACE CONTACT.
C
C   RECORD 10.3   (FOR ELASTOHYDRODYNAMIC TRACTION MODELS ONLY,
C   -----------   KTC.GT.0 ON RECORD 3.2)
C
C   JREC  RECORD IDENTIFIER.
C   HC    CRITICAL FILM THICKNESS (L) FOR LUBRICANT MODEL BREAKDOWN AT
C         ROLLING ELEMENT/RACE INTERFACE.
C   STR   STARVATION PARAMETER.  RATIO OF THE LUBRICANT MANISCUS DIS-
C         TANCE FROM THE EDGE OF CONTACT TO THE CONTACT HALF WIDTH.
C
C   RECORD 10.4.1   (ONLY FOR SPECIAL ELASTOHYDRODYNAMIC TRACTION MODEL
C   -------------   COEFFICIENTS, KTC=8 ON RECORD 3.2)
C
C   THE PROPERTIES SUPPLIED ON THIS RECORD ARE UNDER LOW PRESSURE INLET
C   CONDITIONS AND THEY ARE USED TO COMPUTE THE ELASTOHYDRODYNAMIC FILM
C   THICKNESS OF THE LUBRICANT AT THE ROLLING ELEMENT/RACE INTERFACE.
C
C   JREC  RECORD IDENTIFIER.
C   TO    REFERENCE TEMPERATURE (TM) OF LUBRICANT.
C   XMUO  LUBRICANT VISCOCITY (F*T/L**2) AT REFERENCE TEMPERATURE.
C   GAM   PRESSURE-VISCOCITY COEFFICIENT (L**2/F).
C   BET   TEMPERATURE-VISCOCITY COEFFICIENT (TM).
C   FK    THERMAL CONDUCTIVITY (F/S/TM).
C
C   RECORD 10.4.2   (ONLY FOR SPECIAL ELASTOHYDRODYNAMIC TRACTION MODEL
C   -------------   COEFFICIENTS, KTC=8 ON RECORD 3.2)
C
C   THE PROPERTIES ON THIS RECORD DEFINE THE LUBRICANT BEHAVIOR IN THE
C   HIGH PRESSURE CONTACT REGION AND THEY ARE USED TO COMPUTE THE TRAC-
C   TION IN THE ELASTOHYDRODYNAMIC CONTACT ZONE AT THE ROLLING
C   ELEMENT/RACE INTERFACE.
C
C   JREC  RECORD IDENTIFIER.
C   XMUT  VECTOR OF LENGTH TWO CONTAINING THE VISCOCITY PARAMETERS FOR
C         THE TRACTION MODELS AT OUTER AND INNER RACES (F*T/L**2).
C   GAMMAT VECTOR OF LENGTH TWO CONTAINING THE PRESSURE-VISCOCITY COEFF-
C         ICIENTS (L**2/F) IN THE TRACTION MODELS FOR OUTER AND INNER
C         RACES.
C   BETAT VECTOR OF LENGTH TWO CONTAINING THE TEMPERATURE-VISCOCITY
C         COEFFICIENTS (1/TM) IN THE TRACTION MODELS FOR OUTER AND
C         INNER RACES.
C
C   RECORD 10.5.1   (ONLY IF KBT.GT.0 ON RECORD 3.2)
C   -------------
C
C   TRACTION PARAMETERS FOR THE ROLLING ELEMENT/CAGE INTERACTION IN THE
C   CAGE POCKETS ARE SPECIFIED ON THIS RECORD (NOTES 4 AND 5).
C
C   JREC  RECORD IDENTIFIER.
C   FCFP1 TRACTION COEFFICIENT AT ZERO SLIP UNDER OPTION 1.
C         COEFFICIENT -A- UNDER OPTION 2.
C   FCFP2 MAXIMUM TRACTION COEFFICIENT UNDER OPTION 1.
C         COEFFICIENT -B- (T/L) UNDER OPTION 2.
C   FCFP3 TRACTION COEFFICIENT AT INFINITE SLIP UNDER OPTION 1.
C         COEFFICIENT -C- (T/L) UNDER OPTION 2.
C   FCFP4 SLIP (L/T) AT MAXIMUM TRACTION UNDER OPTION 1.
C         MAXIMUM PERMISSIBLE TRACTION COEFFICIENT WITH A NEGATIVE SIGN
C         (FCFP4.LT.0.) UNDER OPTION 2.
C   HCBP  CRITICAL FILM THICKNESS (L) DEFINING METAL CONTACT IN CAGE
C         POCKETS.
C
C   RECORD 10.5.2   (ONLY IF KBT.GT.0 AND EITHER IRG1 OR IRG2 = 1 ON
C   -------------   RECORD 3.2)
```

```
C
C  THIS RECORD CONTAINS THE TRACTION PARAMETERS FOR THE CAGE/RACE
C  INTERACTION (NOTES 4 AND 5).
C
C  JREC  RECORD IDENTIFIER.
C  FCFR1  TRACTION COEFFICIENT AT ZERO SLIP UNDER OPTION 1.
C         COEFFICIENT -A- UNDER OPTION 2.
C  FCFR2  MAXIMUM TRACTION COEFFICIENT UNDER OPTION 1.
C         COEFFICIENT -B- UNDER OPTION 2.
C  FCFR3  TRACTION COEFFICIENT AT INFINITE SLIP UNDER OPTION 1.
C         COEFFICIENT -C- (T/L) UNDER OPTION 2.
C  FCFR4  SLIP (L/T) AT MAXIMUM TRACTION UNDER OPTION 1.
C         MAXIMUM PERMISSIBLE TRACTION COEFFICIENT WITH A NEGATIVE SIGN
C         (FCFR4.LT.0.) UNDER OPTION 2.
C  HCCR   CRITICAL FILM THICKNESS (L) DEFINING METAL CONTACT AT THE
C         CAGE/RACE INTERFACE.
C
C  RECORD 10.6    (ONLY IF KTC =0 AND EITHER IHBC OR IHRC = 1 ON
C  -----------     RECORD 3.2)
C
C  JREC   RECORD IDENTIFIER.
C  VISP   EFFECTIVE LUBRICANT VISCOCITY (F*T/L**2) FOR HYDRODYNAMIC
C         INTERACTION IN CAGE POCKETS.
C  VISR   EFFECTIVE LUBRICANT VISCOCITY (F*T/L**2) FOR HYDRODYNAMIC
C         INTERACTION AT THE CAGE/RACE INTERFACE.
C
C  RECORD 10.7    (ONLY IF CHURNING IS INCLUDED, ICHR.GT.0 ON RECORD 3.2)
C  -----------
C
C  JREC   RECORD IDENTIFIER.
C  CRO    EFFECTIVE LUBRICANT DENSITY (MA/L**3) FOR CHURNING EFFECTS.
C  VISC   EFFECTIVE LUBRICANT VISCOCITY (F*T/L**2) FOR CHURNING, WHEN
C         ICHR=1 OR A VISCOCITY MULTIPLIER WHEN ICHR=2 ON RECORD 3.2.
C         THE VISCOCITY FROM THE TRACTION MODEL (IF KTC.GT.0) WILL BE
C         MULTIPLIED BY VISC TO DETERMINE EFFECTIVE VISCOCITY FOR THE
C         CHURNING EFFECTS, WHEN ICHR=2.
C
C  RECORD 11    (ONLY IF MODE.GE.0 ON RECORD 1)
C  ---------
C
C  JREC   RECORD IDENTIFIER.
C  G      A VECTOR OF LENGTH THREE CONTAINING THE COMPONENTS OF ACCELE-
C         RATION DUE TO GRAVITY VECTOR (L/T**2) IN INERTIAL FRAME.
C
C  RECORDS 12.1 TO 12.N    (OPTIONAL INPUTS TO USER SUBROUTINES)
C  --------------------
C
C  ANY NUMBER OF DATA RECORDS REQUIRED BY THE USER PROGRAMMABLE SUB=
C  ROUTINES ADRX1, ADRX2, ADRX3, ADRX4, ADRX5 AND/OR ADRX6.
C
C  RECORD 13.1 TO 13.N    (OPTIONAL, ONLY IF JPLT.GT.0 ON RECORD 1)
C  -------------------
C
C  THESE DATA RECORDS ARE OPTIONAL TO THE PLOT ROUTINES.  IN THE
C  ABSENCE OF THESE RECORDS ALL THE POSSIBLE PLOTS WILL BE PROCESSED
C  WITH THE DEFAULT PARAMETERS LISTED BELOW IN THE DESCRIPTION OF THE
C  OPTIONAL PLOT INPUT RECORD 13.K.1.  IF EITHER A SELECTION OF
C  PARTICULAR PLOTS OR A CHANGE IN DEFAULT PARAMETERS IS DESIRED, THEN
C  THE FOLLOWING TWO DATA RECORDS MUST BE SUPPLIED FOR EACH PLOT FILE
C  IN THE ORDER SOL7, SOL1,SOL2,.....,SOL6.  THUS RECORDS 13.1.1 AND
C  13.1.2 CORRESPOND TO SOL7, RECORDS 13.2.1 AND 13.2.2 CORRESPOND TO
C  FILE SOL1, RECORDS 13.3.1 AND 13.3.2 CORRESPOND TO SOL2, AND SO ON.
C  THE SELECTION  OF PARTICULAR PLOTS CAN ALSO BE MADE BY THE DISSPLA
C  POST PROCESSOR AFTER GENERATING ALL THE PLOTS IN THE NEUTRAL PICTURE
C  FILE BY EXECUTING ADORE WITH THE DEFAULT PLOT INPUTS.
C
C      RECORD 13.K.1
C      -------------
C
C
C  JREC   RECORD IDENTIFIER.
C  XMAGI  PLOT MAGNIFICATION IN THE X DIRECTION (DEFAULT = 1.25).
```

```
C  YMAGI   PLOT MAGNIFICATION IN THE Y DIRECTION (DEFAULT = 1.25).         ADORE
C  JROT    PLOT ROTATION BY 90 DEGREES FROM NORMAL CONFIGURATION --        ADORE
C           0 = NO ROTATION.                                               ADORE
C           1 = PLOT ROTATED BY 90 DEGREES.                                ADORE
C  JRI     STARTING STEP NUMBER FOR PLOTTING (DEFAULT = 0).                ADORE
C  JRF     FINAL STEP NUMBER FOR PLOTTING (DEFAULT = 2473).                ADORE
C  ITP     FREQUENCY OF DATA POINTS TO BE PLOTTED.  ITP=1 WILL PLOTS       ADORE
C          ALL POINTS, ITP=2 WILL PLOT EVERY OTHER POINT AND SO ON.        ADORE
C          THE TOTAL NUMBER OF PLOT POINTS MUST NOT EXCEED 2473            ADORE
C          (DEFAULT = 1).                                                  ADORE
C  JAI     INDEX OF THE VARIABLE (X-COMP=1, Y-COMP=2, Z-COMP=3) TO BE      ADORE
C          PLOTTED ON X=AXIS FOR THE MASS CENTER ORBIT PLOT (DEFAULT =2).  ADORE
C  JAJ     INDEX OF THE VARIABLE TO BE PLOTTED ON Y-AXIS FOR THE MASS      ADORE
C          CENTER PLOT (DEFAULT = 3).                                      ADORE
C  NP      NUMBER OF PLOTS DESIRED FOR THE PARTICULAR PLOTS FILES.         ADORE
C          DEFAULT IS ALL POSSIBLE PLOTS.                                  ADORE
C                                                                          ADORE
C      RECORD 13.K.2                                                       ADORE
C      -------------                                                       ADORE
C                                                                          ADORE
C  JREC    RECORD IDENTIFIER                                               ADORE
C  JP      A VECTOR OF LENGTH NP CONTAINING THE REQUIRED PLOT NUMBERS      ADORE
C          FOR THE PARTICULAR PLOT FILE.                                   ADORE
C                                                                          ADORE
C                                                                          ADORE
C                                                                          ADORE
C  USER PROGRAMMABLE SUBROUTINES                                           ADORE
C  =============================                                           ADORE
C                                                                          ADORE
C  FOR THE SIMULATION OF HIGHLY SPECIALIZED AND SOPHISTICATED APPLI-       ADORE
C  CATIONS ADORE OFFERS SIX USER PROGRAMMABLE SUBROUTINES WHERE THE        ADORE
C  USER CAN PROVIDE AN APPROPRIATE CODE TO CUSTOMIZED ADORE FOR THE        ADORE
C  PARTICULAR APPLICATION.  THE VARIOUS SUBROUTINES ARE --                 ADORE
C                                                                          ADORE
C  ADRX1   THIS SUBROUTINE PROVIDES THE FLEXIBILITY TO PRESCRIBE TIME      ADORE
C          VARYING LOADS AND/OR ACCELERATION IMPOSED ON THE BEARING        ADORE
C          RACES.  IF THE BEARING AS A WHOLE IS MOVING IN SPACE, ADRX1     ADORE
C          ALLOWS THE USER TO PRESCRIBE VELOCITIES AND ACCELERATIONS       ADORE
C          IMPOSED ON THE NORMALLY SPACE FIXED FRAME OF REFERENCE.         ADORE
C          THUS ALL THE CENTRIFUGAL AND CORIOLIS COMPONENTS CAN BE         ADORE
C          PROPERLY ACCOUNTED FOR.  ADRX1 ALSO PROVIDES THE USER WITH      ADORE
C          THE OPTION OF VARYING THE ACCELERATION DUE TO GRAVITY           ADORE
C          GRAVITY VECTOR, CHANGING THE VARIOUS TOLERANCES AND RELAXING    ADORE
C          THE SYMETRY CONSIDERATIONS WHICH ADORE NORMALLY EMPLOYS.        ADORE
C                                                                          ADORE
C  ADRX2   ANY ARBITRARY LOAD-DEFLECTION RELATION FOR THE ROLLER TO RACE   ADORE
C          FLANGE INTERACTION CAN BE PRESCRIBED IN THIS SUBROUTINE.        ADORE
C          NORMALLY ADORE EMPLOYS A SIMULATED HERTZIAN CONTACT.            ADORE
C                                                                          ADORE
C  ADRX3   THE NORMALLY USED HERTZIAN CONTACT AT THE ROLLING ELEMENT/      ADORE
C          CAGE INTERACTION CAN BE REPLACED BY ANY ARBITRARY SPRING        ADORE
C          RATE OR LOAD-DEFLECTION RELATION IN THIS SUBROUTINE.            ADORE
C                                                                          ADORE
C  ADRX4   SIMILAR TO ADRX3, THIS SUBROUTINE ALLOWS THE USER TO PRES-      ADORE
C          CRIBE ANY ARBITRARY LOAD-DEFLECTION RELATION FOR THE CAGE/      ADORE
C          RACE CONTACT WHICH IS NORMALLY TREATED AS A CONVENTIONAL        ADORE
C          LINE CONTACT.                                                   ADORE
C                                                                          ADORE
C  ADRX5   THE VARIATION IN THE RADIUS OF THE ROLLER AS A FUNCTION OF      ADORE
C          THE AXIAL AND ANGULAR COORDINATE MAY BE PRESCRIBED IN THIS      ADORE
C          SUBROUTINE.  THUS ROLLER BEARINGS WITH ARBITRARY GEOMETRICAL    ADORE
C          VARIATIONS IN ROLLER GEOMETRY CAN BE TREATED.                   ADORE
C                                                                          ADORE
C  ADRX6   VARIATIONS IN THE GEOMETRY OF THE RACES CAN BE PRESCRIBED IN    ADORE
C          THIS SUBROUTINE.  FOR CYLINDRICAL AND TAPERED ROLLER BEARINGS   ADORE
C          THE RADIUS OF THE RACEWAY IS PRESCRIBED AS A FUNCTION OF THE    ADORE
C          AXIAL AND ANGULAR COORDINATES.  FOR BALL, SPHERICAL ROLLER      ADORE
C          AND SPHERICAL TAPERED ROLLER BEARINGS BOTH THE VARIATIONS IN    ADORE
C          THE LOCATION OF THE CENTER OF CURVATURE AND RADIUS OF CURVA-    ADORE
C          TURE OF THE RACE GROOVE MAY BE PRESCRIBED.                      ADORE
C                                                                          ADORE
C  NECESSARY DOCUMENTATION FOR PROGRAMMING EACH OF THESE SUBROUTINES       ADORE
```

```
C  IS INCLUDED WITH THE SOURCE LISTING OF THE SUBROUTINES.
C
C
C  PRINT OUTPUT
C  ============
C
C  THE AMOUNT OF PRINT OUTPUT FROM ADORE CAN BE GREATLY CONTROLLED
C  BY THE USER.  THE FIRST PART OF THE OUTPUT, WHICH IS ALWAYS PRINTED,
C  CONSISTS OF THE INPUT DATA CONTAINING THE BEARING GEOMETRY,
C  MATERIAL PROPERTIES, INERTIAL PARAMETERS, LUBRICATION PARAMETERS,
C  INITIAL OPERATING CONDITIONS, THE VARIOUS SCALE FACTORS AND ANY
C  OUTPUT PRODUCED BY THE THE USER PROGRAMMABLE SUBROUTINES.
C  FOLLOWING THIS OUTPUT ADORE PRINTS THE STIFFNESS-SPEED TABLE, IF
C  COMPUTED; OR A ONE PAGE OUTPUT FOR THE QUASI-STATIC SOLUTION
C  IF ADORE IS RUN WITH MODE.LT.0 ON RECORD 1.  FOR A DYNAMIC
C  SOLUTION (MODE.GE.0), THE PRINT OUTPUT AT EACH TIME STEP IS DIVIDED
C  IN FOUR SECTIONS WITH CONSIST OF THE FOLLOWING --
C
C  1.   ROLLING ELEMENT PARAMETERS.
C          1A. LOAD DISCRIBUTION ALONG ROLLER NO. 1
C          1B. RACE FLANGE INTERACTION.
C          1C. ROLLER END AND RACE FLANGE WEAR DISTRIBUTION.
C  2.   RACE AND CAGE PARAMETERS.
C  3.   APPLIED PARAMETERS.
C  4.   TIME STEP SUMMARY.
C
C  NOTE THAT 1A, 1B AND 1C ARE RELEVANT FOR ROLLER BEARINGS ONLY.
C
C
C  PLOT OUTPUT
C  ===========
C
C  SOLUTIONS FOR A MAXIMUM OF SIX BEARING ELEMENTS MAY BE STORED IN
C  DATA FILES FOR PLOT OUTPUT.  THE NUMBER OF PLOTS DEPENDS ON THE
C  TYPE OF THE ELEMENT AND THE BEARING.  THE TOTAL OUTPUT IS DIVIDED
C  INTO SUBSETS AND THE MAXIMUM NUMBER OF PLOTS IN EACH SUBSET IS
C  AS FOLLOWS --
C
C  1.   POWER DISSIPATION AND LIFE    (3 PLOTS)
C       --------------------------
C
C          PLOT 1  BEARING LIFE AND POWER LOSS.
C          PLOT 2  APPLIED MOMENTS ON THE RACES.
C          PLOT 3  TIME AVERAGE WEAR RATES.
C
C  2.   ROLLING ELEMENT    (8 PLOTS)
C       ---------------
C
C          PLOT 1  MASS CENTER ACCELERATIONS.
C          PLOT 2  MASS CENTER POSITIONS.
C          PLOT 3  CONTACT LOADS, ANGLES AND ORBITAL VELOCITY.
C          PLOT 4  ANGULAR VELOCITY AND SPIN/ROLL RATIOS.
C          PLOT 5  ROLLING ELEMENT/RACE SLIP, HEAT GENERATION AND LUBRI-
C                  CANT FILM THICKNESS.
C          PLOT 6  ROLLING ELEMENT/CAGE LOAD, GEOMETRIC INTERACTION AND
C                  CONTACT ANGLES.
C          PLOT 7  ROLLER/OUTER RACE FLANGE INTERACTION.
C          PLOT 8  ROLLER/INNER RACE FLANGE INTERACTION.
C
C  3.   OUTER OR INNER RACE    (6 PLOTS)
C       -------------------
C
C          PLOT 1  MASS CENTER WHIRL AND RACE ANGULAR VELOCITIES.
C          PLOT 2  APPLIED FORCES ON THE RACE.
C          PLOT 3  APPLIED MOMENTS ON THE RACE.
C          PLOT 4  RACE MASS CENTER ACCELERATION.
C          PLOT 5  RACE MASS CENTER ORBIT.
C          PLOT 6  RACE MASS CENTER POSITIONS.
C
C  4.   CAGE OR CAGE SEGMENT    (5 + 2*(SN/3) PLOTS, NS = NO. OF POCKETS)
C       --------------------
C
```

```
C        PLOT 1  MASS CENTER WHIRL AND CAGE ANGULAR VELOCITIES.
C        PLOT 2  CAGE/RACE INTERACTION AT THE LAND LOCATED ON THE
C                NEGATIVE X-AXIS (LAND NO. 1).
C        PLOT 3  CAGE/RACE INTERACTION AT THE LAND LOCATED ON THE
C                POSITIVE X-AXIS (LAND NO. 2).
C        PLOT 4  CAGE MASS CENTER ACCELERATIONS.
C        PLOT 5  CAGE MASS CENTER ORBIT.
C
C      THESE PLOTS ARE FOLLOWED BY (NS/3) PLOTS FOR THE FORCES IN NS
C      POCKETS OF THE CAGE OR CAGE SEGMENT (SOLUTIONS FOR THREE POCK-
C      ETS ARE SHOWN ON ONE PLOT) AND (NS/3) PLOTS FOR THE GEOMETRIC
C      INTERACTION IN THE NS POCKETS.  NOTE THAT IF NS IS NOT DIVISIBLE
C      BY 3 THEN THE QUANTITY (NS/3) IS ACTUALLY (NS/3 +1).
C
C
C  DATA FILES
C  ==========
C
C  ADORE EMPLOYS A NUMBER OF SEQUENTIAL DATA FILES FOR THE INPUT/OUTPUT
C  OF THE BEARING DATA AND THE VARIOUS SOLUTIONS.  THE VARIOUS NAMES
C  AND DEVICE CODES USED ARE --
C
C  FILE NAME  DEVICE CODE  DESCRIPTION
C  ---------  -----------  -----------
C
C  INPUT       5           BASIC INPUT FILE
C  OUTPUT      6           BASIC OUTPUT FILE.
C  MASTER      7           MASTER DATA FILE CONTAINING THE BEARING
C                          INPUT DATA AND THE LAST SOLUTION VECTOR.
C  FINAL       8           LAST SOLUTION VECTOR OR INITIAL CONDITION
C                          VECTOR WHEN ICOD=1 ON RECORD 3.3.
C  SOL1       11           PLOT SOLUTIONS FOR ELEMENT NO. 1.
C  SOL2       12           PLOT SOLUTIONS FOR ELEMENT NO. 2.
C  SOL3       13           PLOT SOLUTIONS FOR ELEMENT NO. 3.
C  SOL4       14           PLOT SOLUTIONS FOR ELEMENT NO. 4.
C  SOL5       15           PLOT SOLUTIONS FOR ELEMENT NO. 5.
C  SOL6       16           PLOT SOLUTIONS FOR ELEMENT NO. 6.
C  SOL7       17           POWER DISSIPATION AND LIFE SOLUTIONS.
C
C
C  EXTERNALS CALLED
C  ================
C
C  EXCEPT FOR THE GRAPHIC SOFTWARE, ALL THE EXTERNAL SUBROUTINES
C  EITHER CALLED BY ADORE OR CALLED BY ANY OF THE SUBROUTINES ARE
C  INCLUDED IN THE ADORE SOURCE FILE.  THE FOLLOWING IS A LIST OF THE
C  VARIOUS SUBROUTINES --
C
C
C    ADRA1    INPUT DATA
C    ADRA2    INITIAL COMPUTATIONS
C    ADRA3    PRINT OUTPUT
C    ADRA4    QUASI-STATIC EQUILIBRIUM SOLUTION
C    ADRA5    INITIAL ANGULAR VELOCITIES
C
C    ADRB1    COMPUTATION OF DERIVATIVES
C    ADRB2    COORDINATE FRAME TRANSFORMATION MATRIX
C    ADRB3    MODIFIED ARCTAN
C    ADRB4    ANGULAR ACCELERATION TRANSFORMATION
C    ADRB5    ANGULAR VELOCITY TRANSFORMATION
C
C    ADRC1    ROLLING ELEMENT/RACE NORMAL LOAD
C    ADRC2    CENTRIFUGAL AND THERMAL EXPANSION
C    ADRC3    ROLLER/FLANGE INTERACTION
C    ADRC31   GEOMETRICAL FUNCTION FOR ROLLER/FLANGE CONTACT
C    ADRC4    POINT CONTACT ELASTICITY SOLUTION
C    ADRC41   GEOMETRICAL FUNCTION FOR POINT CONTACT
C    ADRC5    ROLLER/RACE DEFLECTION
C    ADRC6    LINE CONTACT ELASTICITY SOLUTION
C
C    ADRD1    ROLLING ELEMENT/RACE TRACTION
C    ADRD2    TRACTION ON ELEMENTARY AREA
```

```
C    ADRD3    HYPOTHETICAL TRACTION/SLIP RELATION                     ADORE    1139
C    ADRD4    COMPUTATION OF COEFFICIENTS FOR HYPOTHETICAL MODEL      ADORE    1140
C    ADRD41   FUNCTION FOR TRACTION COEFFICIENT COMPUTATION           ADORE    1141
C    ADRD5    ELASTOHYDRODYNAMIC TRACTION MODELS                      ADORE    1142
C    ADRD6    COMPUTATION OF LOCAL SLIP FOR ROLLING ELEMENTS          ADORE    1143
C    ADRD7    EFFECTIVE RHEOLOGICAL PARAMETERS                        ADORE    1144
C    ADRD8    THERMAL AND SIDE LEAKAGE FACTORS                        ADORE    1145
C                                                                     ADORE    1146
C    ADRE1    ROLLING ELEMENT/CAGE INTERACTION                        ADORE    1147
C    ADRE2    RACE/CAGE INTERACTION                                   ADORE    1148
C    ADRE21   GEOMETRICAL FUNCTION FOR RACE/CAGE CONTACT              ADORE    1149
C    ADRE3    HYDRODYNAMIC SOLUTIONS                                  ADORE    1150
C    ADRE4    SHORT BEARING SOLUTION                                  ADORE    1151
C    ADRE5    LONG BEARING SOLUTION                                   ADORE    1152
C    ADRE51   FUNCTION FOR EXIT COORDINATE IN LONG BEARING SOLUTION   ADORE    1153
C    ADRE6    SLIP VELOCITY FOR INTERACTING SLIDING ELEMENTS          ADORE    1154
C    ADRE7    GENERALIZED FRICTION FORCE VECTOR                       ADORE    1155
C                                                                     ADORE    1156
C    ADRF1    FATIGUE LIFE                                            ADORE    1157
C    ADRF2    CHURNING AND DRAG LOSSES                                ADORE    1158
C                                                                     ADORE    1159
C    ADRG1    RUNGA-KUTTA-FEHLBERG METHOD                             ADORE    1160
C    ADRG2    PREDICTOR-CORRECTOR METHOD                              ADORE    1161
C    ADRG21   COEFFICIENTS FOR PREDICTOR-CORRECTOR METHOD             ADORE    1162
C    ADRG3    ROOT OF A NONLINEAR FUNCTION                            ADORE    1163
C    ADRG4    VECTOR CROSS PRODUCT                                    ADORE    1164
C    ADRG5    LINEAR INTERPOLATION                                    ADORE    1165
C    ADRG6    TEN POINT GAUSSIAN QUADRATURE                           ADORE    1166
C    ADRG7    ELLIPTIC INTEGRALS                                      ADORE    1167
C                                                                     ADORE    1168
C    ADRP1    PLOT OUTPUT                                             ADORE    1169
C    ADRP2    PLOT AXIS BOUNDS                                        ADORE    1170
C                                                                     ADORE    1171
C    ADRX1    OPTIONAL APPLIED LOAD AND SPEED FUNCTIONS               ADORE    1172
C    ADRX2    OPTIONAL ROLLER/FLANGE SPRING RATE                      ADORE    1173
C    ADRX3    OPTIONAL ROLLING ELEMENT/CAGE SPRING RATE               ADORE    1174
C    ADRX4    OPTIONAL RACE/CAGE SPRING RATE                          ADORE    1175
C    ADRX5    OPTIONAL ROLLER GEOMETRY                                ADORE    1176
C    ADRX6    OPTIONAL RACE GEOMETRY                                  ADORE    1177
C                                                                     ADORE    1178
C                                                                     ADORE    1179
C  GRAPHIC SOFTWARE                                                   ADORE    1180
C  ================                                                   ADORE    1181
C                                                                     ADORE    1182
C  ADORE USES THE GRAPHICS PACKAGE DISSPLA, VERSION 8.2, TO GENERATE  ADORE    1183
C  ALL THE PLOT OUTPUT.  DISSPLA HAS BEEN DEVELOPED AND IT IS         ADORE    1184
C  SUPPORTED BY ISSCO, 4186 SORRENTO VALLEY BLVD., SAN DIEGO,         ADORE    1185
C  CALIFORNIA 92121.                                                  ADORE    1186
C                                                                     ADORE    1187
C                                                                     ADORE    1188
C  NOTES                                                              ADORE    1189
C  =====                                                              ADORE    1190
C                                                                     ADORE    1191
C  1.  THE LENGTH OF A RUN IS DEFINED EITHER BY THE MAXIMUM NUMBER    ADORE    1192
C      OF STEPS ON RECORD 1 OR THE FINAL TIME SPECIFIED ON RECORD 2,  ADORE    1193
C      WHICHEVER IS ENCOUNTERED FIRST.  FOR COMPUTATIONAL CONVENIENCE ADORE    1194
C      THE FINAL TIME AND SIZE OF TIME STEP ARE SPECIFIED IN DIMENSION- ADORE  1195
C      LESS TERMS.  SINCE THE TIME SCALE IS NOT KNOWN APRIORI TO      ADORE    1196
C      THE USER, IT WILL BE NECESSARY TO MAKE AN INITIAL RUN AND      ADORE    1197
C      IDENTIFY THE RELEVANT TIME SCALE FROM THE OUTPUT.  A REALISTIC ADORE    1198
C      VALUE FOR THE MAXIMUM STEP SIZE CAN THEN BE ESTIMATED BY       ADORE    1199
C      KNOWING THIS TIME SCALE AND THE DESIRED HIGHEST FREQUENCY IN   ADORE    1200
C      THE DYNAMIC SOLUTIONS.                                         ADORE    1201
C                                                                     ADORE    1202
C  2.  SINCE ADORE CONSIDERS ONLY A SINGLE ROW OF ROLLING ELEMENTS, THE ADORE  1203
C      TREATMENT OF SPHERICAL ROLLER BEARINGS IS RESTRICTED TO A      ADORE    1204
C      SINGLE ROW RADIALLY LOADED BEARING ONLY.  A MORE RIGOROUS      ADORE    1205
C      DEVELOPMENT FOR A TWO ROW SPHERICAL ROLLER BEARING IS ANTI-    ADORE    1206
C      CIPATED IN THE FUTURE.                                         ADORE    1207
C                                                                     ADORE    1208
C  3.  FOR A SEGMENTED CAGE, IT IS NECESSARY THAT ALL SEGMENTS BE     ADORE    1209
C      IDENTICAL TO EACH OTHER AND THE SEGMENTATION TAKES PLACE EITHER ADORE   1210
```

```
C       IN THE CENTER OF THE POCKETS OR IN THE CENTER OF THE WALL          ADORE
C       BETWEEN POCKETS.  THE GEOMETRY OF A SEGMENTED CAGE IS PRES-        ADORE
C       CRIBED SIMPLY AS IF IT WERE A FULL ONE PIECE CAGE.  SEGMENTATION   ADORE
C       IS INTRODUCED BY SPECIFYING THE NUMBER OF SEGMENTS AND THE         ADORE
C       ANGULAR WIDTH OF CUT (DEGREES) USED TO SEGMENT THE CAGE.  NO       ADORE
C       HYDRODYNAMIC EFFECTS (BOTH IN THE CAGE POCKET AND AT THE           ADORE
C       CAGE/RACE INTERACTION) MAY BE CONSIDERED WITH A SEGMENTED CAGE.    ADORE
C                                                                          ADORE
C   4.  IN LIEU OF A LUBRICANT TRACTION MODEL, TRACTION AT THE ROLLING     ADORE
C       ELEMENT/RACE INTERFACE MAY BE DEFINED BY A TRACTION-SLIP           ADORE
C       RELATION OF THE FORM --                                            ADORE
C                                                                          ADORE
C            K = (A + B*V)*EXP(C*V) + D                                    ADORE
C                                                                          ADORE
C       WHERE K IS THE TRACTION COEFFICIENT AT A SLIP VELOCITY, V          ADORE
C       (L/T), AND A, B, C, D ARE PREDETERMINED COEFFICIENTS.  NOTE        ADORE
C       THAT B AND C WILL HAVE UNITS (T/L).  THE TRACTION AT THE           ADORE
C       ROLLING ELEMENT/CAGE AND CAGE/RACE INTERACTIONS IS ALWAYS          ADORE
C       DEFINED BY SUCH A RELATION.                                        ADORE
C                                                                          ADORE
C   5.  IN ORDER TO EASE THE DETERMINATION OF THE COEFFICIENTS A, B, C,    ADORE
C       D, ADORE OFFERS TWO OPTIONS --                                     ADORE
C                                                                          ADORE
C         OPTION 1 - UNDER THIS OPTION THE COEFFICIENTS ARE COMPUTED       ADORE
C                    BY ADORE BY THE FOUR CONDITIONS (TRACTION COEFFI-     ADORE
C                    CIENT AT ZERO SLIP, MAXIMUM TRACTION COEFFICIENT,     ADORE
C                    TRACTION COEFFICIENT AT INFINITE SLIP AND THE SLIP    ADORE
C                    VELOCITY (L/T) AT WHICH THE TRACTION IS A MAXIMUM)    ADORE
C                    WHICH CAN BE EASILY EXTRACTED FROM AN EXPERIMENTAL    ADORE
C                    TRACTION-SLIP CURVE.  THIS OPTION IS TRIGGERED BY     ADORE
C                    A POSITIVE AND NONZERO VALUE OF THE FOURTH CONDI-     ADORE
C                    TION (SLIP VELOCITY).                                 ADORE
C                                                                          ADORE
C         OPTION 2 - FOR THE PURPOSE OF MODELLING VERY SIMPLE TRACTION-    ADORE
C                    SLIP RELATIONS, THE ACTUAL VALUES OF THE COEFFI-      ADORE
C                    CIENTS A, B AND C CAN BE DIRECTLY PRESCRIBED AS THE   ADORE
C                    FIRST THREE INPUTS FOR THE TRACTION CURVE UNDER       ADORE
C                    THIS OPTION.  THE FOURTH INPUT IS NOW SET EQUAL TO    ADORE
C                    THE MAXIMUM PERMISSIBLE TRACTION COEFFICIENT WITH A   ADORE
C                    NEGATIVE SIGN.  ADORE USES THE NEGATIVE SIGN TO       ADORE
C                    IDENTIFY THE DATA WITH THIS OPTION.  THE ACTUAL       ADORE
C                    VALUE OF THE COEFFICIENT D IS ZERO UNDER THIS         ADORE
C                    OPTION.                                               ADORE
C                                                                          ADORE
C   6.  FOR LUBRICANTS ON WHICH EXTENSIVE TRACTION DATA HAS BEEN GENER-    ADORE
C       ATED, IT MAY BE POSSIBLE TO FIT THE DATA TO THE TYPE OF TRACTION   ADORE
C       MODEL USED IN ADORE AND EXTRACT APPROPRIATE VISCOCITY-PRESSURE-    ADORE
C       TEMPERATURE RELATION.  UNDER SUCH A CASE THE USER MAY SET THE      ADORE
C       TRACTION CODE, KTC = 8 ON RECORD 3.2 AND PRESCRIBE APPROPRIATE     ADORE
C       LUBRICANT PARAMETERS ON RECORDS 10.4.K.                            ADORE
C                                                                          ADORE
C   7.  THE BEARING ELEMENTS (ROLLING ELEMENTS, CAGE AND THE RACES) ARE    ADORE
C       NUMBERED IN THE FOLLOWING SEQUENCE --                              ADORE
C                                                                          ADORE
C           1 TO NB      NB ROLLING ELEMENTS.                              ADORE
C        NB+1 TO NB+KS   KS CAGE SEGMENTS.                                 ADORE
C           NB+KS+1      OUTER RACE.                                       ADORE
C           NB+KS+2      INNER RACE.                                       ADORE
C                                                                          ADORE
C       FOR A CAGELESS BEARING THE OUTER AND INNER RACES ARE NUMBERED      ADORE
C       AS NB+1 AND NB+2 RESPECTIVELY.                                     ADORE
C                                                                          ADORE
C                                                                          ADORE
C    PRADEEP K. GUPTA                                                      ADORE
C                                                                          ADORE
C    VERSION 1.0    1 JUNE 1984                                            ADORE
C--------------------------------------------------------------------------ADORE
C                                                                          ADORE
      COMMON X(541),Y(541),Z(541),XT(541),YT(541),ZT(541),EX(541),         ADORE
     1FX(541),ZX(541),ZST(541,2),YST(541,8),FST(36,40),XIR(4,43),          ADORE
     2QLIN(5,40,2),TIBS(3,3,40),TCPS(3,3,80),CBS(3,2,40),CRS(3,2,40),      ADORE
     3QBF(2,2,40),ABF(2,2,40),BBF(2,2,40),QVFS(2,2,40),SVFS(2,2,40),       ADORE
```

```
4FLDS(2,2,40),ALI(2,2,40),VSBS(4,40),TRBS(4,40),HST(4,40),
5BDR(3,40),BPF(3,80),OMGBS(3,40),RPS(3,80),BC(3,40),BGCS(3,40),
6QVBS(2,40),SVBS(2,40),REA(2,40),SRS(2,40),RR(2,40),AI(2,40),
7BI(2,40),QI(2,40),DEL(2,40),BLEN(2,40),BANG(2,40),QCL(2,40),
8ACL(2,40),QBL(40,2),BPHM(80),BPSV(80),RCLS(40),WLS(40),BRAD(40),
9BRAF(40),CFR(40),GYM(40),D(40),SXX(40),PBN(2,40),SGP(20),
1CF(3,20),CM(3,20),CAI(2,20),OMCS(3,20),CRF(2,4,20),CRHM(2,20),
2CRSV(2,20),TICS(3,3,20),CSZ(3,20),CCZ(3,20),FJS(2,2,40),
3FTH(2,2,40),ZM(300),ZD(300)
 COMMON/BLK102/FL,SRAD,HRAD,DPLY,STC,R(2,2),SF(2),RCAG(2),
1CPL(2),STS(2),IBRG
 COMMON/BLK104/CCDR(2),BPR3,BPR4,ICHR
 COMMON/BLK105/CLB,ADR,CDR,CAR(3),BAR(2),CCL(2),CCR(2),
1RAI(2,2),OMGRS(3,2)
 COMMON/BLK107/CRO,CMU,OME,TE,COM(2),CR(2),CLC(2),CTR(2)
 COMMON/BLK115/NRR,NWR,PI,TOLR(3)
 COMMON/BLK117/KS,ISEG,IBS(2,40),IBP(2,40),ISG(80),NBS(40)
 COMMON/BLK119/RBSS(3),RWSS(2)
 COMMON/BLK120/NRJ,KAM,KA
 COMMON/BLK121/KLP,PQS(2),EFL,SUM,BB,DQN
 COMMON/BLK125/TIRS(3,3,2),RMCS(3,2),VRS(3,2),OMEGA(3,2)
 COMMON/BLK127/NJB(40),ISY,IB
 COMMON/BLK128/LNL,LNF,LNV,LNA,LNM,LNQ,LNP,LNS,LNW,IVER,ISPEC
 COMMON/BLK129/XC(9),XS(9),XJ(8),HL(2),HA(2),ERR,TOL,
1ISC,JSC,MS,NM,N,NQQ,IC,JC,IM,ITCM
 COMMON/BLK130/FY(6),T(3,3),KBT,KBTA,MODE,MDE
 COMMON/BLK131/DCL,AXPS,RPOS,APOS,DINT,DMIN,DMAX,TF,DMR,OMG(2,2)
 COMMON/BLK132/FCF(4,4),OMGR(2),LU
 COMMON/BLK134/ID,NRI,NWFF,NDQ,KSTEP,LSTEP,MAX,NBW,
1NPR(6),IBW(6),NWG(6),IBX(40)
 COMMON/BLK135/PQBR(2),PQBC(2),PQCR(2),LPBR,LPBC,LPCR
 COMMON/BLK136/JPT,IPT,JPLT
 COMMON/BLK137/ANG,OMS,DVEL,DACC,DMOM,DAAC,DPRS
 COMMON/BLK138/RO,TT,QO,QMX
 COMMON/BLK142/IFLNS,IFLL(2)
 COMMON/BLK153/NQ,NBB,ICOD,IROT,IMET,ICAL,IRF(3,2),IRM(2)
 COMMON/BLK154/SVB(2),QVB(2),VSB(4),TRB(4)
 COMMON/BLK158/TIC(3,3),CGC(3),CMC(3),VC(3),OMC(3)
 COMMON/BLK159/IRGS(2)
 COMMON/BLK161/WB,WC,WR,IDX(4)
 COMMON/BLK162/DF,DP,ALFA,BRN,CRN,RCUR(2),RGR(2),RGRC(2)
 COMMON/BLK163/FCL(2),RANG(2),FANG(2,2),FLH(2,2),CRAD(2),
1ERAD(2),TRFS(3,3,4),RFS(3,4),CONS(6,4),RS(4),FHLS(2,2),ARS(2,2),
2THSS(2,2),CURC(3),ECURC(2),IFLN(2,2)
 COMMON/BLK164/JPRT,JPR,IBJ,JBJ,IBT
 COMMON/BLK165/NB,ILC(2,40)
 COMMON/BLK171/GAMA(2,2),QF(2),UF(2),APL(2),RPM(2),JF(2),IFL
 COMMON/BLK173/RBS(3),RBST(3),RRS(3,2),RRST(3,2),
1CB(3,2),CRR(3,2),OMB(3),OMEGR(3,2)
 COMMON/BLK174/RJN,RJM(2),EL(2,2),EM(2,2),IP(2),JP
 COMMON/BLK175/TJ,APF(3,2),APM(3,2),ARD(3,2),OMTD(3,2),G(3)
 DIMENSION LPUS(7,2),LPU(8),T1(3,3),XRE(12),YRE(12),CRE(12,2)
 DIMENSION NPS(40),LLS(6),R1(3)
 CHARACTER*36 ISPEC,ISPC
 CHARACTER*30 ITP1,ITP2,ITP3(2),TC(3)
 CHARACTER*10 LNL,LNF,LNV,LNA,LNM,LNQ,LNP,LNS,LNW,IVER
 CHARACTER*10 JPU(8),LFOR,LSCR,IBLK
 CHARACTER*3 IEND,JEND
 EXTERNAL ADRB1
 DATA ITP1,ITP2,ITP3,IBLK/'POWER DISSIPATION AND LIFE$',
1'CAGE$','OUTER RACE$','INNER RACE$','  '/
 DATA TC,JEND/'MAXIMUM STEP COUNT','TRUNCATION LIMIT COUNT',
1'MAXIMUM TIME COUNT','END'/
 DATA LPUS/4,5,6,9,10,8,9,3,3,5,8,7,5,8/
 DATA NR,NW,NWE,NWF,NWH,NWG/5,6,7,8,17,11,12,13,14,15,16/
 DATA XRE/0.10,1.,10.,100.,1000.,1.0E+4,1.0E+5,2.0E+5,3.0E+5,
14.0E+5,5.0E+5,1.0E+6/
 DATA CRE/275.,30.,4.2,1.2,0.48,0.40,0.45,0.40,0.10,0.09,0.09,
10.09,60.,10.,3.,1.8,1.,1.2,1.2,1.2,0.90,0.65,0.30,0.30/
 DATA NPRJ,ITOLM,NZS/13,5,9/
 IVER='ADORE-1.0'
 PI=4.*ATAN(1.)
```

```
      NWR=NW
      NRR=NR
      NRI=NWE
      NWFF=NWF
C
C-----------------------------------------------------------
C     ASSIGNED CONSTANTS
C
      TOLR(1)=1.0E-10
      TOLR(2)=1.0E-07
      TOLR(3)=1.0E-06
      NM=8
      ITCM=3
      JSC=1
      LPBR=1
      LPBC=1
      LPCR=1
      KSTEP=2
C
C-----------------------------------------------------------
C
      ISC=0
      ISOL=0
      IC=0
      JC=0
      MS=0
      ITER=0
      ICAL=0
      ERR=0.
      CALL ADRB2(0.,0.,0.,T)
      DO 9 I=1,2
      DO 9 J=1,3
      RMCS(J,I)=0.
      OMEGR(J,I)=0.
      VRS(J,I)=0.
      OMEGA(J,I)=0.
      APF(J,I)=0.
      APM(J,I)=0.
      ARD(J,I)=0.
      OMTD(J,I)=0.
      DO 9 K=1,3
      TIRS(K,J,I)=T(K,J)
    9 CONTINUE
      CALL ADRA1
      IF(JPLT.GT.1) GO TO 41
      CALL ADRA2
      NQQ=NQ
      TOL=TOLR(3)
      XB1=FY(4)
      DO 299 I=1,NBW
      LLS(I)=0
      IF(IBW(I).LE.NB) GO TO 299
      KI=IBW(I)-NB
      IF(KI.GT.ISEG) GO TO 299
      JI=KI-1
      IF(JI.LE.0) GO TO 299
      LL=0
      DO 339 J=1,JI
      DO 339 K=1,NBS(J)
      LL=LL+1
  339 CONTINUE
      LLS(I)=LL
  299 CONTINUE
      IDX(1)=NB
      NZM=NZS
      IF(KBTA.GT.1) GO TO 84
      IDX(2)=IDX(1)+NB
      IF(KS.GT.0) IDX(2)=IDX(2)+KS
      GO TO 27
   84 IF(IFLNS.EQ.0) GO TO 98
      NZM=NZM+4
      IDX(1)=IDX(1)+2*NB
```

```
   98 IDX(2)=IDX(1)+2*NB
   27 IDX(3)=IDX(2)+4*ISEG
      IDX(4)=IDX(3)+NZM
      JPR=JPRT
      WB=WB*DVEL*QO
      WC=WC*DVEL*QO
      WR=WR*DVEL*QO
      APL(1)=QF(1)/QO
      APL(2)=QF(2)/QO
C
C     QUASI-STATIC COMPUTATIONS AND INITIAL CONDITIONS
C     ------------------------------------------------
C
      IQT=0
      GAMAR=GAMA(1,2)
      DO 119 I=1,2
      CALL ADRB2(0.,GAMA(1,I),GAMA(2,I),T1)
      DO 119 J=1,3
      DO 119 K=1,3
      TIRS(K,J,I)=T1(K,J)
  119 CONTINUE
      IF(MODE+1) 18,52,53
   18 MDE=-1
   52 IQT=1
      GO TO 40
   53 IF(ICOD-1) 40,37,44
   40 DCR=RWSS(2)
      IF(RCUR(1)-1.0E+6) 2,42,42
   42 DCR=0.
    2 OMC(1)=0.
      OMEGA(1,1)=OMGR(1)
      OMEGA(1,2)=OMGR(2)
      DO 159 I=1,2
      J=2*I-1
      IF(JF(I)) 48,17,48
   17 RMCS(J,2)=FCL(I)+APL(I)*FY(3)
      RMCS(I,1)=RMCS(J,2)
      RMCS(J,2)=RMCS(J,2)+(I-1)*0.50*DCL-I*0.50*DCR*GAMAR**(3-I)
      GO TO 159
   48 RMCS(J,2)=UF(I)/RO
      RMCS(J,2)=RMCS(J,2)-(2-I)*RWSS(2)
      RMCS(I,1)=RMCS(J,2)
  159 CONTINUE
      BC1=0.50*RMCS(1,1)
      BC2=0.50*RMCS(2,1)
      RMCS(1,1)=0.
      RMCS(2,1)=0.
      DCL1=0.
      IF(RWSS(2).NE.0.) GO TO 13
      DCL1=0.250*DCL
   13 DC1=GAMAR*(RBSS(2)+RBSS(3))
      DC2=GAMAR*RBSS(1)
      IF(BRAD(1)*RCUR(1)-1.0E+6) 67,66,66
   66 DC1=0.
      DC2=0.
   67 AC=0.50*(RANG(1)+RANG(2))
      DO 39 I=1,NB
      AC1=BC(3,I)
      CBCI=COS(AC1)
      BC(1,I)=BC1+DC1*CBCI
      BC(2,I)=BC2+DCL1-DC2*CBCI
      AC2=AC+GAMAR*CBCI
      AC3=-GAMAR*SIN(AC1)
      CALL ADRB2(AC1,AC2,AC3,T1)
      K=(I-1)*12+1
      X(K+7)=AC1
      X(K+8)=AC2
      X(K+9)=AC3
      DO 39 J=1,3
      DO 39 K=1,3
      TIBS(K,J,I)=T1(K,J)
   39 CONTINUE
```

```
      IFL=IFLN(2,1)+IFLN(2,2)
      KLP=LPBR
      PQS(1)=PQBR(1)
      PQS(2)=PQBR(2)
      DOM1=0.
      DOM2=0.
      IF(NDQ-1) 20,23,24
   20 IF(MODE.GE.0) GO TO 520
      NDQ=1
      OMG(1,1)=OMGR(1)
      OMG(1,2)=OMGR(2)
      GO TO 22
   24 DOM1=(OMG(2,1)-OMG(1,1))/(NDQ-1)
      DOM2=(OMG(2,2)-OMG(1,2))/(NDQ-1)
C
C     STIFFNESS COMPUTATIONS
C     ----------------------
C
   23 WRITE(NW,265) DMR,LNL,LNL,LNS,LNS
   22 JF1=JF(1)
      JF2=JF(2)
      OMEGA(1,1)=OMG(1,1)
      OMEGA(1,2)=OMG(1,2)
      CALL ADRC2
      DO 619 I=1,NDQ
      IF(I.LT.NDQ) GO TO 532
      IF(MODE.GE.0) GO TO 532
      OMEGA(1,1)=OMGR(1)
      OMEGA(1,2)=OMGR(2)
      GO TO 533
  532 OMEGA(1,1)=OMG(1,1)+(I-1)*DOM1
      OMEGA(1,2)=OMG(1,2)+(I-1)*DOM2
  533 CALL ADRC2
      OMGA=OMEGA(1,2)-OMEGA(1,1)
      OMG2=OMEGA(1,2)*OMS
      OMG1=OMEGA(1,1)*OMS
      OMEGR(3,1)=TIRS(1,3,1)*OMEGA(1,1)
      OMEGR(1,1)=TIRS(1,1,1)*OMEGA(1,1)
      OMEGR(3,2)=TIRS(1,3,2)*OMEGA(1,2)
      OMEGR(1,2)=TIRS(1,1,2)*OMEGA(1,2)
      IP(1)=1
      IP(2)=2
      IF(OMGA) 8,8,10
    8 IP(1)=2
      IP(2)=1
   10 FL=0.
      CALL ADRA4
      IF(OMGA-TOLR(1)) 510,510,21
   21 QBMX=0.
      DO 709 L=1,IB
      JJ=0
      DO 709 J=1,2
      IF(QBMX-QI(J,L)) 6,1,1
    6 QBMX=QI(J,L)
      JP=ILC(J,L)
      AB=AI(J,L)/BI(J,L)
    1 CALI=COS(ALI(J,1,L))
      SALI=SIN(ALI(J,1,L))
      Z(JJ*NB+L)=CALI
      Z((JJ+1)*NB+L)=SALI
      Z((JJ+2)*NB+L)=Y((J-1)*NB+L)
      JJ=JJ+3
  709 CONTINUE
      JP=JP+1
      IF(JP.EQ.2) GO TO 33
      IF(AB-10.) 33,7,7
    7 JP=2
   33 CALL ADRF1(FL)
      FL=FL/(60.*OMGA*OMS)
  510 JF(1)=1
      JF(2)=1
      LL=0
```

```
      DEL1=RMCS(1,2)*RO
      DEL2=RMCS(3,2)
      DO 19 J=1,NB
      FX(J)=BC(1,J)
      FX(NB+J)=BC(2,J)
      FX(2*NB+J)=X(J*12-3)
      FX(3*NB+J)=SXX(J)
      SXX(J)=1.
      DO 19 K=1,3
      DO 19 L=1,3
      LL=LL+1
      YT(LL)=TIBS(L,K,J)
   19 CONTINUE
      APL1=APL(1)
      APL2=APL(2)
      RMC3=RMCS(3,2)
      RMC2=RMCS(2,2)
      DELL=DEL2
      IF(DELL) 515,516,515
  516 DELL=ABS(RMCS(1,2))
  515 DMRR=0.010*DMR*DELL
      RMCS(3,2)=DEL2+DMRR
      IB1=IB
      IB=NB
      CALL ADRA4
      DZ=QO/(RO*DMRR)
      DELZZ=(APL(2)-APL2)*COS(FY(2))*DZ
      RMCS(3,2)=DEL2
      RMCS(2,2)=RMC2+DMRR
      CALL ADRA4
      DELYY=-APL(2)*SIN(FY(2))*DZ
      CFY=COS(FY(2))
      IF(ABS(CFY)-TOLR(1)) 31,31,32
   31 CFY=0.
   32 DELZY=(APL(2)*CFY-APL2)*DZ
      IB=IB1
      LL=0
      DO 29 J=1,NB
      BC(1,J)=FX(J)
      BC(2,J)=FX(NB+J)
      X(J*12-3)=FX(2*NB+J)
      SXX(J)=FX(3*NB+J)
      DO 29 K=1,3
      DO 29 L=1,3
      LL=LL+1
      TIBS(L,K,J)=YT(LL)
   29 CONTINUE
      RMCS(2,2)=RMC2
      DEL2=DEL2*RO
      JF(1)=JF1
      JF(2)=JF2
      APL(1)=APL1
      APL(2)=APL2
      RMCS(3,2)=RMC3
      APL1=APL1*QO
      APL2=APL2*QO
      IF(MODE.GT.0) GO TO 26
      IF(NDQ.EQ.1) GO TO 619
   26 IF(FL) 517,517,518
  517 WRITE(NW,266) I,OMG1,OMG2,DEL1,DEL2,DELZZ,DELYY
      GO TO 619
  518 WRITE(NW,266) I,OMG1,OMG2,DEL1,DEL2,DELZZ,DELYY,FL
  619 CONTINUE
      IF(MODE.GT.0) GO TO 25
      GAMAR=GAMA(1,2)-GAMA(1,1)
      WRITE(NW,100) ISPEC,LNF,APL1,LNF,APL2,GAMAR,OMG1,OMG2,LNL,DEL1,
     1LNL,DEL2,LNS,DELZZ,LNS,DELYY
      IF(FL) 25,25,28
   28 WRITE(NW,101) FL
   25 OMEGA(1,1)=OMGR(1)
      OMEGA(1,2)=OMGR(2)
  520 CALL ADRC2
```

```
      OMGA=OMEGA(1,2)-OMEGA(1,1)
      IP(1)=1
      IP(2)=2
      IF(OMGA) 16,16,513
   16 IP(1)=2
      IP(2)=1
  513 OMEGR(3,1)=TIRS(1,3,1)*OMEGA(1,1)
      OMEGR(1,1)=TIRS(1,1,1)*OMEGA(1,1)
      OMEGR(3,2)=TIRS(1,3,2)*OMEGA(1,2)
      OMEGR(1,2)=TIRS(1,1,2)*OMEGA(1,2)
      CALL ADRA4
      XCC=0.
      K=-11
C
C    SET-UP INITIAL CONDITIONS
C    -------------------------
C
      SS=RBSS(2)+RBSS(3)
      THBTA=0.
      DO 69 I=1,NB
      K=K+12
      SSI=SIN(BC(3,I))
      CSI=COS(BC(3,I))
      IF(KBTA.GT.1) GO TO 534
      X(K+8)=ADRB3(-X(K+10),X(K+12))-1.5*PI
      X(K+10)=-SQRT(X(K+10)**2+X(K+12)**2)
      X(K+11)=0.
      X(K+12)=0.
  534 OM1=X(K+10)
      OM2=X(K+11)
      OM3=X(K+12)
      SGMA=SIN(X(K+9))
      CGMA=COS(X(K+9))
      SBTA=SIN(X(K+8))
      CBTA=COS(X(K+8))
      X(K+10)=(OM1*CGMA-OM2*SGMA)/CBTA
      X(K+11)=OM1*SGMA+OM2*CGMA
      X(K+12)=(-OM1*CGMA+OM2*SGMA)*SBTA/CBTA+OM3
      CALL ADRB2(X(K+7),X(K+8),X(K+9),T1)
      DO 79 J=1,3
      R1(J)=0.
      DO 79 L=1,3
      R1(J)=R1(J)+T1(L,J)*BGCS(L,I)
   79 CONTINUE
      X(K+1)=BC(1,I)-R1(1)
      X(K+2)=BC(2,I)+R1(2)*SSI-R1(3)*CSI
      X(K+3)=BC(3,I)
      XCC=XCC+BC(1,I)
      THBTA=THBTA+X(K+6)
   69 CONTINUE
      IF(KBT.LE.0) GO TO 3
      RCC=0.50*RMCS(3,2)*(IRGS(1)+IRGS(2)-2)+RPOS
      IF(KS.NE.0) RCC=-CGC(3)+ABS(RPOS)
      THBTA=THBTA/NB
      XCC=(XCC/NB)+AXPS+XB1
      OMCI=THBTA
      DO 59 I=1,ISEG
      K=K+12
      X(K+1)=XCC
      SSGP=SIN(SGP(I))
      CSGP=COS(SGP(I))
      X(K+2)=-RCC*SSGP
      X(K+3)=RCC*CSGP
      X(K+5)=-RCC*THBTA*CSGP
      X(K+6)=-RCC*THBTA*SSGP
      X(K+7)=SGP(I)+APOS
      X(K+10)=OMCI
   59 CONTINUE
    3 FY(4)=ABS(RMCS(3,2))
      DO 539 I=1,2
      K=K+12
      IF(IROT.NE.I) GO TO 56
```

```
      X(K+5)=-FY(4)*OMGR(I)
   56 X(K+8)=GAMA(1,I)
      X(K+9)=GAMA(2,I)
      X(K+10)=OMGR(I)*COS(GAMA(2,I))/COS(GAMA(1,I))
      X(K+11)=OMGR(I)*SIN(GAMA(2,I))
      X(K+12)=-OMGR(I)*TAN(GAMA(1,I))*COS(GAMA(2,I))
  539 CONTINUE
      X(K+1)=RMCS(1,2)
      X(K+3)=RMCS(3,2)
   37 IQ=-1
      IR=-1
      IRS=0
      IBWI=0
      JPU(1)=LNL
      JPU(2)=LNF
      JPU(3)=LNV
      JPU(4)=LNA
      JPU(5)=LNQ
      JPU(6)=LNM
      JPU(7)=LNW
      JPU(8)=IBLK
      LFOR=IBLK
      DO 49 I=1,7
      LSCR=JPU(I)
      LPU(I)=LPUS(I,LU)
      IF(LPU(I).EQ.10) GO TO 49
      JPU(I)=IBLK
      L=INDEX(LSCR,'(')-1
      IF(L.LE.0) GO TO 528
      WRITE(LFOR,102) L,LPU(I)
      GO TO 530
  528 WRITE(LFOR,106) LPU(I)
  530 READ(LSCR,LFOR) JPU(I)
   49 CONTINUE
      LPU(8)=0
      NS=1
      NPS(1)=KBT
      WRITE(NWH,112) IVER,ISPEC
      WRITE(NWH,118) ITP1
      WRITE(NWH,115) NPRJ,NB,NS,ISEG,IBWI,IFLL,LPU,(NPS(J),J=1,NS)
      WRITE(NWH,111) JPU
      IF(NBW) 78,78,34
   34 DO 249 I=1,NBW
      NS=1
      NPS(1)=KBT
      NRI=NWG(I)
      WRITE(NRI,112) IVER,ISPEC
      IF(IBW(I).GT.NB) GO TO 54
      IF(IBS(2,IBW(I)).GT.0) NPR(I)=NPR(I)+3
      IF(IFLL(1).EQ.0) GO TO 523
  522 NPR(I)=NPR(I)+12
      GO TO 524
  523 IF(IFLL(2).NE.0) GO TO 522
  524 WRITE(NRI,121) IBW(I)
      GO TO 508
   54 KK=IBW(I)-NB-ISEG
      IF(KK.GT.0) GO TO 55
      KI=IBW(I)-NB
      NS=NBS(KI)
      NPSI=0
      LL=LLS(I)
      DO 229 J=1,NS
      LL=LL+1
      JL=ISG(LL)
      NPS(J)=0
      DO 229 K=1,2
      IF(IBS(K,JL).NE.KI) GO TO 229
      NPS(J)=NPS(J)+1
      NPSI=NPSI+1
  229 CONTINUE
      NPR(I)=NPR(I)+2*NPSI
      IF(ISEG.GT.1) GO TO 504
```

```
      WRITE(NRI,118) ITP2
      GO TO 508
  504 IF(KI.LE.9) WRITE(NRI,116) KI
      IF(KI.GT.9) WRITE(NRI,117) KI
      GO TO 508
   55 WRITE(NRI,118) ITP3(KK)
  508 WRITE(NRI,115) NPR(I),NB,NS,ISEG,IBW(I),IFLL,LPU,(NPS(J),J=1,NS)
      WRITE(NRI,111) JPU
  249 CONTINUE
      GO TO 78
C
C     READ INITIAL CONDITIONS FROM FILE CODE NWE
C     -----------------------------------------
C
   44 READ(NWE,103) IR,ISOL,X(1),DTI,ERR
      ZT(IDX(3)+6)=ERR
      IF(DINT.LE.0.) DINT=DTI
      READ(NWE,113) (X(J),J=2,NQ)
      READ(NWE,113) (ZM(J),J=1,IDX(4))
      IF(ISOL.EQ.0) GO TO 85
      READ(NWE,113) XS,XJ,((YST(J,K),J=1,NQ),K=1,NM)
      IC=1
   85 REWIND NWE
      NBW1=NBW+1
      IRS=IR
      DO 209 I=1,NBW1
      IF(I.GT.NBW) GO TO 87
      NRI=NWG(I)
      GO TO 91
   87 NRI=NWH
   91 READ(NRI,112) IVER,ISPC
      IF(ISPC.NE.ISPEC) GO TO 96
      READ(NRI,118) ITP2
      READ(NRI,115) NPRI,NB,NS,ISEG,IBWI,IFLL,LPU,(NPS(J),J=1,NS)
      READ(NRI,111) JPU
      IF(I.GT.NBW) GO TO 97
      NPR(I)=NPRI
   97 READ(NRI,125,END=80) IQ,Z(1)
      READ(NRI,122) (Z(J),J=1,NPRI)
      IF(IQ.NE.IR) GO TO 97
  209 CONTINUE
      IQ=IQ-1
   78 ITOL=0
      OMEGA(1,1)=OMGR(1)
      OMEGA(1,2)=OMGR(2)
      CALL ADRC2
      MDE=0
      IB=NB
      DO 179 I=1,NB
      SXX(I)=1.
  179 CONTINUE
      IF(IROT) 61,61,62
   62 FY(4)=ABS(SQRT(X((NBB-2)*12+3)**2+X((NBB-2)*12+4)**2)
     1          -SQRT(X((NBB-1)*12+3)**2+X((NBB-1)*12+4)**2))
     2          *OMGR(IROT)**2
   61 MDF=MDE
      IF(ICHR.LE.0) GO TO 76
      DO 519 I=1,12
      XRE(I)=ALOG(XRE(I))
      YRE(I)=ALOG(CRE(I,ID))
  519 CONTINUE
      RENX=X(7)*DP*DF*RO*RO/(2.*TT*CMU)
      RENX=ALOG(ABS(RENX))
      CDR=ADRG5(RENX,XRE,YRE,12)
      CDR=EXP(CDR)
   76 IBT=1
      JPRT=JPR
      DO 349 I=1,2
      DO 349 J=1,NRJ
      DO 349 K=1,5
      QLIN(K,J,I)=0.
  349 CONTINUE
```

```
      IF(MODE-1) 5,4,521
    4 JF(1)=1
      JF(2)=1
      IF(ISY.NE.0) GO TO 521
      IF(NJB(NB).EQ.NJB(NB-1)) MDF=MDF+1
  521 MDF=MDF+1
    5 CALL ADRB1
      IF(MDF.EQ.0) GO TO 77
      DO 509 I=1,NB
      K=(I-1)*12+1
      Y(K+4)=0.
      Y(K+5)=0.
  509 CONTINUE
      IF(IROT.GT.0) GO TO 77
      KI=(NBB-2)*12+1
      Y(KI+4)=0.
      Y(KI+5)=0.
      Y(KI+16)=0.
      Y(KI+17)=0.
   77 MDE=MDF
      NST=0
      DT=DINT
      HL(1)=DMIN
      HL(2)=DMAX
      HA(1)=DT
      HA(2)=DT
      IM=IMET
      IF(JPT.EQ.-2) WRITE(NW,133) ISPEC,LNQ,LNW
      IF(IM.LT.10) GO TO 43
      N=IMET-10
      IM=4
      IF(N.GT.8) N=8
      JSC=2*N
C
C     DYNAMIC SOLUTIONS
C     -----------------
C
   43 IQ=IQ+1
      IBT=0
      K=-11
      QBMX=0.
      DO 99 I=1,IDX(4)
      ZT(I)=0.
   99 CONTINUE
      TI=X(1)*TT
      BPR3=CCDR(2)
      BPR4=BPR3
      DO 699 I=1,NB
      K1=1+12*(I-1)
      JJ=0
      ZT(IDX(3)+5)=ZT(IDX(3)+5)+X(K1+6)
      RRI=OMGBS(2,I)**2+OMGBS(3,I)**2
      OMGA=SQRT(RRI+OMGBS(1,I)**2)*OMS
      RRI=SQRT(RRI)
      OMC2=OMGBS(2,I)
      IF(OMGBS(1,I).EQ.0.) GO TO 525
      OMGBS(2,I)=ADRB3(-RRI,OMGBS(1,I))*ANG-180.
      OMGBS(3,I)=ADRB3(OMC2,OMGBS(3,I))*ANG
  525 OMGBS(1,I)=OMGA
      ZT(I)=(QVBS(1,I)+QVBS(2,I))*WB
      IF(IFLNS.EQ.0) GO TO 503
      ZT(NB+I)=(QVFS(1,1,I)+QVFS(1,2,I))*WB
      ZT(2*NB+I)=(QVFS(2,1,I)+QVFS(2,2,I))*WB
      ZT(IDX(3)+10)=ZT(IDX(3)+10)+QVFS(1,1,I)*WR
      ZT(IDX(3)+11)=ZT(IDX(3)+11)+QVFS(2,1,I)*WR
      ZT(IDX(3)+12)=ZT(IDX(3)+12)+QVFS(1,2,I)*WR
      ZT(IDX(3)+13)=ZT(IDX(3)+13)+QVFS(2,2,I)*WR
      BPR4=BPR4+SVFS(1,1,I)+SVFS(2,1,I)+SVFS(1,2,I)+SVFS(2,2,I)
  503 ZT(IDX(3)+1)=ZT(IDX(3)+1)+QVBS(1,I)*WR
      ZT(IDX(3)+2)=ZT(IDX(3)+2)+QVBS(2,I)*WR
      BPR3=BPR3+BDR(3,I)
      BPR4=BPR4+BDR(3,I)
```

```
      LL=0
      DO 699 J=1,2
      IK=(J-1)*NB+I
      BPR4=BPR4+SVBS(J,I)+ABS(BPF(2,IK)*BPSV(IK))
      IF(QBMX-QI(J,I)) 45,46,46
   45 QBMX=QI(J,I)
      JP=ILC(J,I)
      AB=AI(J,I)/BI(J,I)
   46 CALI=COS(ALI(J,1,I))
      SALI=SIN(ALI(J,1,I))
      Z(JJ*NB+I)=CALI
      Z((JJ+1)*NB+I)=SALI
      Z((JJ+2)*NB+I)=SRS(J,I)
      JJ=JJ+3
      REA(J,I)=REA(J,I)*ANG
      BPF(3,(J-1)*NB+I)=BPF(3,(J-1)*NB+I)*ANG
  699 CONTINUE
      JP=JP+1
      IF(JP.EQ.2) GO TO 47
      IF(AB-10.) 47,63,63
   63 JP=2
   47 FL=0.
      OMGA1=1.
      OMGA=OMEGA(1,2)-OMEGA(1,1)
      OMGA1=OMGA
      IF(OMGA) 11,14,512
   11 OMGA=-OMGA
  512 IF(OMGA-TOLR(1)) 14,14,12
   12 CALL ADRF1(FL)
      FL=FL/(60.*OMGA*OMS)
   14 DO 259 I=1,2
      RRI=SQRT(TIRS(1,2,I)**2+TIRS(1,3,I)**2)
      RAI(1,I)=ADRB3(-RRI,TIRS(1,1,I))*ANG
      RAI(2,I)=ADRB3(TIRS(1,2,I),TIRS(1,3,I))*ANG
      RRI=OMGRS(2,I)**2+OMGRS(3,I)**2
      OMGA=OMGRS(1,I)**2+RRI
      IF(OMGA.EQ.0.) GO TO 259
      OMGA=SQRT(OMGA)*OMS
      RRI=SQRT(RRI)
      OMC2=OMGRS(2,I)
      IF(OMGRS(1,I).EQ.0.) GO TO 526
      OMGRS(2,I)=ADRB3(-RRI,OMGRS(1,I))*ANG
      OMGRS(3,I)=ADRB3(OMC2,OMGRS(3,I))*ANG
  526 OMGRS(1,I)=OMGA
  259 CONTINUE
      IF(KBT.LE.0) GO TO 86
      LL=0
      JK=IDX(1)
      DO 289 I=1,ISEG
      RRI=SQRT(TICS(1,2,I)**2+TICS(1,3,I)**2)
      CAI(1,I)=ADRB3(-RRI,TICS(1,1,I))*ANG
      CAI(2,I)=ADRB3(TICS(1,2,I),TICS(1,3,I))*ANG
      OMC2=OMCS(2,I)
      RRI=OMCS(2,I)**2+OMCS(3,I)**2
      OMGA=SQRT(OMCS(1,I)**2+RRI)
      RRI=SQRT(RRI)
      IF(OMCS(1,I).EQ.0.) GO TO 527
      OMCS(2,I)=ADRB3(-RRI,OMCS(1,I))
      OMCS(3,I)=ADRB3(OMC2,OMCS(3,I))
  527 OMCS(1,I)=OMGA*OMS
      ZT(IDX(2)+2*ISEG+I)=OMGA/OMGA1
      OBV=0.
      K1=(NB+I-1)*12+1
      RRK=X(K1+2)**2+X(K1+3)**2
      IF(RRK) 502,502,501
  501 OBV=(X(K1+2)*X(K1+6)-X(K1+3)*X(K1+5))/RRK
  502 ZT(IDX(2)+3*ISEG+I)=OBV/OMGA1
      ZT(IDX(3)+7)=ZT(IDX(3)+7)+OMGA
      ZT(IDX(3)+8)=ZT(IDX(3)+8)+OBV
      DO 109 J=1,NBS(I)
      LL=LL+1
      JL=ISG(LL)
```

```
      DO 109 L=1,2
      IF(IBS(L,JL).NE.I) GO TO 109
      JK=JK+1
      IK=NB*(L-1)+JL
      ZT(JK)=BPF(1,IK)*BPSV(IK)*WC
      ZT(IDX(3)+9)=ZT(IDX(3)+9)+ZT(JK)
  109 CONTINUE
      ZT(IDX(2)+I)=CRF(1,1,I)*CRSV(1,I)*WC
      ZT(IDX(2)+ISEG+I)=CRF(2,1,I)*CRSV(2,I)*WC
      ZT(IDX(3)+9)=ZT(IDX(3)+9)+ZT(IDX(2)+I)+ZT(IDX(2)+ISEG+I)
      DO 289 J=1,2
      BPR4=BPR4+ABS(CRF(J,2,I)*CRSV(J,I))
      CRF(J,3,I)=CRF(J,3,I)*ANG
  289 CONTINUE
      ZT(IDX(3)+7)=ZT(IDX(3)+7)/(OMGA1*ISEG)
      ZT(IDX(3)+8)=ZT(IDX(3)+8)/(OMGA1*ISEG)
      ZT(IDX(3)+9)=ZT(IDX(3)+9)/ISEG
   86 IF(BPR4.EQ.0.) GO TO 30
      BPR3=BPR3/BPR4
      BPR4=BPR4*Q0*DVEL
   30 ZT(IDX(3)+3)=FL
      ZT(IDX(3)+4)=BPR4
      ZT(IDX(3)+5)=ZT(IDX(3)+5)/(NB*OMGA1)
      ZT(IDX(3)+6)=ERR
      IF(IQ.EQ.IR) GO TO 51
      IF(IQ.GT.0) GO TO 88
      DO 309 I=1,IDX(4)
      ZD(I)=ZT(I)
      ZM(I)=ZT(I)
  309 CONTINUE
      GO TO 83
   51 DO 319 I=1,IDX(4)
      ZD(I)=ZT(I)
  319 CONTINUE
      GO TO 95
   88 DO 169 I=1,IDX(4)
      ZM(I)=(ZM(I)*(X(1)-DT)+0.50*DT*(ZT(I)+ZD(I)))/X(1)
      ZD(I)=ZT(I)
  169 CONTINUE
C
C     WRITE SOLUTIONS IN DATA FILES
C     -----------------------------
C
   83 IF((IQ-IRS).EQ.MAX) GO TO 505
      IF(MOD((IQ-IRS),LSTEP).NE.0) GO TO 95
  505 ZX(1)=APM(1,1)*DMOM
      ZX(2)=APM(1,2)*DMOM
      ZX(3)=APM(2,1)*DMOM
      ZX(4)=APM(2,2)*DMOM
      ZX(5)=APM(3,1)*DMOM
      ZX(6)=APM(3,2)*DMOM
      WRITE(NWH,225) IQ,TI,X(4)
      WRITE(NWH,222) BPR4,BPR3,FL,(ZX(J),J=1,6),ZM(1),ZM(IDX(3)+1),
     1ZM(IDX(3)+2),ZM(IDX(3)+9)
      IF(NBW) 94,95,94
   94 DO 239 I=1,NBW
      NRI=NWG(I)
      IBWI=IBW(I)
      NPRI=NPR(I)
      K1=12*(IBWI-1)+1
      IF(IBWI.GT.NB) GO TO 90
C
C     RE DATA
C     -------
C
      ANGP=X(K1+3)*ANG
      WRITE(NRI,225) IQ,TI,ANGP
      ZX(3)=Y(K1+6)*DAAC
      ZX(6)=X(K1+3)*ANG
      ZX(11)=X(K1+6)*OMS
      ZX(12)=OMGBS(1,I)
      NSB=25
```

```
      ZX(23)=BPF(1,IBWI)*QO
      IF(IBS(2,IBWI).LE.0) GO TO 511
      ZX(24)=BPF(1,NB+IBWI)*QO
      ZX(25)=BPHM(IBWI)*RO
      ZX(26)=BPHM(IBWI+NB)*RO
      ZX(27)=BPF(3,IBWI)
      ZX(28)=BPF(3,NB+IBWI)
      NSB=28
      GO TO 507
  511 ZX(24)=BPHM(IBWI)*RO
      ZX(25)=BPF(3,IBWI)
  507 DO 329 J=1,2
      ZX(J)=Y(K1+3+J)*DACC
      ZX(3+J)=X(K1+J)*RO
      ZX(6+J)=QI(J,IBWI)*QO
      ZX(8+J)=ALI(J,1,IBWI)*ANG
      ZX(12+J)=OMGBS(J+1,I)
      ZX(14+J)=Z((3*J-1)*NB+IBWI)
      ZX(16+J)=VSBS(2*J,IBWI)*RO/TT
      ZX(18+J)=SVBS(J,IBWI)*QO*DVEL
      ZX(20+J)=HST(J,IBWI)*HST(J+2,IBWI)
      IF(NPRI.LE.NSB) GO TO 329
      DO 279 K=1,2
      JK=(K-1)*6
      ZX(NSB+JK+J)=QBF(J,K,IBWI)*QO
      ZX(NSB+JK+J+2)=SVFS(J,K,IBWI)*QO*DVEL
      ZX(NSB+JK+J+4)=-FLDS(J,K,IBWI)*RO
  279 CONTINUE
  329 CONTINUE
      ZX(10)=ZX(10)-180.
      WRITE(NRI,222) (ZX(J),J=1,NPRI)
      GO TO 239
C
C     CAGE DATA
C     ---------
C
   90 KK=IBWI-NB
      IF(KBT.LE.0) GO TO 15
      KI=KK
      KK=KK-ISEG
      IF(KI.GT.ISEG) GO TO 15
      WRITE(NRI,225) IQ,TI,X(4)
      ZX(13)=Y(K1+6)*DACC
      ZX(16)=X(K1+3)*RO
      ZX(1)=ZT(IDX(2)+3*ISEG+KI)
      ZX(2)=ZT(IDX(2)+2*ISEG+KI)
      JJ=4
      DO 129 J=1,2
      ZX(2+J)=OMCS(J+1,KI)
      ZX(JJ+1)=CRF(J,1,KI)*QO
      ZX(JJ+2)=CRHM(J,KI)*RO
      ZX(JJ+3)=CRF(J,2,KI)*CRSV(J,KI)*QO*DVEL
      JJ=JJ+3
      ZX(10+J)=Y(K1+3+J)*DACC
      ZX(13+J)=X(K1+J)*RO
  129 CONTINUE
      K=0
      NPSI=NPRI/2+8
      LL=LLS(I)
      DO 139 J=1,NBS(KI)
      LL=LL+1
      JL=ISG(LL)
      DO 139 L=1,2
      IF(IBS(L,JL).NE.KI) GO TO 139
      IK=(L-1)*NB+JL
      K=K+1
      ZX(16+K)=BPF(1,IK)*QO
      ZX(NPSI+K)=BPHM(IK)/RCLS(JL)
  139 CONTINUE
      WRITE(NRI,222) (ZX(J),J=1,NPRI)
      GO TO 239
C
```

```
C    RACE DATA
C    ---------
C
   15 ZX(1)=0.
      WRITE(NRI,225) IQ,TI,X(4)
      RRK=X(K1+2)**2+X(K1+3)**2
      IF(RRK) 36,36,35
   35 ZX(1)=(X(K1+2)*X(K1+6)-X(K1+3)*X(K1+5))*OMS/RRK
   36 DO 149 J=1,3
      ZX(1+J)=OMGRS(J,KK)
      ZX(4+J)=APF(J,KK)*QO
      ZX(7+J)=APM(J,KK)*RO*QO
      ZX(10+J)=Y(K1+3+J)*DACC
      ZX(13+J)=X(K1+J)*RO
  149 CONTINUE
      WRITE(NRI,222) (ZX(J),J=1,NPRI)
  239 CONTINUE
C
C    INTEGRATION STEP
C    ----------------
C
   95 IF(IQT.EQ.1) GO TO 70
      IF((IQ-IRS).GE.MAX) GO TO 71
      IF(ITOL.GE.ITOLM) GO TO 72
      IF(X(1)-TF) 75,73,73
   71 ITER=1
      GO TO 74
   72 ITER=2
      GO TO 74
   73 ITER=3
   74 IQT=1
   70 CALL ADRA3(IQ)
      GO TO 50
   75 IF(MOD((IQ-IRS),KSTEP).NE.0) GO TO 65
      CALL ADRA3(IQ)
   65 JPRT=0
      IF(IMET.LT.10) GO TO 64
      IF(ISOL.EQ.1) GO TO 68
      NST=NST+1
      XS(NST)=X(1)
      DO 399 I=1,NQ
      YST(I,NST)=Y(I)
  399 CONTINUE
      IF(NST.LT.NM) GO TO 64
      XS(NM+1)=XS(NM)+DT
      ISOL=1
   68 CALL ADRG2(ADRB1,IE)
      DT=XS(NM+1)-XS(NM)
      GO TO 82
   64 CALL ADRG1(ADRB1,DT,IE)
   82 IF(IE.NE.0) GO TO 531
      ITOL=0
      GO TO 43
  531 IQ1=IQ+1
      WRITE(NW,123) IQ1,X(1),DT,ERR
      ITOL=ITOL+1
      GO TO 43
C
C    WRITE CLOSING RECORD
C    --------------------
C
   50 IF(MODE.LT.0) GO TO 81
      REWIND NWE
   92 READ(NWE,105) IEND
      IF(IEND.NE.JEND) GO TO 92
      WRITE(NWE,104) IQ,ISOL,X(1),DT,ERR
      WRITE(NWE,213) (X(J),J=2,NQ)
      WRITE(NWE,213) (ZM(J),J=1,IDX(4))
      IF(ISOL.EQ.0) GO TO 93
      WRITE(NWE,213) XS,XJ,((YST(J,K),J=1,NQ),K=1,NM)
   93 J1=1
      WRITE(NWF,260) J1,J1,X(1)
```

```
      DO 549 I=2,NQ,3
      J1=I
      J2=J1+2
      WRITE(NWF,260) J1,J2,(X(J),J=J1,J2)
  549 CONTINUE
      WRITE(NW,305) TC(ITER)
      WRITE(NW,132) HA,DT,ZM(IDX(3)+6),ICAL
      GO TO 38
   80 WRITE(NW,300)
      NRI=NRI-10
      WRITE(NW,130) NRI,IR,NRI,IQ
      GO TO 81
   96 WRITE(NW,300)
      NRI=NRI-10
      WRITE(NW,131) NRI,ISPEC,NRI,ISPC
      GO TO 81
C
C     PLOT DATA IF REQUIRED
C     ---------------------
C
   38 IF(JPLT.LE.0) GO TO 81
   41 CALL ADRP1
   81 STOP
  100 FORMAT(1H1//44X,33HQUASI-STATIC EQUILIBRIUM SOLUTION/
     144X,33(1H=)///20X,A36//20X,
     210HAXIAL LOAD,14X,A6,21X,1PE10.3/20X,
     311HRADIAL LOAD,13X,A6,21X,E10.3/20X,
     418HREL MISALIGNMENT-Y,7X,5H(RAD),21X,E10.3/20X,
     516HOUTER RACE SPEED,9X,5H(RPM),21X,E10.3/20X,
     516HINNER RACE SPEED,9X,5H(RPM),21X,E10.3/20X,
     720HREL AXIAL DEFLECTION,6X,A4,21X,E10.3/20X,
     821HREL RADIAL DEFLECTION,5X,A4,21X,E10.3/20X,
     919HRADIAL STIFFNESS ZZ,3X,A8,21X,E10.3/20X,
     119HRADIAL STIFFNESS YY,3X,A8,21X,E10.3)
  101 FORMAT(20X,12HFATIGUE LIFE,11X,7H(HOURS),21X,1PE10.3)
  102 FORMAT(1H(,I1,2HX,,1HA,I1,1H))
  103 FORMAT(2X,I12,I11,4E23.15)
  104 FORMAT(2X,I12,I11,1P4E23.15)
  105 FORMAT(1X,A3)
  106 FORMAT(2H(A,I1,1H))
  111 FORMAT(2X,12A10)
  112 FORMAT(2X,A10,5X,A36)
  113 FORMAT(2X,5E23.15)
  115 FORMAT(1X,I3,60I2)
  116 FORMAT(2X,16HCAGE SEGMENT NO.,I2,1H$)
  117 FORMAT(2X,16HCAGE SEGMENT NO.,I3,1H$)
  118 FORMAT(2X,A30)
  121 FORMAT(2X,19HROLLING ELEMENT NO.,I2,1H$)
  122 FORMAT(2X,7E16.7)
  123 FORMAT(//1X,31H***TRUNCATION CHECK AT STEP NO ,I6,3X,
     121HDIMENSIONLESS TIME = ,1PE12.5,3X,12HSTEP SIZE = ,E12.5,
     23X,12HMAX ERROR = ,E12.5)
  125 FORMAT(2X,I16,6E16.7)
  130 FORMAT(5X,49HINCOMPATIBLE STARTING SOLUTIONS IN FILES -MASTER-,
     19H AND -SIM,I1,1H-//5X,33HSTEP NO FOR LAST SOLUTION IN FILE,
     213H -MASTER-    =,I6/5X,33HSTEP NO FOR LAST SOLUTION IN FILE,
     35H -SIM,I1,7H-      =,I6///)
  131 FORMAT(5X,40HBEARING SPECS IN FILES -MASTER- AND -SIM,
     1I1,14H- DO NOT MATCH//5X,31HSPEC CODE IN FILE -MASTER-    = ,
     2A36/5X,22HSPEC CODE IN FILE -SIM,I1,8H-      = ,A36///)
  132 FORMAT(/10X,25HMINIMUM STEP SIZE        = ,1PE12.5/10X,
     125HMAXIMUM STEP SIZE        = ,E12.5/10X,
     225HLAST STEP SIZE           = ,E12.5/10X,
     325HMAX AVE TRUNCATION       = ,E12.5/10X,
     425HTOTAL DERIVATIVE CALLS = ,I12)
  133 FORMAT(1H1/5X,20H4. TIME STEP SUMMARY,10X,A36/5X,20(1H-)///
     141X,21(1H.),23HTIME AVERAGE PARAMETERS,21(1H.)/3X,
     24HSTEP,7X,26HTIME OUTER RACE INNER RACE,11H AV FATIGUE,
     355H        POWER RE ORBITAL CAGE OMEGA CAGE WHIRL         CAGE/
     45X,2HNO,11X,2(3X,8HROTATION),7X,4HLIFE,7X,4HLOSS,2X,
     59HVEL RATIO,2(6X,5HRATIO),11H  WEAR RATE/15X,3H(S),
     62(6X,5H(DEG)),4X,8H(HOURS) ,A10,34X,A10/)
```

```
  213 FORMAT(2X,1P5E23.15)
  222 FORMAT(2X,1P7E16.7)
  225 FORMAT(2X,I16,1P6E16.7)
  260 FORMAT(2(I4,2X),1P3E23.15)
  265 FORMAT(1H1//44X,30HBEARING STIFFNESS COMPUTATIONS/44X,30(1H-)///
     126X,23HDEFLECTION INCREMENT = ,1PE12.5,
     231H  PERCENT OF NOMINAL DEFLECTION///
     317X,23HRACE ANGULAR VELOCITIES,4X,20HRELATIVE DEFLECTIONS,
     42(3X,9HSTIFFNESS),5X,7HFATIGUE/23X,5HOUTER,7X,5HINNER,7X,
     55HAXIAL,6X,6HRADIAL,10X,2HZZ,10X,2HYY,8X,4HLIFE/
     616X,2(7X,5H(RPM)),2(8X,A4),2(4X,A8),5X,7H(HOURS)/)
  266 FORMAT(11X,I2,3H.  ,1P9E12.3)
  300 FORMAT(//5X,31HEXECUTION TERMINATING IN -DREB-///)
  305 FORMAT(//5X,19HEXECUTION COMPLETED/5X,
     125HEXIT FROM -ADORE- DUE TO ,A30//5X,
     222HSTATISTICS OF THIS RUN/5X,22(1H-))
      END


      SUBROUTINE ADRX1
C
C-----------------------------------------------------------------------
C
C    ADRX1   OPTIONAL APPLIED LOAD AND SPEED FUNCTIONS
C    -----   -----------------------------------------
C
C  ANY ARBITRARY VARIATION IN APPLIED LOADS OR RACE MOTION CAN BE
C  ACCOMPLISHED IN THIS OPTIONAL SUBROUTINE.
C
C  GENERAL INSTRUCTIONS
C  --------------------
C
C  1.  ANY READ/WRITE INSTRUCTIONS ARE EXECUTED IN THE FIRST TWO CALLS,
C      WHICH ARE MADE WITH ICM(1) EQUAL TO -1 AND 0 RESPECTIVELY.  IN
C      ALL SUBSEQUENT CALLS ICM(1)=1 AND THE CODE SHOULD BE FREE OF ANY
C      READ/WRITE INSTRUCTIONS.
C
C  2.  THE READ AND WRITE DEVICE CODES ARE NR AND NW RESPECTIVLEY.
C
C  3.  IN ORDER TO ACTIVATE THE USE OF THIS SUBROUTINE THE SWITCH
C      JCM(1) SHOULD BE SET EQUAL TO 1 IN THE FIRST CALL.
C
C  4.  THE COORDINATE FRAME (X,Y,Z) IS LOCATED AT THE BEARING CENTER
C      AND MOST QUANTITIES ARE PRESCRIBED IN THIS FRAME, WHICH IS
C      NORMALLY CONSIDERED AS AN INERTIAL FRAME OF REFERENCE (SPACE
C      FIXED).  HOWEVER, IF THE BEARING AS A WHOLE IS MOVING IN SPACE
C      THEN THE (X,Y,Z) SYSTEM WILL NOT BE FIXED IN SPACE AND IT WILL
C      BE NECESSARY TO TRANSFORM THE VARIOUS QUANTITIES APPROPRIATELY.
C
C  5.  ALL VARIABLES ARE IN DIMENSIONLESS FORM AND THE VALUES FOR THE
C      LENGTH, FORCE AND TIME SCALES ARE CONTAINED IN THE VARIABLES RO,
C      QO AND TT RESPECTIVELY.  THESE FUNDAMENTAL SCALES MUST BE USED
C      TO MAKE APPROPRIATE ASSIGNMENT TO THE VARIOUS VARIABLES.
C
C  6.  THE FOLLOWING VARIABLES ARE AVAILABLE AS INPUTS TO THIS SUB-
C      ROUTINE AND IF NECESSARY THEY MAY BE USED FOR COMPUTING THE
C      VARIOUS QUANTITIES.  MOST OF THE VARIABLES ARE VECTORS WITH
C      THREE COMPONENTS (ARRAYS WITH DIMENSION 3).  THE TWO DIMENSION-
C      AL ARRAYS (DIMENSIONED AS 3,2) CORRESPOND TO THE TWO RACES WITH
C      THE FIRST DIMENSION INDICATING THE VECTOR COMPONENT WHILE THE
C      SECOND DIMENSION DENOTING THE RACE (OUTER=1, INNER=2).
C
C         TI     CURRENT TIME AT WHICH THE VARIOUS QUANTITIES ARE TO BE
C                PRESCRIBED.
C         RMCD   POSITION VECTOR (X,Y,Z) LOCATING THE OUTER AND INNER
C                RACE MASS CENTER RELATIVE TO THE BEARING CENTER.
C         VRS    TRANSLATIONAL VELOCITY VECTORS (X,Y,Z) FOR THE RACE MASS
C                CENTERS.
C         OMGD   ANGULAR VELOCITY VECTORS OF THE RACES IN THE RACE FIXED
C                COORDINATE FRAMES.
C         PI     VALUE OF 3.14159265.
C         TOLR   VECTOR OF LENGTH THREE CONTAINING THE FOLLOWING --
```

```
C                TOLR(1) = TOLERANCE ON ABSOLUTE ZERO (1.0E-10).         ADRX1
C                TOLR(2) = TOLERANCE ON RELATIVE ZERO (1.0E-07).         ADRX1
C                TOLR(3) = TRUNCATION LIMIT AT EACH TIME STEP (1.0E-06)  ADRX1
C              THE VALUES IN PARENTHESIS INDICATE THE DEFAULT VALUES     ADRX1
C              ASSIGNED IN ADORE.  ANY OF THESE TOLERANCE VALUES MAY     ADRX1
C              BE CHANGED AT THE FIRST CALL TO ADRX1 IF DESIRED.         ADRX1
C                                                                        ADRX1
C  7.  THE SYMETRY SWITCH, ISM, MUST BE SET EQUAL TO 1 WHEN EXERCISING   ADRX1
C      ANY VARIABLE LOAD OR SPEED OPTIONS OR ACTIVATING THE ARBITRARY    ADRX1
C      GEOMETRY OPTIONS IN SUBROUTINES ADRX5 AND ADRX6.                  ADRX1
C                                                                        ADRX1
C  8.  FOR ROLLER BEARINGS, IF THE NUMBER OF GRID POINTS ALONG THE       ADRX1
C      ROLLER LENGTH FOR COMPUTING THE ROLLER/RACE INTERACTION IS        ADRX1
C      DESIRED TO BE DIFFERENT FROM THE DEFAULT VALUE OF 10, THEN THE    ADRX1
C      VARIABLE NRG MUST BE SET EQUAL TO THE NUMBER OF GRID POINTS       ADRX1
C      DESIRED (IT MUST BE AN EVEN NUMBER).                              ADRX1
C                                                                        ADRX1
C  TIME VARYING LOADS                                                    ADRX1
C  ------------------                                                    ADRX1
C                                                                        ADRX1
C  TIME VARYING LOADS ARE SIMULATED BY ASSIGNING THE APPROPRIATE         ADRX1
C  VALUES TO THE VARIABLES APF AND APM --                                ADRX1
C                                                                        ADRX1
C    APM   APPLIED FORCE VECTORS ON THE RACES IN THE X,Y,Z SYSTEM.       ADRX1
C    APM   APPLIED MOMENT VECTORS IN THE RACE FIXED FRAMES.              ADRX1
C                                                                        ADRX1
C  WHEN PRESCRIBING ARBITRARY FORCES OR MOMENTS, APPROPRIATE             ADRX1
C  COMPONENTS OF THE SWITCHES, IRF1, IRF2, IRM1 AND IRM2, MUST BE        ADRX1
C  SET TO ZERO AND THE RELEVANT MASSES AND MOMENTS OF INERTIA MUST       ADRX1
C  BE PRESCRIBED ON RECORDS 6.2.K, WITH IRXM=1 ON RECORD 3.2, OF         ADRX1
C  ADORE INPUT.                                                          ADRX1
C                                                                        ADRX1
C  ARBITRARY ACCELERATIONS                                               ADRX1
C  -----------------------                                               ADRX1
C                                                                        ADRX1
C  BOTH THE MASS CENTER AND ANGULAR ACCELERATIONS MAY BE ARBITRARILY     ADRX1
C  PRESCRIBED.                                                           ADRX1
C                                                                        ADRX1
C    ARD   MASS CENTER ACCELERATIONS OF THE OUTER AND INNER RACES IN     ADRX1
C          (X,Y,Z) COORDINATE SYSTEM.                                    ADRX1
C    OMTD  ANGULAR ACCELERATION VECTORS OF THE OUTER AND INNER RACES IN  ADRX1
C          THE RACE FIXED COORDINATE FRAMES.                             ADRX1
C    G     ACCELERATION DUE TO GRAVITY VECTOR IN THE (X,Y,Z) FRAME.      ADRX1
C                                                                        ADRX1
C  MOVING COORDINATE FRAMES                                              ADRX1
C  ------------------------                                              ADRX1
C                                                                        ADRX1
C  IF THE BEARING AS A WHOLE MOVES IN SPACE THEN THE MASS CENTER MOTION  ADRX1
C  OF THE VARIOUS ELEMENTS WILL BE CONSIDERED IN A MOVING COORDINATE     ADRX1
C  FRAME.  UNDER THIS OPTION THE SWITCH IVAR MUST BE SET EQUAL TO 1      ADRX1
C  AND THE MOTION OF THE (X,Y,Z) COORDINATE FRAME MUST BE PRESCRIBED IN  ADRX1
C  TERMS OF THE FOLLOWING VECTOR QUANTITIES --                           ADRX1
C                                                                        ADRX1
C    R     POSITION VECTOR LOCATING THE ORIGIN OF THE (X,Y,Z) FRAME OR   ADRX1
C          THE BEARING CENTER WITH RESPECT TO A SPACE FIXED POINT.       ADRX1
C    OM    ANGULAR VELOCITY OF THE (X,Y,Z) FRAME IN THE (X,Y,Z) SYSTEM.  ADRX1
C    OMT   ANGULAR ACCELERATION OF THE (X,Y,Z) FRAME IN THE (X,Y,Z)      ADRX1
C          SYSTEM.                                                       ADRX1
C    AC    ACCELERATION OF BEARING CENTER IN AN INERTIAL FRAME.          ADRX1
C                                                                        ADRX1
C----------------------------------------------------------------------- ADRX1
C                                                                        ADRX1
      COMMON/BLK100/R(3),OM(3),OMT(3),AC(3),IVAR,NRG,ISM                 ADRX1
      COMMON/BLK101/R0,Q0,TT,ICM(6),JCM(6)                               ADRX1
      COMMON/BLK115/NR,NW,PI,TOLR(3)                                     ADRX1
      COMMON/BLK170/RMCD(3,2),VRD(3,2),OMGD(3,2)                         ADRX1
      COMMON/BLK175/TI,APF(3,2),APM(3,2),ARD(3,2),OMTD(3,2),G(3)         ADRX1
      IF(ICM(1)) 1,2,3                                                   ADRX1
    1 JCM(1)=0                                                           ADRX1
      NRG=0                                                              ADRX1
      ISM=0                                                              ADRX1
      IVAR=0                                                             ADRX1
```

```
C
C--------------------------------
C    INSERT ANY INPUT DATA USING INPUT DEVICE =NR AND
C    OUTPUT DEVICE =NW
C--------------------------------
C
      RETURN
    2 CONTINUE
C
C--------------------------------
C    INSERT ANY OUTPUT TO BE DOCUMENTED WITH THE MAIN OUTPUT
C    USING THE OUTPUT DEVICE CODE =NW
C--------------------------------
C
      RETURN
    3 CONTINUE
C
C--------------------------------
C    INSERT CODING FOR APPROPRIATE MODEL
C--------------------------------
      RETURN
      END


      FUNCTION ADRX2(X)
C
C----------------------------------------------------------------------
C
C    ADRX2   OPTIONAL ROLLER/FLANGE SPRING RATE
C    -----   ------------------------------
C
C  ANY ARBITRARY LOAD-DEFLECTION RELATION FOR THE ROLLER/FLANGE INTER-
C  ACTION MAY BE PRESCRIBED IN THIS FUNCTION.
C
C    X      GEOMETRIC INTERACTION (INPUT).
C    ADRX2  CONTACT STRESS PER UNIT GEOMETRIC INTERACTION (OUTPUT).
C
C  BOTH QUANTITIES ARE DIMENSIONLESS.  THE FUNDAMENTAL SCALES FOR
C  LENGTH (RO), FORCE (QO) AND TIME (TT) SHOULD BE USED AS APPROPRIATE.
C
C  ANY READ/WRITE INSTRUCTIONS ARE CARRIED OUT IN THE FIRST TWO CALLS,
C  WHICH ARE MADE WITH ICM(2) EQUAL TO -1 AND 0 RESPECTIVELY.  USE
C  DEVICE CODES NR AND NW FOR READ AND WRITE INSTRUCTIONS RESPECTIVELY.
C  FOR ALL SUBSEQUENT CALLS (ICM(2)=1) THE CODE SHOULD BE FREE OF
C  ANY READ/WRITE INSTRUCTIONS.
C
C  IN ORDER TO ACTIVATE THE USE OF THIS FUNCTION THE SWITCH JCM(2)
C  MUST BE SET EQUAL TO 1 IN THE FIRST CALL.
C
C----------------------------------------------------------------------
C
      COMMON/BLK101/RO,QO,TT,ICM(6),JCM(6)
      COMMON/BLK115/NR,NW,PI,TOLR(3)
      ADRX2=0.
      IF(ICM(2)) 1,2,3
    1 JCM(2)=0
C
C--------------------------------
C    INSERT ANY INPUT DATA USING INPUT DEVICE =NR AND
C    OUTPUT DEVICE =NW
C--------------------------------
C
      RETURN
    2 CONTINUE
C
C--------------------------------
C    INSERT ANY OUTPUT TO BE DOCUMENTED WITH THE MAIN OUTPUT
C    USING THE OUTPUT DEVICE CODE =NW
C--------------------------------
C
      RETURN
    3 CONTINUE
```

```
C
C--------------------------------
C    INSERT CODING FOR APPROPRIATE MODEL
C--------------------------------
      RETURN
      END


      FUNCTION ADRX3(X)
C
C-----------------------------------------------------------------------
C
C    ADRX3   OPTIONAL ROLLING ELEMENT/CAGE SPRING RATE
C    -----   ----------------------------------------
C
C  THE PURPOSE OF THIS FUNCTION IS TO PRESCRIBE ANY ARBITRARY LOAD-
C  DEFLECTION RELATION FOR THE ROLLING ELEMENT/CAGE INTERACTION IN THE
C  CAGE POCKET.
C
C    X       GEOMETRIC INTERACTION (INPUT).
C    ADRX3   CONTACT STRESS PER UNIT GEOMETRIC INTERACTION (OUTPUT).
C
C  BOTH QUANTITIES ARE DIMENSIONLESS AND THE FUNDAMENTAL SCALES FOR
C  LENGTH (RO), FORCE (QO) AND TIME (TT) SHOULD BE USED AS APPROPRIATE.
C
C  ANY READ/WRITE INSTRUCTIONS MUST BE PERFORMED IN THE FIRST TWO CALLS
C  WHICH ARE MADE WITH ICM(3) EQUAL TO -1 AND 0 RESPECTIVELY.  THE
C  DEVICE CODE FOR READ AND WRITE ARE NR AND NW RESPECTIVELY.  FOR ALL
C  SUBSEQUENT CALLS (ICM(3)=1) THE CODE SHOULD BE FREE OF ANY READ/
C  WRITE INSTRUCTIONS.
C
C  THE USE OF THIS FUNCTION IS ACTIVATED BY SETTING JCM(3)=1 AT THE
C  FIRST CALL.
C
C-----------------------------------------------------------------------
C
      COMMON/BLK101/RO,QO,TT,ICM(6),JCM(6)
      COMMON/BLK115/NR,NW,PI,TOLR(3)
      ADRX3=0.
      IF(ICM(3)) 1,2,3
    1 JCM(3)=0
C
C--------------------------------
C    INSERT ANY INPUT DATA USING INPUT DEVICE =NR AND
C    OUTPUT DEVICE =NW
C--------------------------------
C
      RETURN
    2 CONTINUE
C
C--------------------------------
C    INSERT ANY OUTPUT TO BE DOCUMENTED WITH THE MAIN OUTPUT
C    USING THE OUTPUT DEVICE CODE =NW
C--------------------------------
C
      RETURN
    3 CONTINUE
C
C--------------------------------
C    INSERT CODING FOR APPROPRIATE MODEL
C--------------------------------
      RETURN
      END


      FUNCTION ADRX4(X)
C
C-----------------------------------------------------------------------
C
C    ADRX4   OPTIONAL RACE/CAGE SPRING RATE
C    -----   ------------------------------
C
```

```
C  AN ARBITRARY LOAD-DEFLECTION RELATION FOR THE RACE/CAGE INTERACTION
C  CAN BE PRESCRIBED IN THIS FUNCTION.
C
C    X      GEOMETRIC INTERACTION (INPUT).
C    ADRX4  CONTACT STRESS PER UNIT GEOMETRIC INTERACTION (OUTPUT).
C
C  BOTH QUANTITIES ARE DIMENSIONLESS AND APPROPRIATE SCALES FOR LENGTH
C  (RO), FORCE (QO) AND TIME (TT) SHOULD BE USED.
C
C  ANY READ/WRITE INSTRUCTIONS MUST BE PERFORMED IN THE FIRST TWO
C  CALLS WHICH ARE MADE WITH ICM(4) EQUAL TO -1 AND 0 RESPECTIVELY.
C  THE DEVICE CODES FOR READ AND WRITE ARE NR AND NW RESPECTIVELY.  FOR
C  ALL SUBSEQUENT CALLS (ICM(4)=1) THE CODE SHOULD BE FREE OF ANY
C  READ/WRITE INSTRUCTIONS.
C
C  THE USE OF THIS FUNCTION IS ACTIVATED BY SETTING JCM(4)=1 IN THE
C  FIRST CALL.
C
C-----------------------------------------------------------------------
C
      COMMON/BLK101/RO,QO,TT,ICM(6),JCM(6)
      COMMON/BLK115/NR,NW,PI,TOLR(3)
      ADRX4=0.
      IF(ICM(4)) 1,2,3
    1 JCM(4)=0
C
C--------------------------------
C    INSERT ANY INPUT DATA USING INPUT DEVICE =NR AND
C    OUTPUT DEVICE =NW
C--------------------------------
C
      RETURN
    2 CONTINUE
C
C--------------------------------
C    INSERT ANY OUTPUT TO BE DOCUMENTED WITH THE MAIN OUTPUT
C    USING THE OUTPUT DEVICE CODE =NW
C--------------------------------
C
      RETURN
    3 CONTINUE
C
C--------------------------------
C    INSERT CODING FOR APPROPRIATE MODEL
C--------------------------------
      RETURN
      END
```

```
      FUNCTION ADRX5(I,XX,PHI,IC)
C
C-----------------------------------------------------------------------
C
C    ADRX5   OPTIONAL ROLLER GEOMETRY
C    -----   ------------------------
C
C  FOR ROLLER BEARINGS ANY ARBITRARY VARIATION IN ROLLER RADIUS AS A
C  FUNCTION OF THE AXIAL AND ANGULAR POSITION ON THE ROLLER SURFACE
C  MAY BE SIMULATED BY THIS FUNCTION.
C
C    I       ROLLER NUMBER (INPUT).
C    XX      AXIAL POSITION RELATIVE TO ROLLER GEOMETRIC CENTER (INPUT).
C    PHI     ANGULAR POSITION (RAD) ABOUT THE ROLLER AXIS (INPUT).
C    IC      CLASS OF RADIUS VARIATION (OUTPUT) --
C              0 = ROLLER RADIUS INDEPENDENT OF PHI.
C              1 = ROLLER RADIUS IS A FUNCTION OF PHI.
C    ADRX5   ROLLER RADIUS (OUTPUT) AT COORDINATES (XX,PHI).
C
C  ALL QUANTITIES ARE DIMENSIONLESS AND THE LENGTH SCALE (RO) SHOULD
C  BE USED TO DETERMINE APPROPRIATE VALUES.
C
C  ANY READ/WRITE INSTRUCTIONS MUST BE EXECUTED IN THE FIRST TWO CALLS,
```

```
C  FOR WHICH ICM(5) IS EQUAL TO -1 AND 0 RESPECTIVELY.  THE DEVICE       ADRX5
C  CODES FOR READ AND WRITE ARE NR AND NW RESPECTIVELY.  FOR ALL SUBSE-  ADRX5
C  QUENT CALLS (ICM(5)=1) THE CODE SHOULD BE FREE OF ANY READ/WRITE INS- ADRX5
C  TRUCTION.                                                             ADRX5
C                                                                        ADRX5
C  THE USE OF THIS FUNCTION IS ACTIVATED BY SETTING JCM(5)=1 AT THE      ADRX5
C  FIRST CALL.                                                           ADRX5
C                                                                        ADRX5
C----------------------------------------------------------------------- ADRX5
C                                                                        ADRX5
      COMMON/BLK101/RO,QO,TT,ICM(6),JCM(6)                               ADRX5
      COMMON/BLK115/NR,NW,PI,TOLR(3)                                     ADRX5
      ADRX5=0.                                                           ADRX5
      IF(ICM(5)) 1,2,3                                                   ADRX5
    1 JCM(5)=0                                                           ADRX5
C                                                                        ADRX5
C-------------------------------                                         ADRX5
C     INSERT ANY INPUT DATA USING INPUT DEVICE =NR AND                   ADRX5
C     OUTPUT DEVICE =NW                                                  ADRX5
C-------------------------------                                         ADRX5
C                                                                        ADRX5
      RETURN                                                             ADRX5
    2 CONTINUE                                                           ADRX5
C                                                                        ADRX5
C-------------------------------                                         ADRX5
C     INSERT ANY OUTPUT TO BE DOCUMENTED WITH THE MAIN OUTPUT            ADRX5
C     USING THE OUTPUT DEVICE CODE =NW                                   ADRX5
C-------------------------------                                         ADRX5
C                                                                        ADRX5
      RETURN                                                             ADRX5
    3 CONTINUE                                                           ADRX5
C                                                                        ADRX5
C-------------------------------                                         ADRX5
C     INSERT CODING FOR APPROPRIATE MODEL                                ADRX5
C-------------------------------                                         ADRX5
      RETURN                                                             ADRX5
      END                                                                ADRX5
```

```
      FUNCTION ADRX6(I,XX,SI)                                            ADRX6
C                                                                        ADRX6
C----------------------------------------------------------------------- ADRX6
C                                                                        ADRX6
C     ADRX6   OPTIONAL RACE GEOMETRY                                     ADRX6
C     -----   ----------------------                                     ADRX6
C                                                                        ADRX6
C  THE FUNCTION ADRX6 MAY BE USED TO SIMULATE THE VARIATION IN RACE      ADRX6
C  GEOMETRY FOR BOTH BALL AND ROLLER BEARINGS.                           ADRX6
C                                                                        ADRX6
C     I       RACE NUMBER, OUTER=1 AND INNER=2 (INPUT).                  ADRX6
C     XX      AXIAL POSITION (INPUT) RELATIVE TO RACE GEOMETRIC CENTER   ADRX6
C             FOR CYLINDRICAL AND TAPERED ROLLER BERINGS OR THE RACE CUR- ADRX6
C             VATURE FACTOR VARIATION (OUTPUT) FOR BALL, SPHERICAL       ADRX6
C             AND SPHERICAL TAPERED ROLLER BEARINGS, AT THE ANGULAR      ADRX6
C             POSITION, SI.  THE CURVATURE FACTOR VARIATION IS DEFINED   ADRX6
C             AS AN INCREMENT TO THE NOMINAL VALUE DEFINED IN THE MAIN   ADRX6
C             INPUT TO ADORE.                                            ADRX6
C     SI      ANGULAR POSITION (RAD) ABOUT THE RACE AXIS (INPUT).        ADRX6
C     ADRX6   RACEWAY RADIUS (OUTPUT) AT THE COORDINATE (XX,SI) FOR      ADRX6
C             CYLINDRICAL AND TAPERED ROLLER BEARINGS OR DEVIATION OF    ADRX6
C             THE RADIUS OF THE CIRCLE DEFINING THE LOCUS OF THE CURVA-  ADRX6
C             TURE CENTERS OF THE RACE GROOVE (OUTPUT) FOR BALL, SPHERI- ADRX6
C             CAL AND SPHERICAL TAPERED ROLLER BEARINGS.  THE DEVIATION  ADRX6
C             OF THE RADIUS OF THE CURVATURE CENTER LOCUS IS DEFINED BY  ADRX6
C             A LENGTH INCREMENT TO THE RADIUS FOR A GEOMETRICALLY PER-  ADRX6
C             FECT RACE.                                                 ADRX6
C                                                                        ADRX6
C  ALL QUANTITIES ARE DIMENSIONLESS AND THE LENGTH SCALE (RO) SHOULD BE  ADRX6
C  USED AS APPROPRIATE.                                                  ADRX6
C                                                                        ADRX6
C  ANY READ/WRITE INSTRUCTIONS MUST BE EXECUTED AT THE FIRST TWO CALLS,  ADRX6
C  WHICH ARE MADE WITH ICM(6) EQUAL TO -1 AND 0 RESPECTIVELY.  THE       ADRX6
```

```
C  DEVICE CODES FOR READ AND WRITE ARE NR AND NW RESPECTIVELY.  FOR ALL  ADRX6
C  SUBSEQUENT CALLS (ICM(6)=1) THE CODE SHOULD BE FREE OF ANY            ADRX6
C  READ/WRITE INSTRUCTIONS.                                              ADRX6
C                                                                        ADRX6
C  THE USE OF THIS FUNCTION IS ACTIVATED BY SETTING JCM(6)=1 AT THE      ADRX6
C  FIRST CALL.                                                           ADRX6
C                                                                        ADRX6
C----------------------------------------------------------------------- ADRX6
C                                                                        ADRX6
      COMMON/BLK101/RO,QO,TT,ICM(6),JCM(6)                               ADRX6
      COMMON/BLK115/NR,NW,PI,TOLR(3)                                     ADRX6
      ADRX6=0.                                                           ADRX6
      IF(ICM(6)) 1,2,3                                                   ADRX6
    1 JCM(6)=0                                                           ADRX6
C                                                                        ADRX6
C-------------------------------                                         ADRX6
C    INSERT ANY INPUT DATA USING INPUT DEVICE =NR AND                    ADRX6
C    OUTPUT DEVICE =NW                                                   ADRX6
C-------------------------------                                         ADRX6
C                                                                        ADRX6
      RETURN                                                             ADRX6
    2 CONTINUE                                                           ADRX6
C                                                                        ADRX6
C-------------------------------                                         ADRX6
C    INSERT ANY OUTPUT TO BE DOCUMENTED WITH THE MAIN OUTPUT             ADRX6
C    USING THE OUTPUT DEVICE CODE =NW                                    ADRX6
C-------------------------------                                         ADRX6
C                                                                        ADRX6
      RETURN                                                             ADRX6
    3 CONTINUE                                                           ADRX6
C                                                                        ADRX6
C-------------------------------                                         ADRX6
C    INSERT CODING FOR APPROPRIATE MODEL                                 ADRX6
C-------------------------------                                         ADRX6
      RETURN                                                             ADRX6
      END                                                                ADRX6
```

Typical Example

The following example considers a 20 mm angular contact ball bearing operating at 120,000 rpm with a thrust load of 2,000 N and a rotating radial load of 400 N. The dynamic performance of the bearing is simulated over 100 time steps using a fifth-order Runge–Kutta–Fehlberg formula for integrating the differential equations of motion. In real time this corresponds to 16.87 msec, or 121.5° of shaft rotation.

Input data as prepared for this example

```
'REC 1',1,2,100,1,1,100,5
'REC 2',0.001,0.0005,1.,1000.
'REC 3.1','20MM BALLBRG 120KRPM ROT RAD LOAD'
'REC 3.2',2,1,0,6,1,1,1,1,3,2,0,0,0,0,0,0,0,0
'REC 3.3',0,0,7,0,2,0,0,0,2,1,7
'REC 4.1',0.020,0.042,0.005,0.055,1.25E-5,1.25E-5
'REC 4.2',290.,290.,290.,290.,290.,290.,290.
'REC 5A',0.008,0.031,24.,0.52,0.56
'REC 7.1',0.035,0.030,0.014,0.00025,0.0025,0.
'REC 7.2.1',0.035,0.0025,0.006,0.00025
'REC 7.2.2',0.035,0.0025,0.006,0.00025
'REC 7.3',0.0002,0.0001,0.,0.0001,0.
'REC 9.1',2000.,400.,0.,0.,0.,0.,0.,120000.
'REC 9.2',0.5,0.,0.,0.,120000.
'REC 10.1',0.,0.20,0.,-0.10
'REC 10.3',1.0E-07,5.0
'REC 10.5.1',0.050,0.,0.,-0.050,2.0E-07
'REC 10.5.2',0.050,0.,0.,-0.050,2.0E-07
'REC 10.7',50.,1.0
'REC 11',0.,0.,0.
```

ADORE Print and Plot Output

The various solutions are printed at step numbers 0 and 100, which respectively correspond to the beginning and end of this run. Also, the plot output is generated at the end of this run.

```
LISTING OF INPUT DATA RECORDS ---
******************************

REC 1             1     2   100      1      1   100     5
REC 2        1.0000E-03 5.0000E-04 1.0000E+00 1.0000E+03
REC 3.1     20MM BALLBRG 120KRPM ROT RAD LOAD
REC 3.2        2  1  0  6  1  1  1  1  3  2  0  0  0  0  0  0  0  0
REC 3.3        0  0  7  0  2  0  0  0  2  1  7
REC 4.1      2.0000E-02 4.2000E-02 5.0000E-03 5.5000E-02 1.2500E-05 1.2500E-05
REC 4.2      2.9000E+02 2.9000E+02 2.9000E+02 2.9000E+02 2.9000E+02 2.9000E+02 2.9000E+02
REC 5A       8.0000E-03 3.1000E-02 2.4000E+01 5.2000E-01 5.6000E-01
REC 7.1      3.5000E-02 3.0000E-02 1.4000E-02 2.5000E-04 2.5000E-03 0.
REC 7.2.1    3.5000E-02 2.5000E-03 6.0000E-03 2.5000E-04
REC 7.2.2    3.5000E-02 2.5000E-03 6.0000E-03 2.5000E-04
REC 7.3      2.0000E-04 1.0000E-04 0.         1.0000E-04 0.
REC 9.1      2.0000E+03 4.0000E+02 0.         0.         0.         0.         0.         1.2000E+05
REC 9.2      5.0000E-01 0.         0.         0.         1.2000E+05
REC 10.1     0.         2.0000E-01 0.        -1.0000E-01
REC 10.3     1.0000E-07 5.0000E+00
REC 10.5.1   5.0000E-02 0.         0.        -5.0000E-02 2.0000E-07
REC 10.5.2   5.0000E-02 0.         0.        -5.0000E-02 2.0000E-07
REC 10.7     5.0000E+01 1.0000E+00
REC 11       0.         0.         0.

INPUT FROM USER ACCESSIBLE ROUTINES ---
************************************
```

```
          AAAAAAAAA   DDDDDDDDDD    000000000   RRRRRRRRRR   EEEEEEEEEEE
         AAAAAAAAAAA  DDDDDDDDDDD  00000000000  RRRRRRRRRRR  EEEEEEEEEEE
         AA       AA    DD      DD 00        00   RR      RR EE
         AA       AA    DD      DD 00        00   RR      RR EE
         AA       AA    DD      DD 00        00   RRRRRRRRR  EEEEEEE
         AA       AA    DD      DD 00        00   RRRRRRRRR  EEEEEEE
         AAAAAAAAAAA    DD      DD 00        00   RR  RR     EE
         AAAAAAAAAAA    DD      DD 00        00   RR   RR    EE
         AA       AA  DDDDDDDDDDD  00000000000    RR    RR   EEEEEEEEEEE
         AA       AA  DDDDDDDDDD    000000000     RR     RR  EEEEEEEEEEE

         ADVANCED        DYNAMICS       OF       ROLLING        ELEMENTS
         ========        ========       ==       =======        ========


             -A REAL TIME SIMULATION OF ROLLING BEARING PERFORMANCE-
                             (VERSION ADORE-1.0 )


                                     BY
                              PRADEEP K. GUPTA

****************************************************************************************************************************
*                                                                                                                          *
*    BEARING TYPE = BALL                              PROGRAM MODE = 1       SPEC CODE = 20MM BALLBRG 120KRPM ROT RAD LOAD  *
*                                                                                                                          *
****************************************************************************************************************************


BEARING GEOMETRY
----------------

NO OF BALLS      6           OUTER CUR FAC    5.20000E-01     OUTER FIT    (M) 1.25000E-05     SHAFT OD     (M) 2.00000E-02
BALL DIA    (M) 8.00000E-03  INNER CUR FAC    5.60000E-01     INNER FIT    (M) 1.25000E-05     SHAFT ID     (M) 5.00000E-03
PITCH DIA   (M) 3.10000E-02  DIA PLAY     (M) 1.10662E-04                                      BEARING OD   (M) 4.20000E-02
CON ANGLE (DEG) 2.40000E+01  END PLAY     (M) 5.20623E-04                                      HOUSING OD   (M) 5.50000E-02

CAGE OD     (M) 3.50000E-02  CAGE WIDTH   (M) 1.40000E-02     LAND DIA     (M) 3.50000E-02     LAND WIDTH   (M) 2.50000E-03
CAGE ID     (M) 3.00000E-02  POC DIA CL   (M) 2.00000E-04                      3.50000E-02                      2.50000E-03
OUTER CLS   (M) 2.50000E-04  GUIDE LAND TYPE  1      1        LAND CLS     (M) 2.50000E-04     LAND POS     (M) 6.00000E-03
INNER CLS   (M) 2.50000E-03                                                    2.50000E-04                      6.00000E-03
```

MATERIAL PROPERTIES

	DENSITY (KGM/M**3)	ELASTIC MODULUS (PA)	POISSION-S RATIO	COEFF OF THER EXP (1/K)	HARDNESS (PA)	WEAR COEFFICIENT
SHAFT	7.75037E+03	1.99948E+11	2.50000E-01	1.17000E-05		
HOUSING	7.75037E+03	1.99948E+11	2.50000E-01	1.17000E-05		
ROLL ELE	7.75037E+03	1.99948E+11	2.50000E-01	1.17000E-05	7.84532E+09	5.00000E-06
RACE	7.75037E+03	1.99948E+11	2.50000E-01	1.17000E-05	7.84532E+09	5.00000E-06
CAGE	7.75037E+03	1.99948E+11	2.50000E-01	1.80000E-05	7.84532E+08	5.00000E-04

INERTIAL PARAMETERS

	.......MASS (KGM)	..........MOMENT OF INERTIA.......... (KGM*M**2)			.........MASS TO GEO CENTER.......... (M)		
		X-COMP	Y-COMP	Z-COMP	X-COMP	Y-COMP	Z-COMP
RE	2.07774E-03	1.32975E-08	1.32975E-08	1.32975E-08	0.	0.	0.
CAGE	2.15570E-02	5.72607E-06	3.21513E-06	3.21513E-06	0.	0.	0.

LUBRICATION PARAMETERS

	CRITICAL FILM (M)	TRAC COEFF AT ZERO SLIP	MAXIMUM TRAC COEFF	TRAC COEFF AT INF SLIP	SLIP AT MAX TRACTION (M/S)	COEFFICIENT A	COEFFICIENT B (S/M)	COEFFICIENT C (S/M)	COEFFICIENT D
RE/RACE	1.00000E-07					0.	2.00000E-01	0.	-1.00000E-01
RE/CAGE	2.00000E-07					5.00000E-02	0.	0.	-5.00000E-02
CAGE/RACE	2.00000E-07					5.00000E-02	0.	0.	-5.00000E-02

LUBRICANT CODE..... 3 MIL-L-7808 FILM THICKNESS CODE.. 2

	REF TEMP (K)	EFF DENSITY (KGM/M**3)	VISCOCITY (PA*S)	PR-VIS COEFFICIENT (1/PA)	TEMP-VIS COEFF TYPE 1 (1/K)	TEMP-VIS COEFF TYPE 2 (K)	THERMAL CONDUCTIVITY (N/S/K)	STARVATION PARAMETER	REF ROLLING VELOCITY (M/S)
LUB FILM	2.90000E+02		2.77466E-02	1.51885E-08		3.71674E+03	9.65790E-02	5.00000E+00	
TRACTION	2.90000E+02		2.45897E-01	5.22140E-09	3.95485E-02				2.54000E+01
CHURNING	2.90000E+02	5.00000E+01	2.77466E-02						

```
INITIAL OPERATING CONDITIONS
----------------------------
                                                                       OUTER RACE    INNER RACE

APPL AXIAL LOAD     (N)  2.00000E+03      ANGULAR VELOCITY      (RPM)  0.            1.20000E+05
APPL RADIAL LOAD    (N)  4.00000E+02      TEMPERATURE             (K)  2.90000E+02   2.90000E+02
SHAFT TEMP          (K)  2.90000E+02      MISALIGNMENT-Y        (RAD)  0.            0.
HOUSING TEMP        (K)  2.90000E+02      MISALIGNMENT-Z        (RAD)  0.            0.
ROLL ELE TEMP       (K)  2.90000E+02      TRANSLATIONAL CONSTRAINTS      1   1   1     1   1   1
ROOM TEMP           (K)  2.90000E+02      ROTATIONAL CONSTRAINTS         1             1
                                          ORBITING RACE                                XXX

CAGE POS...AXIAL    (M)  0.               GRAVITY VECTOR     (M/S**2)  0.            0.            0.
       ...RADIAL    (M)  1.00000E-04      CAGE TEMPERATURE        (K)  2.90000E+02
       ..ANGULAR (RAD)  0.

SCALE FACTORS AND OUTPUT CONTROLS
---------------------------------

       ......SCALE FACTORS.......         ......STEP SIZE INFO......         NO OF STEPS    100
       LENGTH    (M)  4.00000E-03         MINIMUM        5.00000E-04         DATA CONTROL     2    100
       LOAD      (N)  2.00000E+03         MAXIMUM        1.00000E+00         AUTO PLOTS       1      7      0
       TIME      (S)  6.44630E-05         INITIAL        1.00000E-03                          0      0      0
                                                                             INT METHOD       5

OUTPUT FROM USER ACCESSIBLE ROUTINES ---
************************************
```

BEARING STIFFNESS COMPUTATIONS

DEFLECTION INCREMENT = 5.00000E-01 PERCENT OF NOMINAL DEFLECTION

	RACE ANGULAR VELOCITIES OUTER (RPM)	INNER (RPM)	RELATIVE DEFLECTIONS AXIAL (M)	RADIAL (M)	STIFFNESS ZZ (N/M)	STIFFNESS YY (N/M)	FATIGUE LIFE (HOURS)
1.	0.	0.	3.085E-05	3.240E-06	1.233E+08	1.234E+08	
2.	0.	2.000E+04	3.059E-05	3.237E-06	1.234E+08	1.236E+08	1.466E+01
3.	0.	4.000E+04	2.963E-05	3.266E-06	1.223E+08	1.225E+08	7.599E+00
4.	0.	6.000E+04	2.760E-05	3.399E-06	1.173E+08	1.177E+08	5.304E+00
5.	0.	8.000E+04	2.430E-05	3.647E-06	1.093E+08	1.097E+08	4.144E+00
6.	0.	1.000E+05	1.991E-05	3.931E-06	1.016E+08	1.018E+08	3.389E+00
7.	0.	1.200E+05	1.488E-05	4.147E-06	9.655E+07	9.646E+07	2.803E+00

STEP NO	TAU	TIME (S)	OUTER RACE ROT (DEG)	INNER RACE ROT (DEG)	
0	0.	0.	0.	0.	20MM BALLBRG 120KRPM ROT RAD LOAD

1. ROLLING ELEMENT PARAMETERS

RE NO	...ORBITAL POSITION (DEG)	....CONTACT ANGLE.... (DEG) OUTER RACE	INNER RACE	....CONTACT LOAD..... (N) OUTER RACE	INNER RACE	...CONTACT STRESS.... (PA) OUTER RACE	INNER RACE	..MAJOR HALF WIDTH... (M) OUTER RACE	INNER RACE	..MINOR HALF WIDTH... (M) OUTER RACE	INNER RACE
1	0.	1.463E+01	2.787E+01	1.578E+03	8.527E+02	2.323E+09	2.942E+09	1.517E-03	8.375E-04	2.137E-04	1.652E-04
3	1.200E+02	1.270E+01	2.883E+01	1.368E+03	6.236E+02	2.214E+09	2.648E+09	1.447E-03	7.543E-04	2.040E-04	1.491E-04
5	2.400E+02	1.270E+01	2.883E+01	1.368E+03	6.236E+02	2.214E+09	2.648E+09	1.447E-03	7.543E-04	2.040E-04	1.491E-04

```
RE ...ORBITAL ........ANGULAR VELOCITY........ ...RE ANG POSITION... ......SPIN/ROLL...... ....CONTACT LOSS..... ..TIME AVE
     VELOCITY  AMPLITUDE      THETA        PHI      THETA        PHI                                  (N*M/S)        WEAR RATE
        (RPM)      (RPM)      (DEG)      (DEG)      (DEG)      (DEG) OUTER RACE INNER RACE OUTER RACE INNER RACE   (M**3/S)

 1  4.675E+04  2.288E+05 -1.167E+01  0.          1.167E+01  1.800E+02  2.657E-14  3.447E-01  1.089E+02  2.214E+02  2.789E-12
 3  4.721E+04  2.311E+05 -1.013E+01  0.          1.013E+01  1.800E+02  6.762E-15  3.864E-01  8.116E+01  1.491E+02  2.104E-12
 5  4.721E+04  2.311E+05 -1.013E+01  0.          1.013E+01  1.800E+02  6.762E-15  3.864E-01  8.116E+01  1.491E+02  2.104E-12


RE ....SLIP VELOCITY.... .....TRAC COEFF...... ....ISO LUB FILM..... ...THERMAL RED FAC... ......DRAG ..CHUR MOM ..NET LOSS
NO          (M/S)                                       (M)                                       (N)      (N*M)     (N*M/S)
   OUTER RACE INNER RACE OUTER RACE INNER RACE OUTER RACE INNER RACE OUTER RACE INNER RACE                       (DRAG+CHUR)

 1 -6.560E-01 -5.211E-02 -1.000E-01 -1.000E-01  2.838E-06  2.289E-06  1.020E-01  9.090E-02  2.081E+00  2.163E-03  2.100E+02
 3 -6.013E-01  2.606E-02 -1.000E-01  8.254E-02  2.883E-06  2.328E-06  1.034E-01  9.883E-02  2.123E+00  2.197E-03  2.161E+02
 5 -6.013E-01  2.606E-02 -1.000E-01  8.254E-02  2.883E-06  2.328E-06  1.034E-01  9.883E-02  2.123E+00  2.197E-03  2.161E+02

STEP         TAU             TIME          OUTER RACE ROT  INNER RACE ROT
 NO                           (S)               (DEG)           (DEG)
   0       0.              0.              0.              0.              20MM BALLBRG 120KRPM ROT RAD LOAD
====      ==========      ==========      ==========      ==========      ====================================

2. RACE AND CAGE PARAMETERS
---------------------------


          RE .................RE/CAGE INTERACTION..................
          NO    GEO INT    CONTACT    CONTACT    CONTACT   TIME AVE
                             FORCE      ANGLE       LOSS  WEAR RATE
                    (M)        (N)      (DEG)    (N*M/S)   (M**3/S)

           1  9.596E-05  0.          2.700E+02  0.         0.
           2  1.337E-05  7.472E+00   1.814E+02  7.245E+01  4.576E-10
           3  1.337E-05  7.524E+00   1.787E+02  7.362E+01  4.642E-10
           4  9.577E-05  6.782E-14   9.000E+01  1.999E-25  3.404E-37
           5  1.337E-05  7.524E+00   1.340E+00  7.362E+01  4.642E-10
           6  1.337E-05  7.472E+00   3.586E+02  7.245E+01  4.576E-10
```

LAND NO	............RACE/CAGE (N) NORMAL	(N) TRACTION	FORCES............ (DEG) CON ANGLE	(DEG) ATT ANGLE	RACE/CAGE GEO INT (M)	RACE/CAGE SLIP VEL (M/S)	EFFECTIVE DIA PLAY (M)	CONTACT LOSS (N*M/S)	TIME AVE WEAR RATE (M**3/S)
1	3.376E+01	1.113E+01	3.600E+02	1.551E+02	1.617E-05	8.626E+01	2.323E-04	9.602E+02	1.856E-09
2	3.376E+01	1.113E+01	3.600E+02	1.551E+02	1.617E-05	8.626E+01	2.323E-04	9.602E+02	1.856E-09

	......MASS AXIAL (M)	CENTER POSITION..... RADIAL (M)	ORBITAL (DEG)	...ORBITAL VELOCITY (RPM)	.......ANGULAR AMPLITUDE (RPM)	VELOCITY........ THETA (DEG)	PHI (DEG)	....ANG POSITION..... THETA (DEG)	PHI (DEG)	......HOOP STRESS (PA)	..TIME AVE WEAR RATE (M**3/S)
CAGE	-2.475E-05	1.000E-04	0.	4.706E+04	4.706E+04	0.	0.	0.	0.	5.477E+07	1.844E-09
ORACE	0.	0.	0.	0.	0.	0.	0.	0.	0.	-4.997E+07	4.918E-12
IRACE	1.488E-05	4.147E-06	0.	1.200E+05	1.200E+05	0.	0.	0.	0.	1.540E+08	9.077E-12

STEP NO	TAU	TIME (S)	OUTER RACE ROT (DEG)	INNER RACE ROT (DEG)	
0	0.	0.	0.	0.	20MM BALLBRG 120KRPM ROT RAD LOAD
====	==========	==========	==========	==========	====================================

3. APPLIED PARAMETERS

	........APPLIED (N) COMP-X	FORCES........ (N) COMP-Y	(N) COMP-Z	.......APPLIED (N*M) COMP-X	MOMENTS........ (N*M) COMP-Y	(N*M) COMP-Z
CAGE	-4.012E-01	-3.103E+01	-8.721E+01	-1.532E+00	6.905E-03	0.
ORACE	2.000E+03	2.960E+01	4.290E+02	-4.171E-01	2.914E+00	3.945E-03
IRACE	-1.999E+03	-1.482E+00	-4.000E+02	1.378E-01	-3.027E+00	-1.207E-02

BASIC FATIGUE LIFE	(HOURS)	2.803E+00
INTERNAL CLEARANCE	(M)	7.532E-05
OUTER RACE FIT	(M)	6.250E-06
INNER RACE FIT	(M)	3.097E-08
TOTAL POWER LOSS	(N*M/S)	1.065E+04
CHURNING LOSS FRACTION		6.435E-01

4. TIME STEP SUMMARY

STEP NO	TIME (S)	OUTER RACE ROTATION (DEG)	INNER RACE ROTATION (DEG)	....................TIME FATIGUE LIFE (HOURS)	POWER LOSS (N*M/S)	AVERAGE RE ORBITAL VEL RATIO	PARAMETERS.... CAGE OMEGA RATIO	CAGE WHIRL RATIO	.... CAGE WEAR RATE (M**3/S)
0	0.	0.	0.	2.803E+00	1.065E+04	3.921E-01	3.921E-01	3.921E-01	5.555E-09

```
***TRUNCATION CHECK AT STEP NO      1   DIMENSIONLESS TIME =  5.00000E-04   STEP SIZE =  5.00000E-04   MAX ERROR =  8.15326E-05

   STEP          TAU            TIME       OUTER RACE ROT  INNER RACE ROT
    NO                           (S)            (DEG)           (DEG)
   100        2.617E+00      1.687E-04      0.              1.215E+02      20MM BALLBRG 120KRPM ROT RAD LOAD
   ====       ==========     ==========     ==========      ==========     ==================================


   1. ROLLING ELEMENT PARAMETERS
   ----------------------------


   RE ...ORBITAL ....CONTACT ANGLE.... ....CONTACT LOAD..... ...CONTACT STRESS.... ..MAJOR HALF WIDTH... ..MINOR HALF WIDTH...
   NO   POSITION         (DEG)                   (N)                   (PA)                  (M)                   (M)
           (DEG) OUTER RACE INNER RACE OUTER RACE INNER RACE OUTER RACE INNER RACE OUTER RACE INNER RACE OUTER RACE INNER RACE

    1  4.724E+01  1.382E+01  2.829E+01  1.460E+03  7.355E+02  2.263E+09  2.799E+09  1.478E-03  7.971E-04  2.083E-04  1.574E-04
    3  1.675E+02  1.432E+01  2.804E+01  1.525E+03  8.023E+02  2.297E+09  2.883E+09  1.500E-03  8.206E-04  2.113E-04  1.619E-04
    5  2.878E+02  1.209E+01  2.914E+01  1.305E+03  5.613E+02  2.179E+09  2.556E+09  1.424E-03  7.283E-04  2.008E-04  1.440E-04


   RE ...ORBITAL .......ANGULAR VELOCITY........ ...RE ANG POSITION... ......SPIN/ROLL...... ....CONTACT LOSS..... ..TIME AVE
        VELOCITY  AMPLITUDE      THETA        PHI      THETA        PHI                                (N*M/S)       WEAR RATE
           (RPM)      (RPM)      (DEG)      (DEG)      (DEG)      (DEG) OUTER RACE INNER RACE OUTER RACE INNER RACE  (M**3/S)

    1  4.666E+04  2.281E+05 -1.201E+01  3.589E+02  1.032E+01  1.900E+02 -2.166E-02  3.482E-01  1.081E+02  1.787E+02  2.731E-12
    3  4.665E+04  2.284E+05 -1.065E+01  3.590E+02  9.113E+00  1.909E+02  1.687E-02  3.634E-01  1.126E+02  2.060E+02  2.490E-12
    5  4.713E+04  2.306E+05 -1.020E+01  3.587E+02  8.758E+00  1.897E+02 -1.212E-02  3.916E-01  8.698E+01  1.288E+02  2.039E-12


   RE ....SLIP VELOCITY.... .....TRAC COEFF...... ....ISO LUB FILM..... ...THERMAL RED FAC... ......DRAG ..CHUR MOM ..NET LOSS
   NO         (M/S)                                         (M)                                         (N)      (N*M)    (N*M/S)
      OUTER RACE INNER RACE OUTER RACE INNER RACE OUTER RACE INNER RACE OUTER RACE INNER RACE                   (DRAG+CHUR)

    1 -5.672E-01 -5.495E-01 -8.283E-02 -8.068E-02  2.848E-06  2.309E-06  1.037E-01  9.454E-02  2.074E+00  2.154E-03  2.087E+02
    3 -6.244E-01 -9.074E-01 -8.879E-02 -9.465E-02  2.840E-06  2.293E-06  1.029E-01  9.268E-02  2.073E+00  2.159E-03  2.088E+02
    5 -5.522E-01 -4.657E-01 -8.170E-02 -7.708E-02  2.889E-06  2.341E-06  1.044E-01  1.012E-01  2.116E+00  2.190E-03  2.149E+02
```

STEP NO	TAU	TIME (S)	OUTER RACE ROT (DEG)	INNER RACE ROT (DEG)	
100	2.617E+00	1.687E-04	0.	1.215E+02	20MM BALLBRG 120KRPM ROT RAD LOAD

2. RACE AND CAGE PARAMETERS

RE NO	RE/CAGE INTERACTION GEO INT (M)	CONTACT FORCE (N)	CONTACT ANGLE (DEG)	CONTACT LOSS (N*M/S)	TIME AVE WEAR RATE (M**3/S)
1	3.296E-05	3.203E+00	3.591E+02	5.054E+01	1.025E-10
2	6.907E-05	1.306E+00	1.870E+02	3.472E+01	1.827E-10
3	1.104E-05	8.677E+00	1.822E+02	7.646E+01	9.553E-10
4	5.672E-06	1.415E+01	1.802E+02	9.299E+01	1.665E-09
5	8.870E-05	8.983E-01	1.640E+02	2.913E+01	1.547E-10
6	3.020E-05	3.533E+00	2.071E+00	5.316E+01	5.295E-10

LAND NO	RACE/CAGE FORCES (N) NORMAL	(N) TRACTION	(DEG) CON ANGLE	(DEG) ATT ANGLE	RACE/CAGE GEO INT (M)	RACE/CAGE SLIP VEL (M/S)	EFFECTIVE DIA PLAY (M)	CONTACT LOSS (N*M/S)	TIME AVE WEAR RATE (M**3/S)
1	1.494E+01	9.272E+00	5.563E+01	1.490E+02	2.385E-05	8.572E+01	2.324E-04	7.948E+02	1.236E-09
2	1.554E+01	9.350E+00	5.569E+01	1.494E+02	2.340E-05	8.572E+01	2.324E-04	8.015E+02	1.253E-09

	MASS CENTER POSITION AXIAL (M)	RADIAL (M)	ORBITAL (DEG)	ORBITAL VELOCITY (RPM)	ANGULAR VELOCITY AMPLITUDE (RPM)	THETA (DEG)	PHI (DEG)	ANG POSITION THETA (DEG)	PHI (DEG)	HOOP STRESS (PA)	TIME AVE WEAR RATE (M**3/S)
CAGE	-2.502E-05	9.259E-05	5.566E+01	4.766E+04	4.676E+04	9.192E-05	6.183E+00	2.724E-03	2.445E+02	5.409E+07	3.590E-09
ORACE	0.	0.	0.	0.	0.	0.	0.	0.	0.	-4.997E+07	5.408E-12
IRACE	1.488E-05	4.147E-06	1.215E+02	1.200E+05	1.200E+05	0.	0.	0.	0.	1.540E+08	9.130E-12

```
STEP         TAU            TIME       OUTER RACE ROT  INNER RACE ROT
 NO                          (S)           (DEG)           (DEG)
 100      2.617E+00      1.687E-04      0.             1.215E+02       20MM BALLBRG 120KRPM ROT RAD LOAD
====     ==========     ==========     ==========     ==========       ===================================


3. APPLIED PARAMETERS
---------------------
               .........APPLIED FORCES......... ........APPLIED MOMENTS.........  BASIC FATIGUE LIFE    (HOURS)  2.748E+00
                      (N)        (N)        (N)      (N*M)      (N*M)      (N*M)  INTERNAL CLEARANCE        (M)  7.532E-05
                   COMP-X     COMP-Y     COMP-Z     COMP-X     COMP-Y     COMP-Z  OUTER RACE FIT            (M)  6.250E-06
                                                                                  INNER RACE FIT            (M)  3.097E-08
      CAGE    -2.608E-01  3.736E+01 -2.028E+01 -1.178E+00  7.419E-04  1.181E-02
     ORACE     2.378E+03 -3.485E+02 -1.589E+02  9.895E-03 -1.290E+00  2.659E+00  TOTAL POWER LOSS      (N*M/S)  1.034E+04
     IRACE    -2.045E+03  3.247E+02  1.939E+02 -3.685E-01 -2.983E+00  8.396E-02  CHURNING LOSS FRACTION         6.534E-01


4. TIME STEP SUMMARY
--------------------

                                      .....................TIME AVERAGE PARAMETERS.....................
STEP       TIME OUTER RACE INNER RACE    FATIGUE      POWER RE ORBITAL CAGE OMEGA CAGE WHIRL       CAGE
 NO               ROTATION   ROTATION       LIFE       LOSS  VEL RATIO      RATIO      RATIO  WEAR RATE
            (S)      (DEG)      (DEG)    (HOURS)    (N*M/S)                                   (M**3/S)

 100  1.687E-04  0.         1.215E+02  2.743E+00  1.063E+04  3.910E-01  3.906E-01  4.583E-01  6.079E-09


EXECUTION COMPLETED
EXIT FROM -ADORE- DUE TO MAXIMUM STEP COUNT

STATISTICS OF THIS RUN
----------------------

     MINIMUM STEP SIZE      =  5.00000E-04
     MAXIMUM STEP SIZE      =  4.63430E-02
     LAST STEP SIZE         =  1.62492E-02
     MAX AVE TRUNCATION     =  4.11922E-07
     TOTAL DERIVATIVE CALLS =          836
```

```
OUTPUT FROM PLOT ROUTINE
------------------------

DATA FILE CODE = 17      PLOT NO.  1 COMPLETE
DATA FILE CODE = 17      PLOT NO.  2 COMPLETE
DATA FILE CODE = 17      PLOT NO.  3 COMPLETE
DATA FILE CODE = 11      PLOT NO.  1 COMPLETE
DATA FILE CODE = 11      PLOT NO.  2 COMPLETE
DATA FILE CODE = 11      PLOT NO.  3 COMPLETE
DATA FILE CODE = 11      PLOT NO.  4 COMPLETE
DATA FILE CODE = 11      PLOT NO.  5 COMPLETE
DATA FILE CODE = 11      PLOT NO.  6 COMPLETE
DATA FILE CODE = 12      PLOT NO.  1 COMPLETE
DATA FILE CODE = 12      PLOT NO.  2 COMPLETE
DATA FILE CODE = 12      PLOT NO.  3 COMPLETE
DATA FILE CODE = 12      PLOT NO.  4 COMPLETE
DATA FILE CODE = 12      PLOT NO.  5 COMPLETE
DATA FILE CODE = 12      PLOT NO.  6 COMPLETE
DATA FILE CODE = 12      PLOT NO.  7 COMPLETE
DATA FILE CODE = 12      PLOT NO.  8 COMPLETE
DATA FILE CODE = 12      PLOT NO.  9 COMPLETE
```

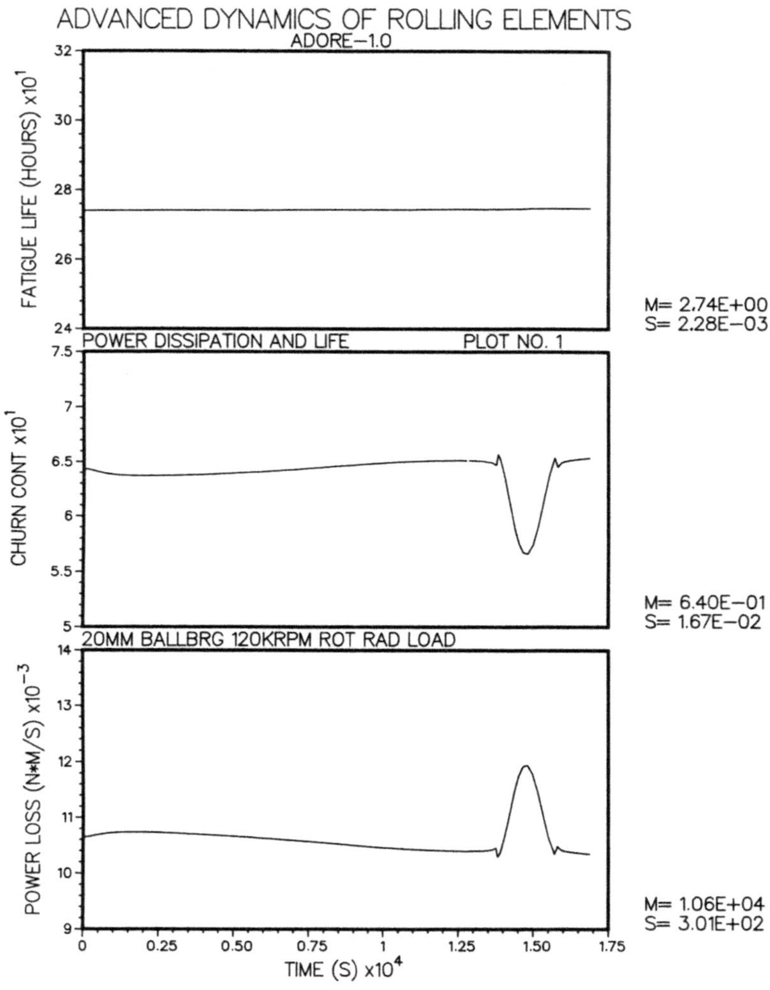
ADVANCED DYNAMICS OF ROLLING ELEMENTS
ADORE-1.0
FATIGUE LIFE (HOURS) x10^1
32
30
28
26
24
M= 2.74E+00
S= 2.28E-03
POWER DISSIPATION AND LIFE
PLOT NO. 1
CHURN CONT x10^1
7.5
7
6.5
6
5.5
5
M= 6.40E-01
S= 1.67E-02
20MM BALLBRG 120KRPM ROT RAD LOAD
POWER LOSS (N*M/S) x10^-3
14
13
12
11
10
9
M= 1.06E+04
S= 3.01E+02
0
0.25
0.50
0.75
1
1.25
1.50
1.75
TIME (S) x10^4

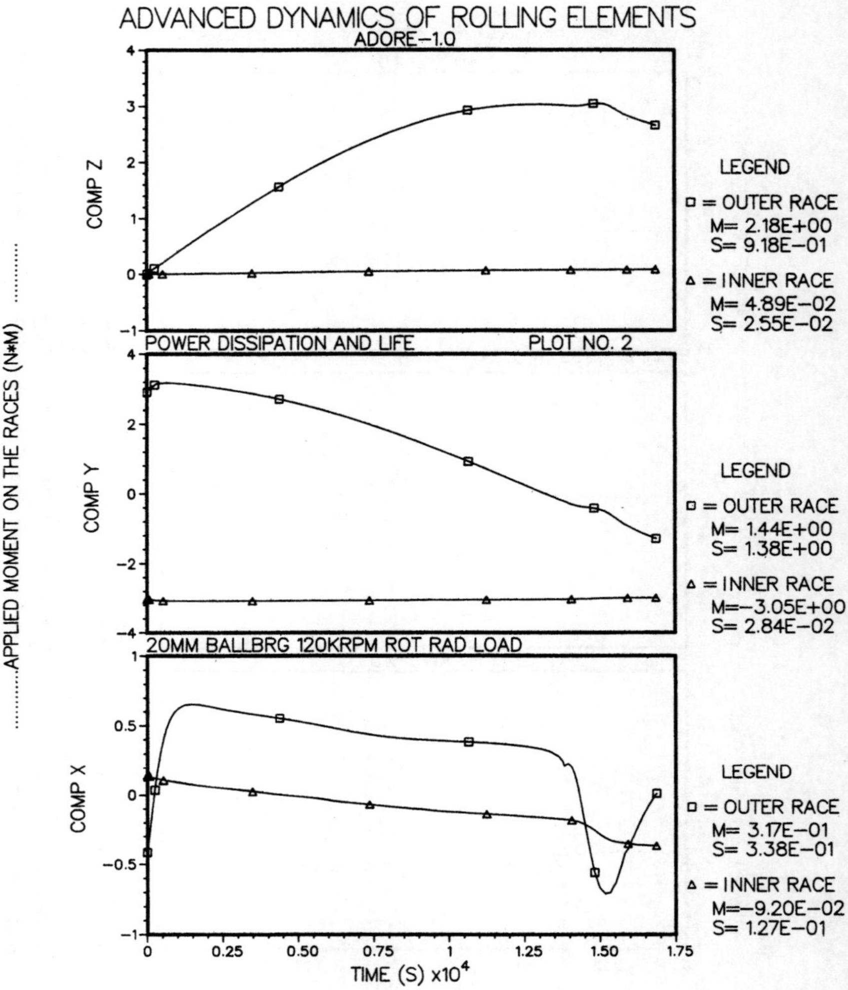
ADVANCED DYNAMICS OF ROLLING ELEMENTS
ADORE-1.0
POWER DISSIPATION AND LIFE
PLOT NO. 2
20MM BALLBRG 120KRPM ROT RAD LOAD
........APPLIED MOMENT ON THE RACES (N*M) ....
COMP Z
COMP Y
COMP X
TIME (S) x10^4
0 0.25 0.50 0.75 1 1.25 1.50 1.75
LEGEND
□ = OUTER RACE
M= 2.18E+00
S= 9.18E−01
Δ = INNER RACE
M= 4.89E−02
S= 2.55E−02
LEGEND
□ = OUTER RACE
M= 1.44E+00
S= 1.38E+00
Δ = INNER RACE
M=−3.05E+00
S= 2.84E−02
LEGEND
□ = OUTER RACE
M= 3.17E−01
S= 3.38E−01
Δ = INNER RACE
M=−9.20E−02
S= 1.27E−01

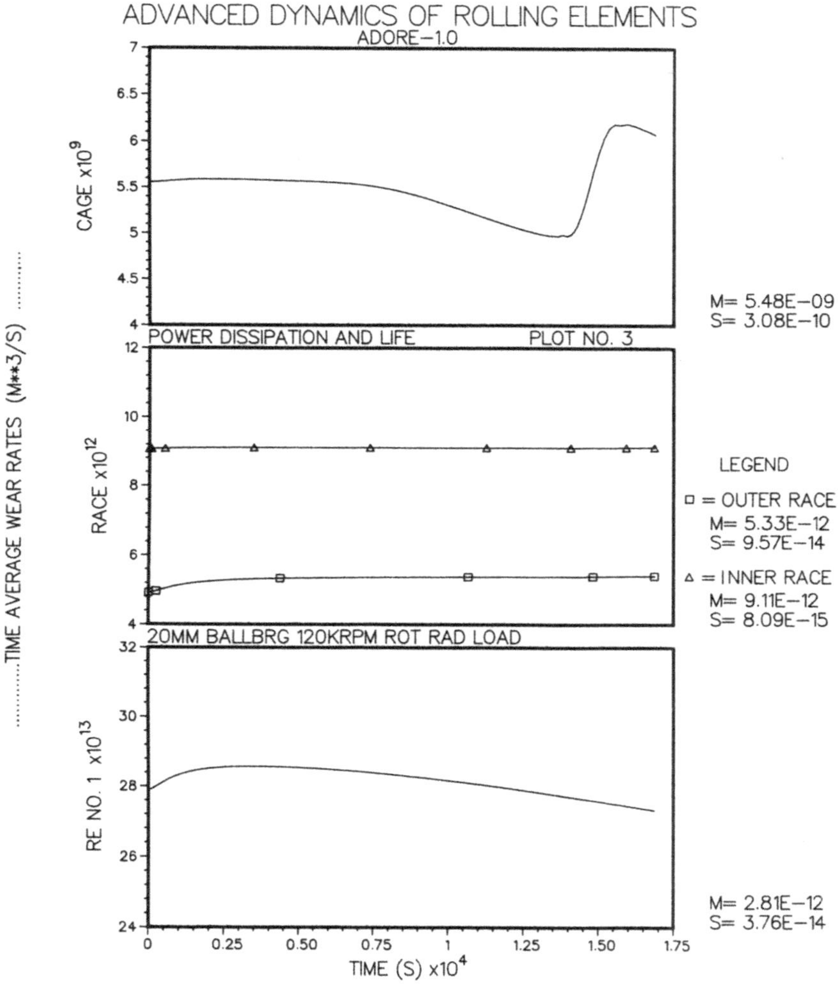
ADVANCED DYNAMICS OF ROLLING ELEMENTS
ADORE-1.0
CAGE x10^9
M= 5.48E-09
S= 3.08E-10
POWER DISSIPATION AND LIFE
PLOT NO. 3
RACE x10^12
LEGEND
□ = OUTER RACE
M= 5.33E-12
S= 9.57E-14
△ = INNER RACE
M= 9.11E-12
S= 8.09E-15
20MM BALLBRG 120KRPM ROT RAD LOAD
RE NO. 1 x10^13
M= 2.81E-12
S= 3.76E-14
TIME (S) x10^4
TIME AVERAGE WEAR RATES (M**3/S)

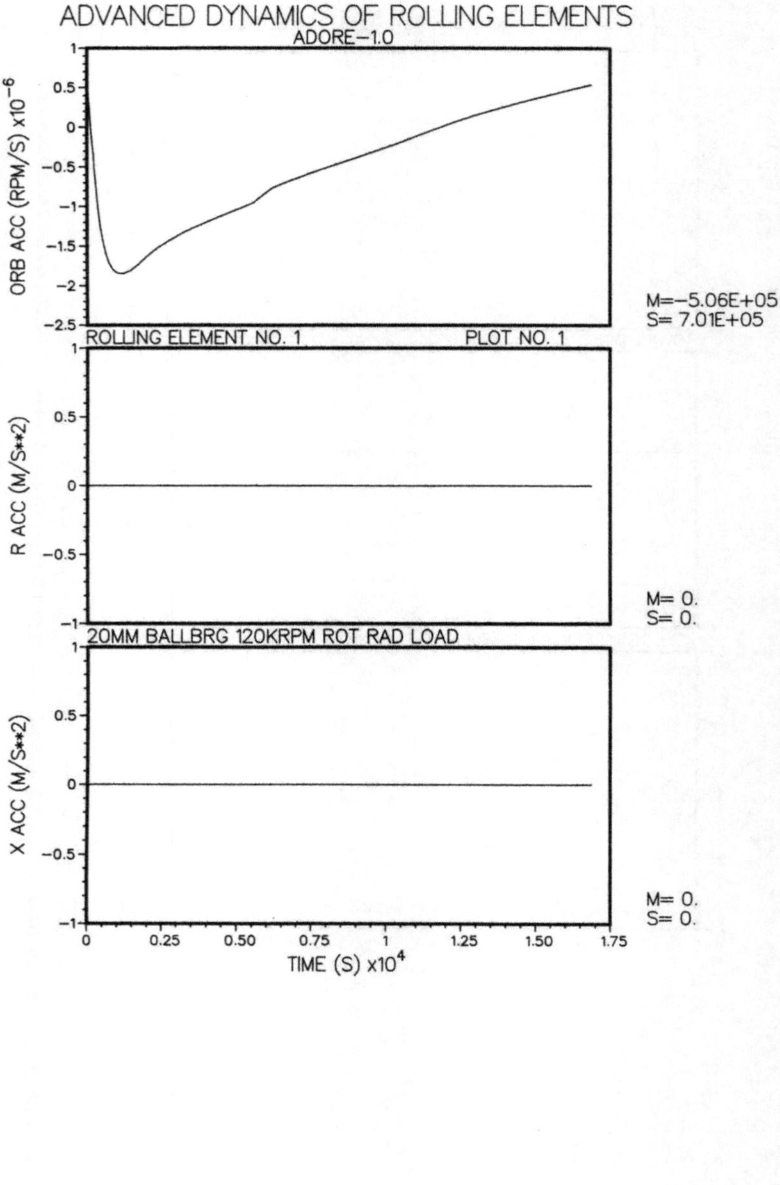
ADVANCED DYNAMICS OF ROLLING ELEMENTS
ADORE-1.0
ORB ACC (RPM/S) x10^-6
1
0.5
0
-0.5
-1
-1.5
-2
-2.5
M=-5.06E+05
S= 7.01E+05
ROLLING ELEMENT NO. 1
PLOT NO. 1
R ACC (M/S**2)
M= 0.
S= 0.
20MM BALLBRG 120KRPM ROT RAD LOAD
X ACC (M/S**2)
M= 0.
S= 0.
0
0.25
0.50
0.75
1
1.25
1.50
1.75
TIME (S) x10^4

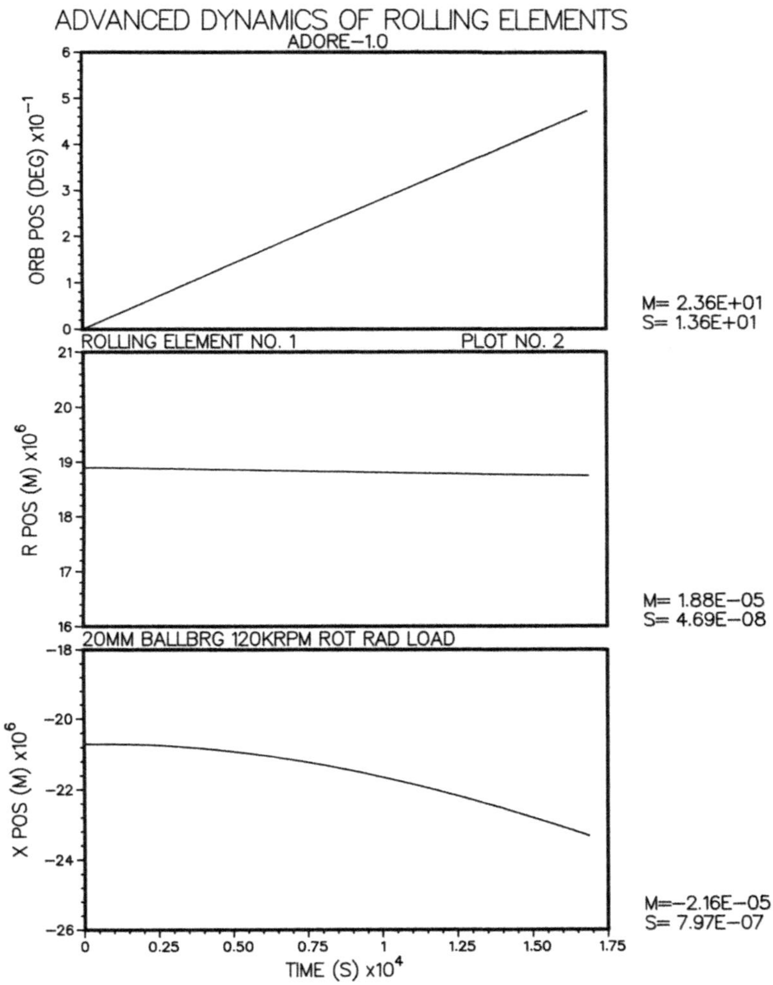
ADVANCED DYNAMICS OF ROLLING ELEMENTS
ADORE-1.0
ORB POS (DEG) x10^-1
M= 2.36E+01
S= 1.36E+01
ROLLING ELEMENT NO. 1
PLOT NO. 2
R POS (M) x10^6
M= 1.88E-05
S= 4.69E-08
20MM BALLBRG 120KRPM ROT RAD LOAD
X POS (M) x10^6
M=-2.16E-05
S= 7.97E-07
TIME (S) x10^4

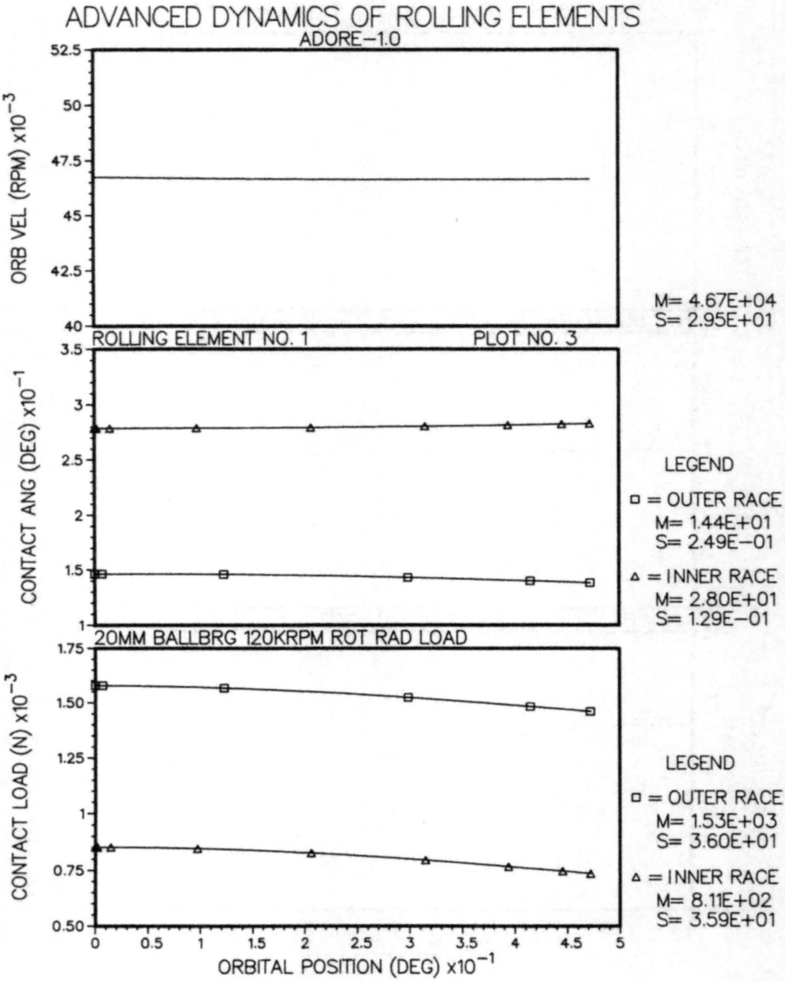

ADVANCED DYNAMICS OF ROLLING ELEMENTS
ADORE-1.0
ORB VEL (RPM) x10^-3
52.5
50
47.5
45
42.5
40
M= 4.67E+04
S= 2.95E+01
ROLLING ELEMENT NO. 1
PLOT NO. 3
CONTACT ANG (DEG) x10^-1
3.5
3
2.5
2
1.5
1
LEGEND
= OUTER RACE
M= 1.44E+01
S= 2.49E-01
= INNER RACE
M= 2.80E+01
S= 1.29E-01
20MM BALLBRG 120KRPM ROT RAD LOAD
CONTACT LOAD (N) x10^-3
1.75
1.50
1.25
1
0.75
0.50
LEGEND
= OUTER RACE
M= 1.53E+03
S= 3.60E+01
= INNER RACE
M= 8.11E+02
S= 3.59E+01
0
0.5
1
1.5
2
2.5
3
3.5
4
4.5
5
ORBITAL POSITION (DEG) x10^-1

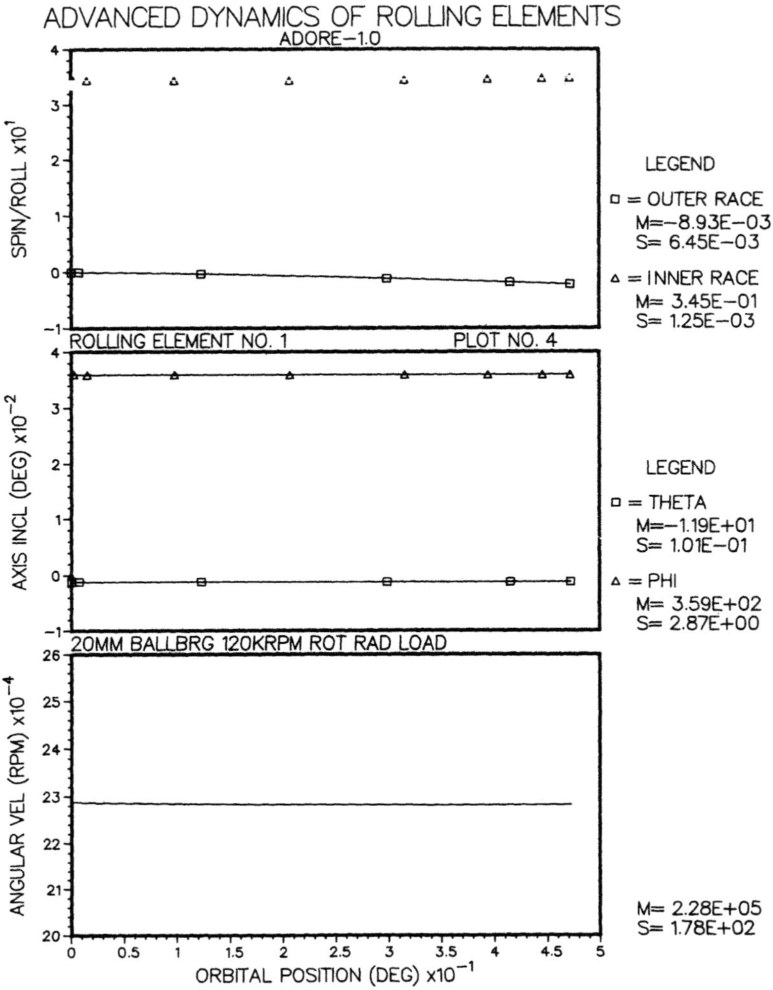
ADVANCED DYNAMICS OF ROLLING ELEMENTS
ADORE-1.0
SPIN/ROLL x10^1
LEGEND
□ = OUTER RACE
M=-8.93E-03
S= 6.45E-03
△ = INNER RACE
M= 3.45E-01
S= 1.25E-03
ROLLING ELEMENT NO. 1
PLOT NO. 4
AXIS INCL (DEG) x10^-2
LEGEND
□ = THETA
M=-1.19E+01
S= 1.01E-01
△ = PHI
M= 3.59E+02
S= 2.87E+00
20MM BALLBRG 120KRPM ROT RAD LOAD
ANGULAR VEL (RPM) x10^-4
M= 2.28E+05
S= 1.78E+02
ORBITAL POSITION (DEG) x10^-1

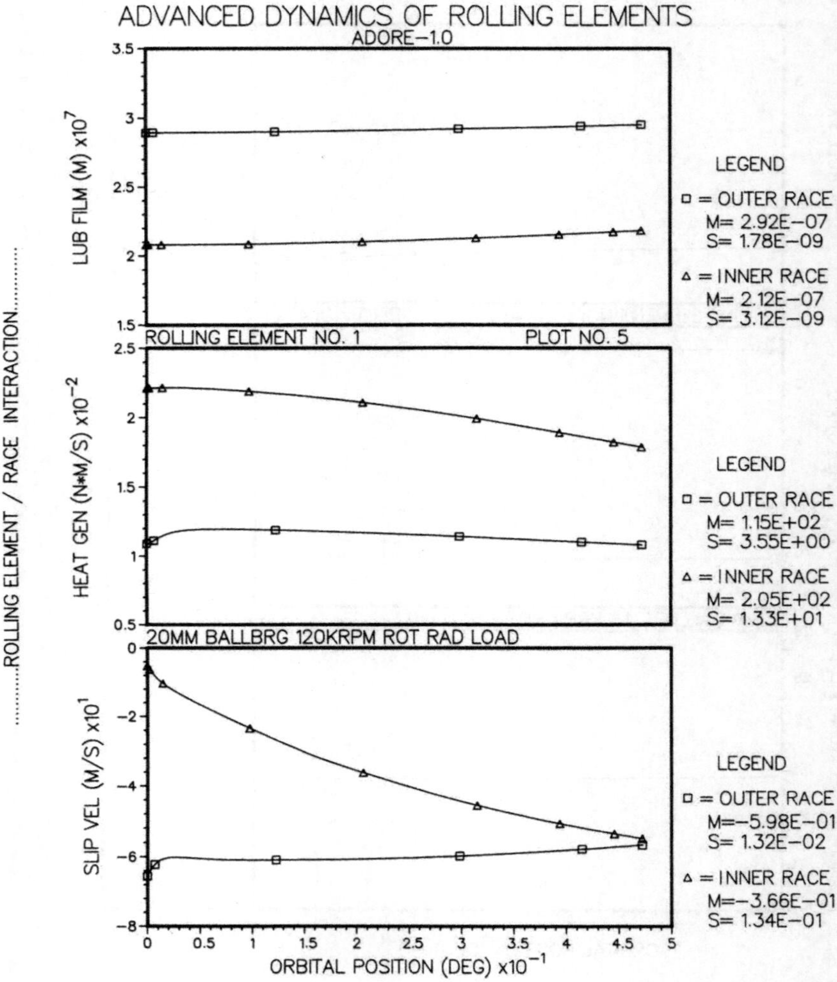
ADVANCED DYNAMICS OF ROLLING ELEMENTS
ADORE-1.0
ROLLING ELEMENT / RACE INTERACTION
LUB FILM (M) x10^7
3.5
3
2.5
2
1.5
LEGEND
□ = OUTER RACE
M= 2.92E-07
S= 1.78E-09
△ = INNER RACE
M= 2.12E-07
S= 3.12E-09
ROLLING ELEMENT NO. 1
PLOT NO. 5
HEAT GEN (N*M/S) x10^-2
2.5
2
1.5
1
0.5
LEGEND
□ = OUTER RACE
M= 1.15E+02
S= 3.55E+00
△ = INNER RACE
M= 2.05E+02
S= 1.33E+01
20MM BALLBRG 120KRPM ROT RAD LOAD
SLIP VEL (M/S) x10^1
0
-2
-4
-6
-8
LEGEND
□ = OUTER RACE
M=-5.98E-01
S= 1.32E-02
△ = INNER RACE
M=-3.66E-01
S= 1.34E-01
0
0.5
1
1.5
2
2.5
3
3.5
4
4.5
5
ORBITAL POSITION (DEG) x10^-1

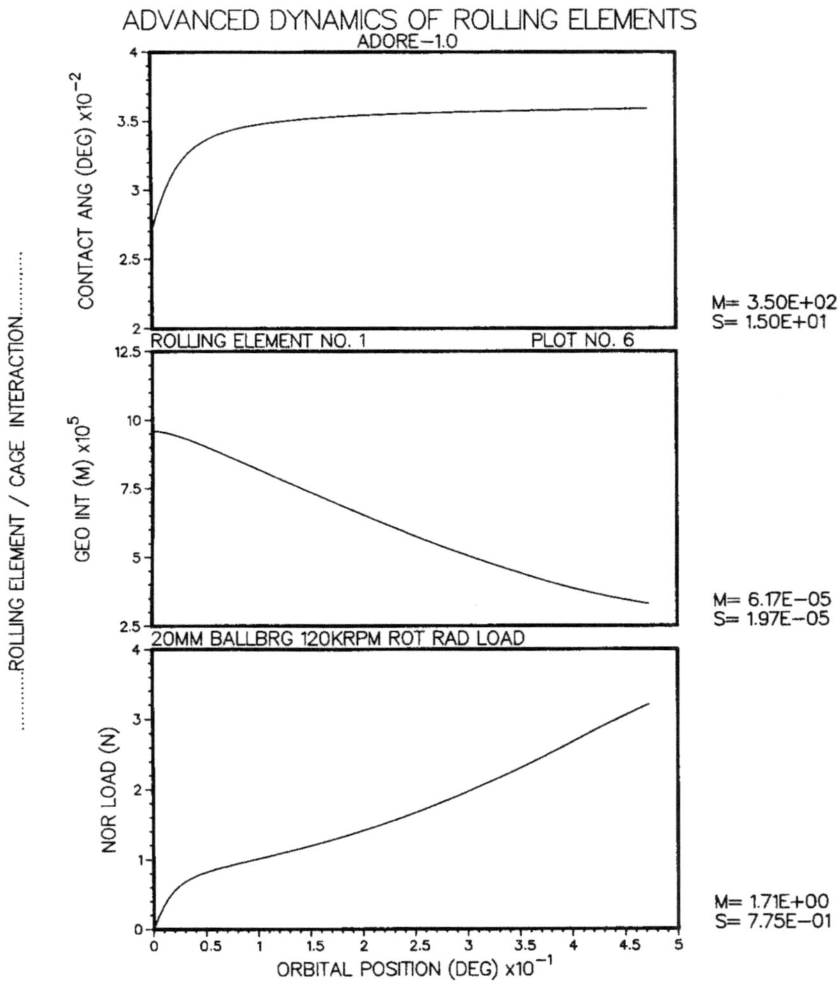
ADVANCED DYNAMICS OF ROLLING ELEMENTS
ADORE-1.0
CONTACT ANG (DEG) x10^-2
M= 3.50E+02
S= 1.50E+01
ROLLING ELEMENT NO. 1
PLOT NO. 6
GEO INT (M) x10^5
M= 6.17E-05
S= 1.97E-05
20MM BALLBRG 120KRPM ROT RAD LOAD
NOR LOAD (N)
M= 1.71E+00
S= 7.75E-01
ORBITAL POSITION (DEG) x10^-1
ROLLING ELEMENT / CAGE INTERACTION

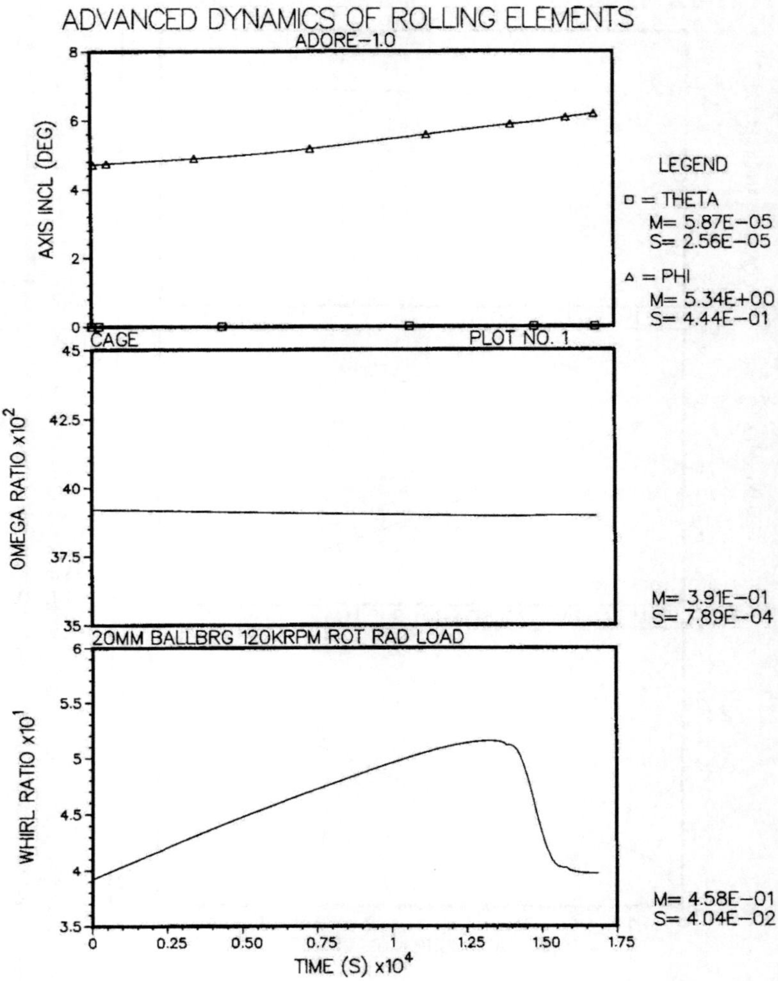
ADVANCED DYNAMICS OF ROLLING ELEMENTS
ADORE-1.0
AXIS INCL (DEG)
LEGEND
□ = THETA
M= 5.87E-05
S= 2.56E-05
△ = PHI
M= 5.34E+00
S= 4.44E-01
CAGE
PLOT NO. 1
OMEGA RATIO x10^2
M= 3.91E-01
S= 7.89E-04
20MM BALLBRG 120KRPM ROT RAD LOAD
WHIRL RATIO x10^1
M= 4.58E-01
S= 4.04E-02
TIME (S) x10^4

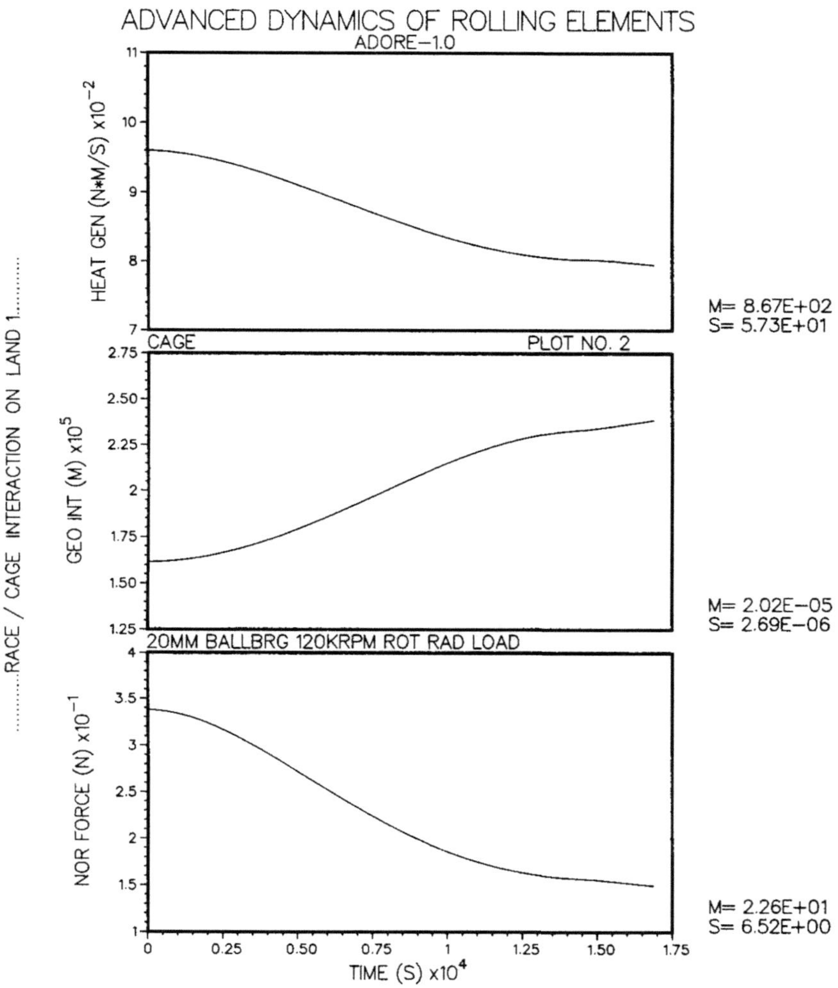
ADVANCED DYNAMICS OF ROLLING ELEMENTS
ADORE-1.0
HEAT GEN (N*M/S) x10^-2
M= 8.67E+02
S= 5.73E+01
CAGE
PLOT NO. 2
GEO INT (M) x10^5
M= 2.02E-05
S= 2.69E-06
20MM BALLBRG 120KRPM ROT RAD LOAD
NOR FORCE (N) x10^-1
M= 2.26E+01
S= 6.52E+00
TIME (S) x10^4
.........RACE / CAGE INTERACTION ON LAND 1.........

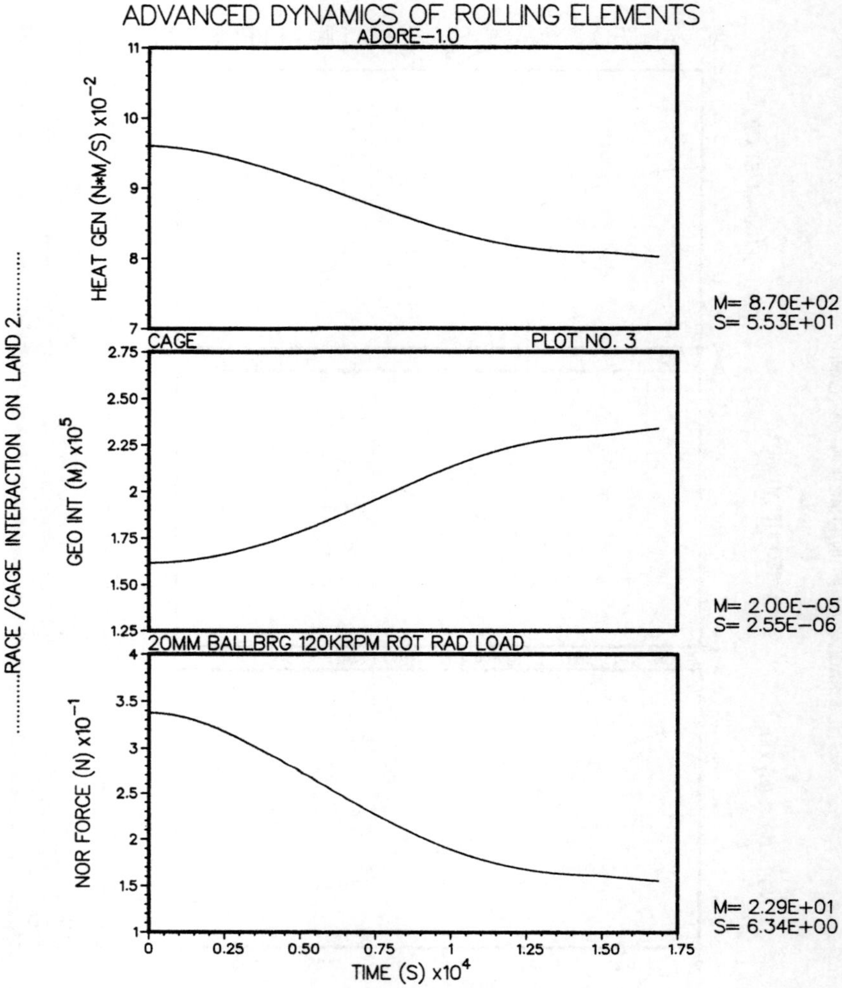
ADVANCED DYNAMICS OF ROLLING ELEMENTS
ADORE-1.0
........RACE /CAGE INTERACTION ON LAND 2........
HEAT GEN (N*M/S) x10^-2
M= 8.70E+02
S= 5.53E+01
CAGE
PLOT NO. 3
GEO INT (M) x10^5
M= 2.00E-05
S= 2.55E-06
20MM BALLBRG 120KRPM ROT RAD LOAD
NOR FORCE (N) x10^-1
M= 2.29E+01
S= 6.34E+00
TIME (S) x10^4
0
0.25
0.50
0.75
1
1.25
1.50
1.75

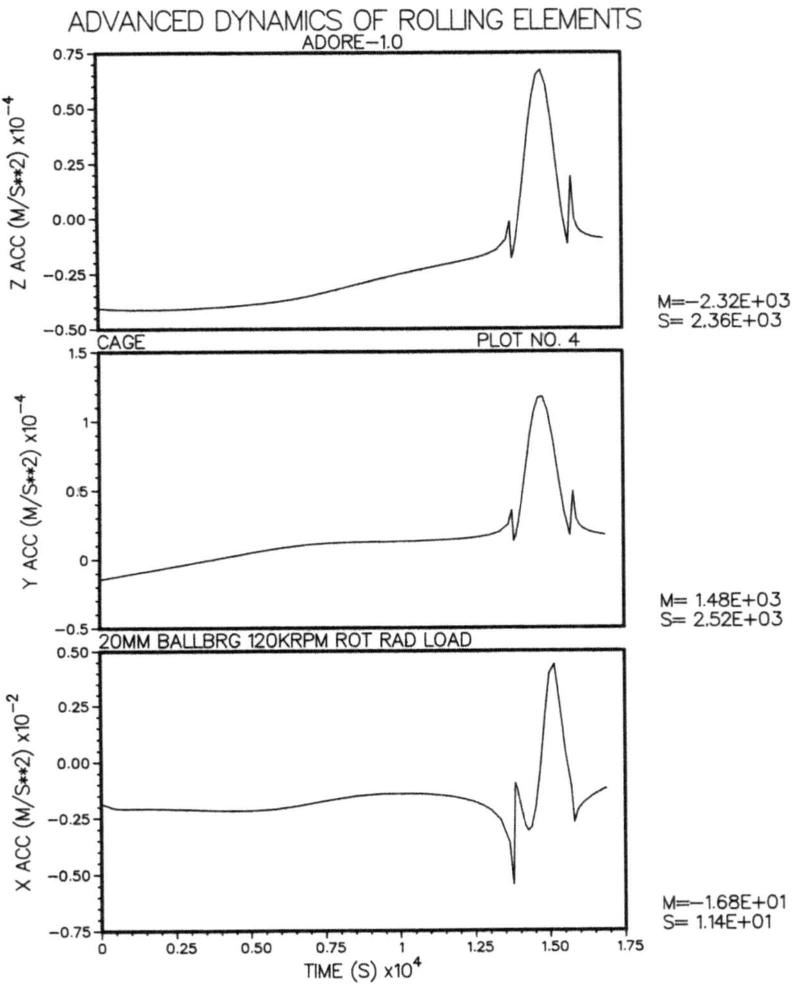
ADVANCED DYNAMICS OF ROLLING ELEMENTS
ADORE-1.0
Z ACC (M/S**2) x10^-4
0.75
0.50
0.25
0.00
-0.25
-0.50
M=-2.32E+03
S= 2.36E+03
CAGE
PLOT NO. 4
Y ACC (M/S**2) x10^-4
1.5
1
0.5
0
-0.5
M= 1.48E+03
S= 2.52E+03
20MM BALLBRG 120KRPM ROT RAD LOAD
X ACC (M/S**2) x10^-2
0.50
0.25
0.00
-0.25
-0.50
-0.75
M=-1.68E+01
S= 1.14E+01
0
0.25
0.50
0.75
1
1.25
1.50
1.75
TIME (S) x10^4

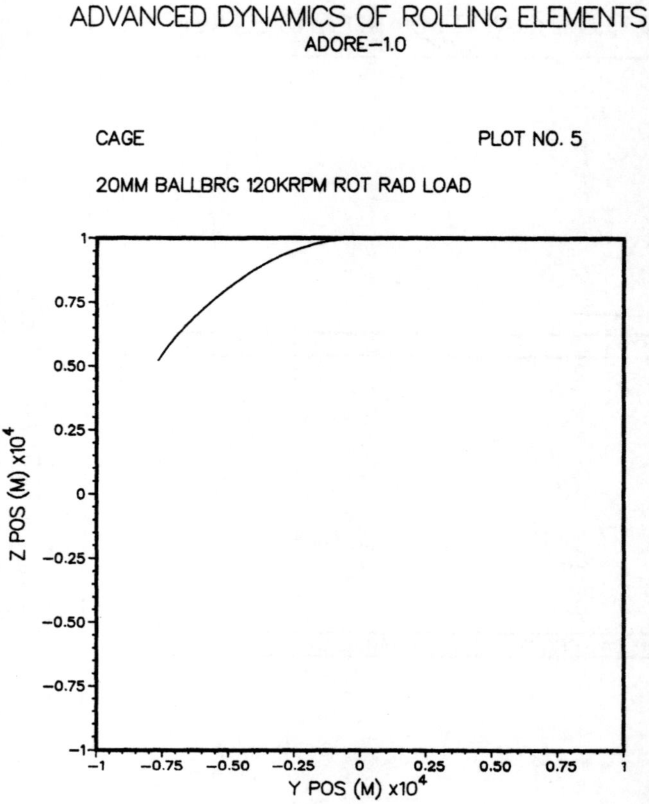
ADVANCED DYNAMICS OF ROLLING ELEMENTS
ADORE-1.0
CAGE
PLOT NO. 5
20MM BALLBRG 120KRPM ROT RAD LOAD
Z POS (M) x10⁴
Y POS (M) x10⁴
1
0.75
0.50
0.25
0
-0.25
-0.50
-0.75
-1
-1
-0.75
-0.50
-0.25
0
0.25
0.50
0.75
1

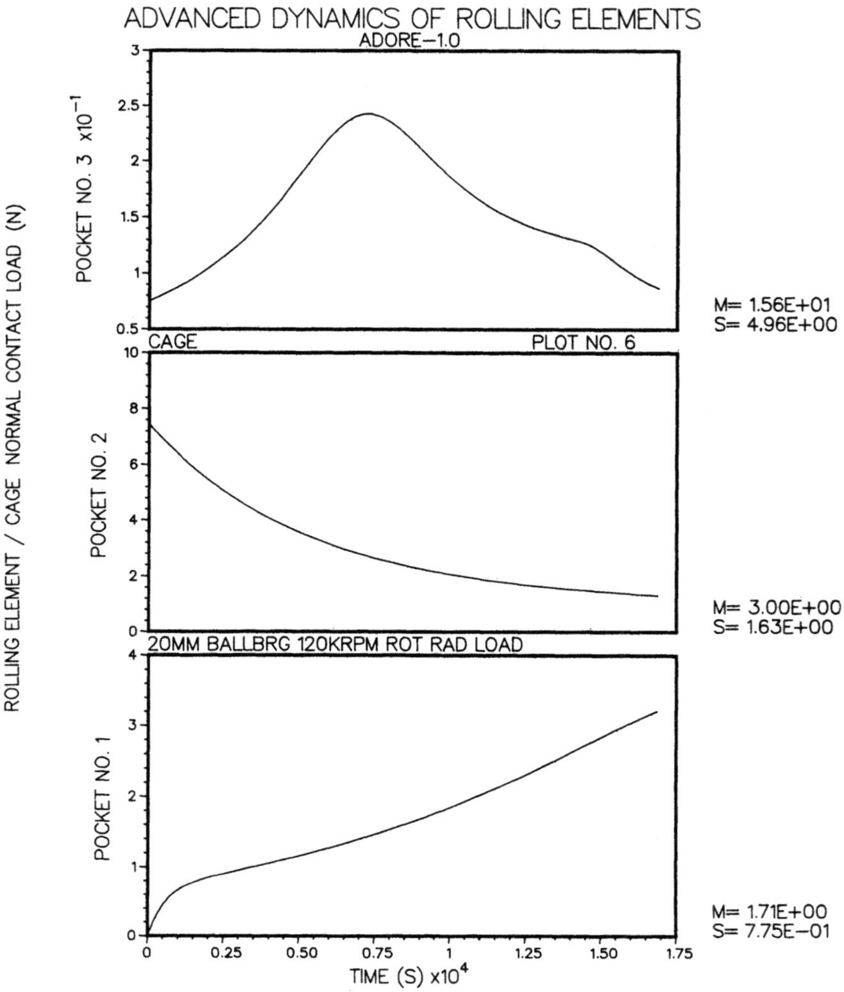
ADVANCED DYNAMICS OF ROLLING ELEMENTS
ADORE-1.0
ROLLING ELEMENT / CAGE NORMAL CONTACT LOAD (N)
POCKET NO. 3 x10⁻¹
M= 1.56E+01
S= 4.96E+00
CAGE
PLOT NO. 6
POCKET NO. 2
M= 3.00E+00
S= 1.63E+00
20MM BALLBRG 120KRPM ROT RAD LOAD
POCKET NO. 1
M= 1.71E+00
S= 7.75E-01
TIME (S) x10⁴

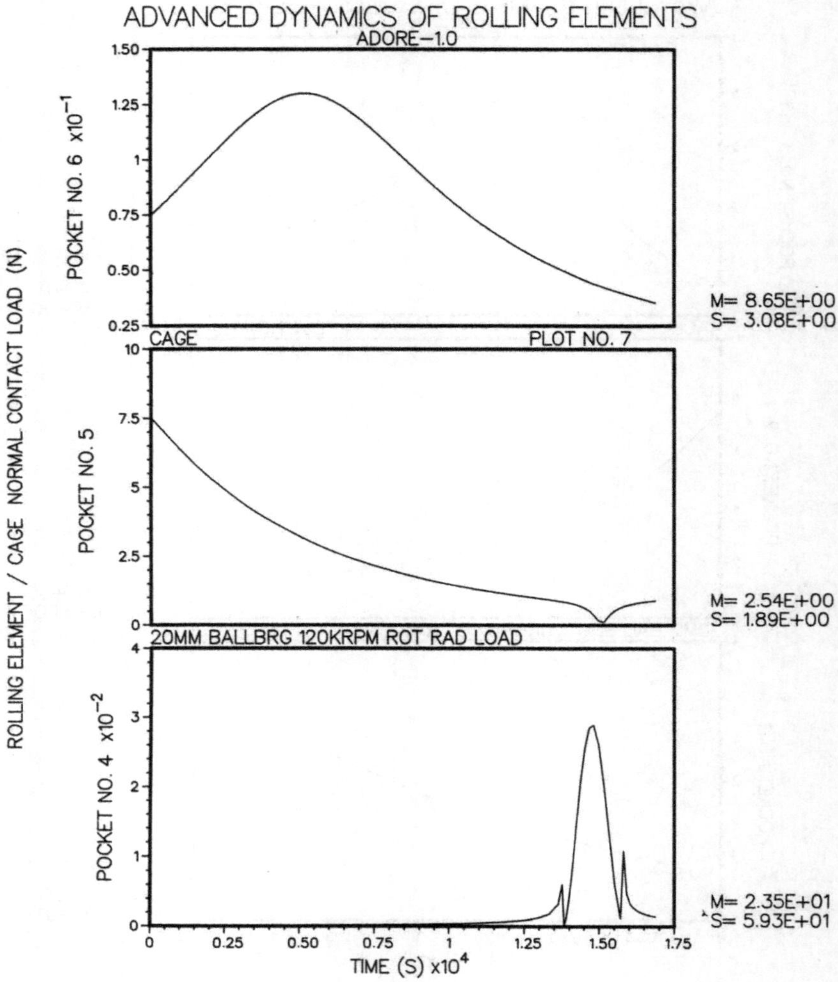
ADVANCED DYNAMICS OF ROLLING ELEMENTS
ADORE-1.0
ROLLING ELEMENT / CAGE NORMAL CONTACT LOAD (N)
POCKET NO. 6 x10^-1
1.50
1.25
1
0.75
0.50
0.25
M= 8.65E+00
S= 3.08E+00
CAGE
PLOT NO. 7
POCKET NO. 5
10
7.5
5
2.5
0
M= 2.54E+00
S= 1.89E+00
20MM BALLBRG 120KRPM ROT RAD LOAD
POCKET NO. 4 x10^-2
4
3
2
1
0
M= 2.35E+01
S= 5.93E+01
0
0.25
0.50
0.75
1
1.25
1.50
1.75
TIME (S) x10^4

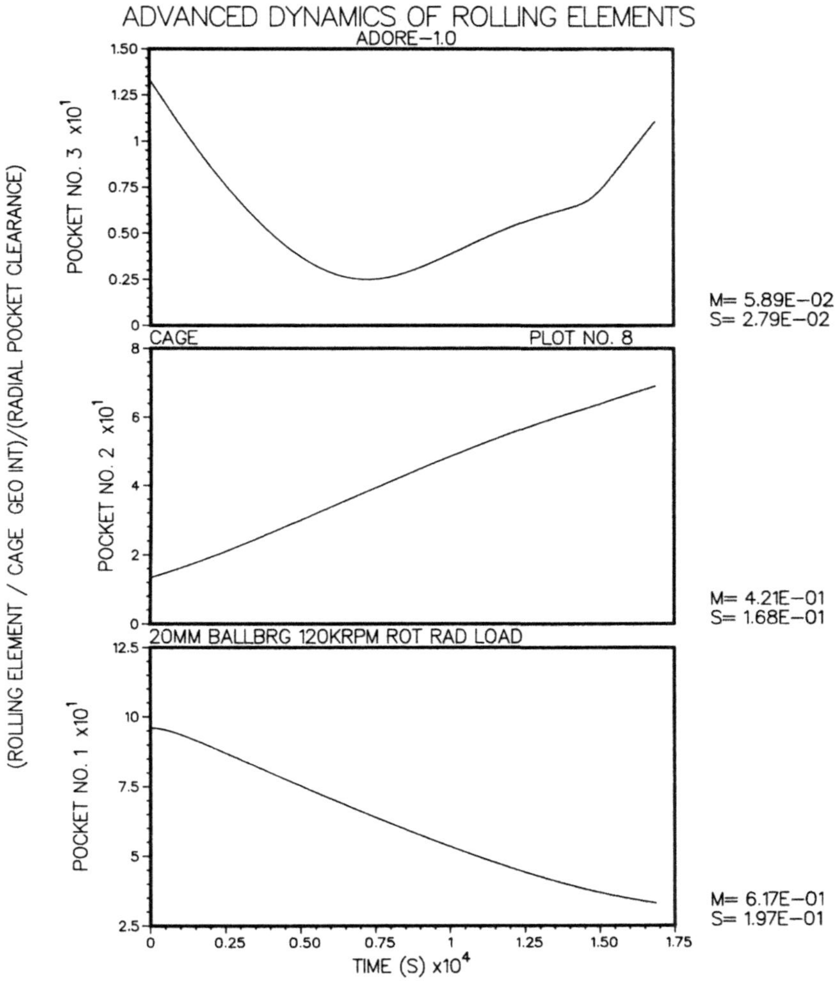
ADVANCED DYNAMICS OF ROLLING ELEMENTS
ADORE-1.0
(ROLLING ELEMENT / CAGE GEO INT)/(RADIAL POCKET CLEARANCE)
POCKET NO. 3 x10¹
1.50
1.25
1
0.75
0.50
0.25
0
M= 5.89E-02
S= 2.79E-02
CAGE
PLOT NO. 8
POCKET NO. 2 x10¹
8
6
4
2
0
M= 4.21E-01
S= 1.68E-01
20MM BALLBRG 120KRPM ROT RAD LOAD
POCKET NO. 1 x10¹
12.5
10
7.5
5
2.5
M= 6.17E-01
S= 1.97E-01
0
0.25
0.50
0.75
1
1.25
1.50
1.75
TIME (S) x10⁴

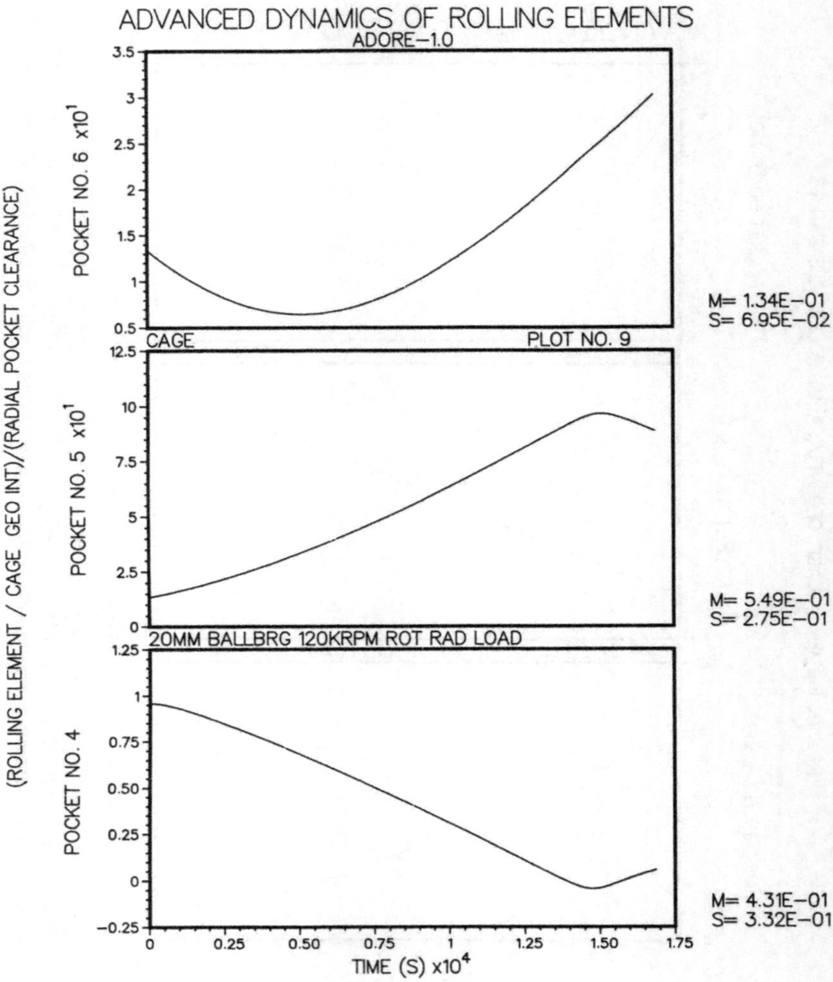
ADVANCED DYNAMICS OF ROLLING ELEMENTS
ADORE-1.0
(ROLLING ELEMENT / CAGE GEO INT)/(RADIAL POCKET CLEARANCE)
POCKET NO. 6 x10[1]
M= 1.34E-01
S= 6.95E-02
CAGE
PLOT NO. 9
POCKET NO. 5 x10[1]
M= 5.49E-01
S= 2.75E-01
20MM BALLBRG 120KRPM ROT RAD LOAD
POCKET NO. 4
M= 4.31E-01
S= 3.32E-01
TIME (S) x10[4]

References

1. Abramowitz, M. and Segun, I.A., *Handbook of Mathematical Functions*, Dover Publications, New York, 1965.
2. Archard, J.F. and Baglin, K.P., Non-dimensional presentation of frictional tractions in elastohydrodynamic lubrication—Part I: Fully-flooded conditions, *Journal of Lubrication Technology, ASME Trans.*, **97**: 398–411, 1975.
3. Archard, J.F. and Baglin, K.P., Non-dimensional presentation of frictional tractions in elastohydrodynamic lubrication—Part II: Starved conditions, *Journal of Lubrication Technology, ASME Trans.*, **97**: 412–423, 1975.
4. Bair, S. and Winer, W.O., A rheological model for elastohydrodynamic contacts based on primary laboratory data, *Journal of Lubrication Technology, ASME Trans.*, **101**: 258–265, 1979.
5. Bair, S. and Winer, W.O., Regimes of traction in concentrated contact lubrication, *Journal of Lubrication Technology, ASME Trans.*, **104**: 382–391, 1982.
6. Bair, S. and Winer, W.O., Some observations in high pressure rheology of lubricants, *Journal of Lubrication Technology, ASME Trans.*, **104**: 357–364, 1982.
7. Bamberger, E.N., Harris, T.A., Kacmarsky, W.M., Moyer, C.A., Parker, R.J., Sherlock, J.J., and Zaretsky, E.V., *Life Adjustment Factors for Ball and Roller Bearings*, an ASME Engineering Design guide, 1971.
8. Bisson, E.E. and Anderson, W.J., *ADVANCED BEARING TECHNOLOGY*, NASA Publication NASA SP-38, National Aeronautics and Space Administration, Washington, D.C., 1965.
9. Brown, P.F., Carrano, M.J., Dobek, L.J., McFadden, R.J., Miner, J.R., and Robinson, J.D., *Main Shaft High-Speed Cylindrical Roller Bearings for Gas Turbine Engines, Parts I to IV*, Report NAPC-PE-60C, prepared for Naval Air Propulsion Center, Trenton, New Jersey, and Air Force Aero Propulsion Laboratory, Wright-Patterson Air Force Base, Ohio, 1980.
10. Brüggemann, H., and Kollman, F.G., A numerical solution of the thermal elastohydrodynamic lubrication in an elliptical contact, *Journal of Lubrication Technology, ASME Trans.*, **104**: 392–400, 1982.
11. Cameron, A., *Principles of Lubrication*, Longmans, London, 1966.
12. Carlson, S.F., and Winer, W.O., The viscous lubrication of rolling and sliding rigid cylinders, *Journal of Lubrication Technology, ASME Trans.*, **97**: 180–186, 1975.

13. Castle, P., and Dowson, D., A theoretical analysis of the starved elastohydrodynamic lubrication problem for cylinder in line contact, *Proc. Instn. Mech. Engineers, London, Second Symposium on Elastohydrodynamic Lubrication*, pp 131–137, 1972.
14. Cheng, H.S., and Sternlicht, B., A numerical solution for the pressure, temperature, and film thickness between two infinitely long, lubricated rolling and sliding cylinders, under heavy loads, *Journal of Basic Engineering, ASME. Trans.*, **87**: 695–707, 1965.
15. Cheng, H.S., A numerical solution of the elastohydrodynamic film thickness in an elliptical contact, *Journal of Lubrication Technology, ASME Trans.*, **92**: 155–162, 1970.
16. Conry, T.F., *Transient Dynamic Analysis of High Speed Lightly Loaded Cylindrical Roller Bearings, Parts I and II*, NASA Technical Reports 3334 and 3335, NASA–Lewis Research Center, Cleveland, Ohio, 1981.
17. Crecelius, W.J., and Pirvics, J., *Computer Program Operation Manual on SHABERTH, A Computer Program for the Analysis of the Steady-State and Transient Thermal Performance of Shaft Bearing Systems*, U.S. Air Force Technical Report AFAPL-TR-76-90, U.S. Aero Propulsion Laboratory, Wright-Patterson Air Force Base, Ohio, 1976.
18. Crook, A.W., Lubrication of rollers, Part III: A theoretical discussion of friction and temperatures in the oil film, *Philosophical Trans., Royal Society of London, Series A*, **254**: 237–258, 1961.
19. Daniels, B.K., Non-Newtonian thermo-viscoelastic EHD traction for combined slip and spin, *ASLE Trans.*, **23**: 53–62, 1980.
20. Den Hartog, J.P., *Advanced Strength of Materials*, McGraw Hill, 1952.
21. Den Hartog, J.P., *Mechanical Vibrations*, McGraw Hill, 1956.
22. Dill, J.F., and Gupta, P.K., Ball bearing response to an unbalanced load, Paper #AIAA-83-1129, presented at the AIAA/SAE/ASME 19*th* Joint Propulsion conference, June 27–29, Seattle, Washington, 1983.
23. Dowson, D., and Higginson, G.R., *Elastohydrodynamic Lubrication*, Pergamon Press, 1966.
24. Dowson, D., Saman, W.Y., and Toyoda, S., A study of starved elastohydrodynamic line contacts, *Proc. Fifth Leeds-Lyon Symposium on Tribology, Elastohydrodynamics and Related Topics*, ed. by D. Dowson, C.M. Taylor, M. Godet, and D. Berthe, Mechanical Engineering Publications, Bury St. Edmuncs, Suffolk, pp 92–103, 1979.
25. Evans, H.P., and Snidle, R.W., The isothermal elastohydrodynamic lubrication of Spheres, *Journal of Lubrication Technology, ASME Trans.*, **103**: 547–557, 1981.
26. Evans, H.P., and Snidle, R.W., Inverse solution of Reynolds' equation of lubrication under point contact elastohydrodynamic conditions, *Journal of Lubrication Technology, ASME Trans.*, **103**: 539–546, 1981.
27. Fehlberg, E., *Classical Fifth-, Sixth-, Seventh-, and Eighth-Order Runge-Kutta Formulas with Step Size Control*, NASA Technical Report, NASA TR R-287, George C. Marshall Space Flight Center, Huntsville, Alabama, October 1968.
28. Fehlberg, E., *Low Order Classical Runge-Kutta Formulas with Stepsize Control and Their Application to Some Heat Transfer Problems*, NASA Technical Report NASA TR R-315, George C. Marshall Space Flight Center, Huntsville, Alabama, July 1969.
29. Forsythe, G.E., Malcolm, M.A., and Moler, C.B., Computer Methods for Mathematical Computations, Prentice Hall, 1977.
30. Gear, C.W., *Numerical Initial Value Problems in Ordinary Differential Equations*, Prentice Hall, 1971.
31. Gentle, C.R., Duckworth, R.R., and Cameron, A., Elastohydrodynamic film thickness at extreme pressures, *Journal of Lubrication Technology, ASME Trans.*, **97**: 383–389, 1975.
32. Greenwood, J.A., and Kauzlarich, J.J., Inlet shear heating in elastohydrodynamic lubrication, *Journal of Lubrication Technology, ASME Trans.*, **95**: 401–416, 1973.
33. Grubin, A.N., Fundamentals of the hydrodynamic theory of lubrication of heavily loaded cylindrical surfaces, Central Scientific Institute for Technology and Mechanical Engineering, Moscow, Book #3, Chapter 2, English Translation #R-3554, Special Libraries Association, Chicago, 1949.
34. Gupta, P.K., Transient ball motion and skid in ball bearings, *Journal of Lubrication Technology, ASME Trans.*, **97**: 261–269, 1975.
35. Gupta, P.K., A Review of Computational Simulations of Roller Bearing Performance, *Computer-Aided Design of Bearings and Seals, Proc. ASME Annual Meeting*, Dec. 1976, pp 19–30.

36. Gupta, P.K., *Dynamics of rolling element bearings, Part VII*, DREB compilation listing, MTI-77TR60, Mechanical Technology Incorporated, Latham, New York, 1977.
37. Gupta, P.K., *Dynamics of rolling element bearings, Part VI*, DREB computer program manual, MTI-77TR55, Mechanical Technology Incorporated, Latham, New York, 1977.
38. Gupta, P.K., Winn, L.W., and Wilcock, D.F., Vibrational characteristics of Ball Bearings, *Journal of Lubrication Technology, ASME Trans.*, 99: 284–289, 1977.
39. Gupta, P.K., Dynamics of rolling element bearings, Parts I, II, III, and IV, *Journal of Lubrication Technology, ASME Trans.*, **101**: 293–326, 1979.
40. Gupta, P.K., *Interactive Graphic Simulation of Rolling Element Bearings—Phase I: Low Frequency Phenomenon and RAPIDREB Development*, U.S. Air Force Technical Report AFWAL-TR-81-4148, Materials Laboratory, Air Force Wright Aeronautical Laboratories, Air Force Systems Command, Wright-Patterson Air Force Base, Ohio, 1981.
41. Gupta, P.K., Flamand, L., Berthe, D., and Godet, M., On the traction behavior of several lubricants, *Journal of Lubrication Technology, ASME Trans.*, **103**: 55–64, 1981.
42. Gupta, P.K., *Interactive Graphic Simulation of Rolling Element Bearings—Phase I: Low-Frequency Phenomenon and RAPIDREB Development*, U.S. Air Force AFWAL-TR-81-4148, Materials Laboratory, Air Force Wright Aeronautical Laboratories, Air Force Systems Command, Wright-Patterson Air Force Base, Ohio, 1981.
43. Gupta, P.K., Simulation of low-frequency components in the dynamic response of a ball bearing, *Advances in Computer-Aided Bearing Design, Proc. 1982 ASLE/ASME Lubrication Conference*, ASME Publication, 1982.
44. Gupta, P.K., Dill, J.F., and Bandow, H.E., Dynamics of rolling element bearings: Experimental validation of the DREB and RAPIDREB computer programs, *Journal of Tribology, ASME Trans.*, to be published.
45. Hald, A., *Statistical Theory with Engineering Applications*, John Wiley and Sons, 1967.
46. Hamrock, B.J., and Dowson, D., *Ball Bearing Lubrication: The Elastohydrodynamics of Elliptical Contacts*, John Wiley and Sons, 1981.
47. Harris, T.A., *Rolling Bearing Analysis*, John Wiley, 1966.
48. Harris, T.A., An analytical method to predict skidding in thrust-loaded angular contact ball bearings, *Journal of Lubrication Technology, ASME Trans.*, **93**: 17–24, 1971.
49. Harris, T.A., Ball motion in thrust-loaded angular contact bearings with coulomb friction, *Journal of Lubrication Technology, ASME Trans.*, 93: 32–38, 1971.
50. Herrebrugh, K., Solving the incompressible and isothermal problem in elastohydrodynamic lubrication through an integral equation, *Journal of Lubrication Technology, ASME Trans.*, **90**: 262–270, 1968.
51. Hildebrand, F.B., *Introduction to Numerical Analysis*, McGraw Hill, 1966.
52. Hindmarsh, A.C., *GEAR: Ordinary Differential Equation Solver*, Lawrence Livermore Laboratory, Report UCID-30001, Rev. 3, 1974.
53. Hindmarsh, A.C., and Byrne, G.D., *EPISODE: An Effective Package for the Integration of Systems of Ordinary Differential Equations*, Lawrence Livermore Laboratory, Report, UCID-30112, Rev. 1, 1977.
54. Houpert, L., Flamand, L., and Berthe, D., Rheological and thermal effects in lubricated EHD contacts, *Journal of Lubrication Technology*, **103**: 526–532, 1981.
55. Jefferis, J.A., and Johnson, K.L., Sliding friction between lubricated rollers, *Proc. Instn. Mech. Engineers, London*, Vol. 182, Pt. 1, #14, 1967–1968, pp 281–291.
56. Johnson, K.L., and Cameron, R., Shear behavior of elastohydrodynamic oil films at high rolling contact pressures, *Proc. Instn. Mech. Engineers, London*, Vol. 182, Pt. 1, #14, 1967–1968, pp 307–330.
57. Johnson, K.L., and Tevaarwerk, J.L., Shear behavior of elastohydrodynamic oil films, *Proc. Royal Society, London, Series A*, **356**: 215–236, 1977.
58. Johnson, K.L., and Greenwood, J.A., Thermal analysis of an eyring fluid in elastohydrodynamic traction, *Wear*, **61**: 353–374, 1980.
59. Jones, A.B., Ball motion and sliding friction in ball bearings, *Journal of Basic Engineering, ASME Trans.*, Series D, vol 81, pp 1–12, 1959 *XX*: 1–12, 1959.
60. Jones, A.B., A general theory for elastically constrained ball and radial roller bearings under arbitrary load and speed conditions, Journal of Basic Engineering, ASME Trans., Series D, vol 82, pp 309–320, 1960.
61. Jones, A.B., The mathematical theory of rolling-element bearings, in *Mechanical Design and Systems Handbook*, McGraw Hill, 1963.

62. Kannel, J.W., Bell, J.C., and Allen, C.M., *A Study of the Influence of Lubricant on High-Speed Rolling-Contact Bearing Performance*, Technical Report ASD-TR-61-643, Part III, Wright-Patterson Air Force Base, Ohio, 1963.
63. Kannel, J.W., and Walowit, J.A., Simplified analysis for tractions between rolling–sliding elastohydrodynamic contacts, *Journal of Lubrication Technology*, **93**: 39–46, 1971.
64. Kannel, J.W., and Dow, T.A., The relation between pressure and temperature in rolling–sliding EHD contact, *ASLE Trans.*, **23**: 262–268, 1980.
65. Kleckner, R.J., Pirvics, J., and Castelli, V., High-speed cylindrical rolling element bearing analysis CYBEAN—analytical formulation, *Journal of Lubrication Technology, ASME Trans.*, **102**: 380–390, 1980.
66. Kleckner, R.J., and Pirvics, J., Spherical roller bearing analysis, *Journal of Lubrication Technology, ASME Trans*, **104**: 99–108, 1982.
67. Kragelskii, I.V., *Friction and Wear*, Butterworths, 1965.
68. Lambert, J.D., *Computational Methods in Ordinary Differential Equations*, John Wiley, 1973.
69. Lund, J.W., The stability of an elastic rotor in journal bearings with flexible damped supports, *Journal of Applied Mechanics, ASME Trans.*, **87**: 911–920, 1965.
70. Lund, J.W., Stability and damped critical speeds of a flexible rotor in fluid film bearings, *Journal of Engineering for Industry, ASME Trans.*, **94**: 531–538, 1972.
71. Lundberg, G., Elastic Berihung Zweier Halbraiine, *VDI Forschung*, **10** (No. 5): 320, 1939.
72. Mauriello, J.A., Lagasse, N., Jones, A.B., and Murray, W., *Rolling Element Bearing Retainer Analysis*, USAAMRDL Report 72-45, Eustis Directorate, U.S. Army Air Mobility Research and Development Laboratory, Fort Eustis, Virginia, 1973.
73. McGrew, J.M., Gu, A., Cheng, H.S., Murray, S.F., *Elastohydrodynamic Lubrication—Preliminary Design Manual*, U.S. Air Force Technical Report #AFAPL-TR-70-27, Air Force Aero-Propulsion Laboratory, Air Force Systems Command, Wright-Patterson Air Force Base, Ohio, 1970.
74. *Mobil EHL Guidebook*, Mobil Oil Corporation, New York, New York, 1979.
75. Nagaraj, H.S., and Gupta, P.K., *EHD traction day for a traction fluid*, MTI Technical Report #MTI 78TR143, Mechanical Technology Incorporated, Latham, N.Y., 1978.
76. Palmgren, A., *Ball and Roller Bearing Engineering, 3rd Edition*, SKF Industries Inc., 1959.
77. Pan, CH.T., Wu, E.R., and Krauter, A.I., *Rotor–Bearing Dynamics Technology Design Guide, Part I: Flexible Rotor Dynamics*, U.S. Air Force Technical Report AFAPL-TR-78-6, U.S. Air Force Aero Propulsion Laboratory, Wright-Patterson Air Force Base, Ohio, 1980.
78. Philips, G.J., Bearing performance investigations through speed ratio measurements, *Trans. ASLE*, **22** (No. 4): 307–314, 1979.
79. Pinkus, O., and Sternlicht, B., *Theory of Hydrodynamic Lubrication*, McGraw Hill, 1961.
80. Ragen, M.A., Steady-state and transient thermal analysis of a transmission systems—SKF Computer Program "TRANSIM," SKF Industries Report No. AT79Y001, 1979.
81. Roark, R.J., Formulas for Stress and Strain, McGraw Hill, 1965.
82. Rumbarger, J.H., Filetti, E.G., and Gubernick, D., Gas turbine engine main shaft roller bearing system analysis, *Journal of Lubrication Technology, ASME Trans.*, **95**: 401–416, 1973.
83. Schlichtig, H., *Boundary Layer Theory*, McGraw Hill, 1968, pp 15–19, 93–99, 606–608.
84. Shampine, L.F., and Allen, R.C., *Numerical Computing: An Introduction*, W.B. Saunders, 1973.
85. Sibley, L.B., and Pirvics, J., *Computer Analysis of Rolling Bearings, Computer-Aided Design of Bearings and Seals, Proc. ASME Annual Meeting*, Dec. 1976, pp 95–115.
86. Smith, R.L., Walowit, J.A., Gupta, P.K., and McGrew, J.M., *Research on Elastohydrodynamic Lubrication of High Speed Rolling–Sliding Contacts*, U.S. Air Force Technical Report #AFAPL-TR-71-54, Air Force Aero Propulsion Laboratory, Air Force Systems Command, Wright-Patterson Air Force Base, Ohio, 1971.
87. Smith, R.L., Walowit, J.A., and McGrew, J.M., Elastohydrodynamic traction coefficients of 5P4E polyphenyl ether, *Journal of Lubrication Technology, ASME Trans.*, **97**: 353–362, 1973.
88. Synge, J.L., and Griffith, B.A., Principles of Mechanics, McGraw Hill, 1959.
89. Tallian, T.E., Rolling Bearing Life Modifying Factors for film thickness, surface roughness and friction, *Journal of Lubrication Technology, ASME Trans.*, **103**: 509–520, 1981.
90. Tevaarwerk, J.L., and Johnson, K.L., The influence of fluid rheology on performance of traction drives, *Journal of Lubrication Technology, ASME Trans.*, **101**: 266–274.

91. Timoshenko, S., and Goodier, J.N., *Theory of Elasticity*, McGraw Hill, 1951.
92. Trachman, E.G., and Cheng, H.S., Thermal and non-Newtonian effects on traction in elastohydrodynamic contacts, *Proc. Instn. Mech. Engineers, London, Second Symposium on Elastohydrodynamic Lubrication*, 1972, pp 142–148.
93. Trachman, E.G., A simplified technique for predicting traction in elastohydrodynamic contacts, *ASLE Trans.*, **21**: 33–62.
94. Walowit, J.A., and Smith, R.L., Traction characteristics of MIL-L-7808 oil, *Journal of Lubrication Technology, ASME Trans.*, **98**: 607–612, 1976.
95. Walters, C.T., The dynamics of ball bearings, *Journal of Lubrication Technology, ASME Trans.*, **93**: 1–10, 1971.
96. Wedevan, L.D., Evans, D., and Cameron, A., Optical analysis of ball bearing starvation, *Journal of Lubrication Technology, ASME Trans.*, **93**: 349–363, 1971.
97. Wedevan, L.D., Traction and film thickness measurements under starved elastohydrodynamic conditions, *Journal of Lubrication Technology, ASME Trans.*, **97**: 321–329, 1975.
98. Wolveridge, P.E., Baglin, K.P., and Archard, J.F., The starved lubrications of cylinders in line contact, *Proc. Instn. Mech. Engineers, London, Vol. 185 81/71*, pp 1159–1169, 1970–71.
99. Zaretsky, E.V., Parker, R.J. and Anderson, W.J., Component hardness differences and their effect on bearing fatigue, *Journal of Lubrication Technology, ASME Trans*, **89**: 47–62, 1967.
100. Zorzi, E.S., and Nelson, H.D., Finite element simulation of rotor–bearing systems with internal damping, *Journal of Engineering for Power, ASME Trans.*, **99**: 71–76, 1977.

Author Index

Subject Index

GPSR Compliance

The European Union's (EU) General Product Safety Regulation (GPSR) is a set of rules that requires consumer products to be safe and our obligations to ensure this.

If you have any concerns about our products, you can contact us on ProductSafety@springernature.com

In case Publisher is established outside the EU, the EU authorized representative is:

Springer Nature Customer Service Center GmbH
Europaplatz 3
69115 Heidelberg, Germany

Batch number: 10620143

Printed by Printforce, the Netherlands